Advances in Intelligent Systems and Computing

Volume 1155

The series "Advances in Intelligent Systems and Computing" contains publications on theory, applications, and design methods of Intelligent Systems and Intelligent Computing. Virtually all disciplines such as engineering, natural sciences, computer and information science, ICT, economics, business, e-commerce, environment, healthcare, life science are covered. The list of topics spans all the areas of modern intelligent systems and computing such as: computational intelligence, soft computing including neural networks, fuzzy systems, evolutionary computing and the fusion of these paradigms, social intelligence, ambient intelligence, computational neuroscience, artificial life, virtual worlds and society, cognitive science and systems, Perception and Vision, DNA and immune based systems, self-organizing and adaptive systems, e-Learning and teaching, human-centered and human-centric computing, recommender systems, intelligent control, robotics and mechatronics including human-machine teaming, knowledge-based paradigms, learning paradigms, machine ethics, intelligent data analysis, knowledge management, intelligent agents, intelligent decision making and support, intelligent network security, trust management, interactive entertainment, Web intelligence and multimedia.

The publications within "Advances in Intelligent Systems and Computing" are primarily proceedings of important conferences, symposia and congresses. They cover significant recent developments in the field, both of a foundational and applicable character. An important characteristic feature of the series is the short publication time and world-wide distribution. This permits a rapid and broad dissemination of research results.

** Indexing: The books of this series are submitted to ISI Proceedings, EI-Compendex, DBLP, SCOPUS, Google Scholar and Springerlink **

More information about this series at http://www.springer.com/series/11156

Brijesh Iyer · A. M. Rajurkar ·
Venkat Gudivada
Editors

Applied Computer Vision and Image Processing

Proceedings of ICCET 2020, Volume 1

 Springer

Editors
Brijesh Iyer
Department of Electronics and
Tele-communication Engineering
Dr. Babasaheb Ambedkar
Technological University
Lonere, Maharashtra, India

A. M. Rajurkar
Department of Computer Engineering
MGMs College of Engineering
Nanded, Maharashtra, India

Venkat Gudivada
Department of Computer Science
East Carolina University
Greenville, NC, USA

ISSN 2194-5357   ISSN 2194-5365   (electronic)
Advances in Intelligent Systems and Computing
ISBN 978-981-15-4028-8   ISBN 978-981-15-4029-5   (eBook)
https://doi.org/10.1007/978-981-15-4029-5

This Springer imprint is published by the registered company Springer Nature Singapore Pte Ltd.
The registered company address is: 152 Beach Road, #21-01/04 Gateway East, Singapore 189721, Singapore

Preface

Dr. Babasaheb Ambedkar Technological University, Lonere-402103, is a State Technological University of the Maharashtra State in India. Over the years, the Department of Electronics and Telecommunication Engineering of this University has been organizing faculty and staff development and continuing education programmes.

In the year 2013, the department had taken a new initiative to organize international conferences in the frontier areas of Engineering and computing technologies. The ICCET series (earlier ICCASP) is an outcome of this initiative. The 5th ICCET 2020 is jointly organized by Department of E&TC Engineering of the University and Department of Computer Science & Engineering, MGM College of Engineering, Nanded-India. Keynote lectures, invited talks by eminent professors and panel discussion of the delegates with the academicians and industry personnel are the key features of 5th ICCET 2020.

This volume aims at a collection of scholarly articles in the area of Applied Computer Vision and Image Processing. We have received a great response in terms of the quantity and quality of individual research contributions for consideration. The conference had adopted a "double-blind peer review" process to select the papers with a strict plagiarism verification policy. Hence, the selected papers are the true record of research work for the theme of this volume.

We are thankful to the reviewers, session chairs and rapporteurs for their support. We also thank the authors and the delegates for their contributions and presence. We are extremely grateful to University officials and Organizing Committee of MGMs College of Engineering, Nanded for the support for this activity.

We are pledged to take this conference series to greater heights in the years to come with the aim to put forward the need-based research and innovation.

Thank you one and all.

Lonere, India   Dr. Brijesh Iyer
Nanded, India   Dr. Mrs. A. M. Rajurkar
Greenville, USA   Dr. Venkat Gudivada

Contents

About the Editors

Dr. Brijesh Iyer received his Ph.D. degree in Electronics and Telecommunication Engineering from Indian Institute of Technology, Roorkee, India, in 2015. He is an Associate Professor in the University Department of E & TC Engineering at Dr. Babasaheb Ambedkar Technological University, Lonere (A State Technological University). He is a recipient of INAE research fellowship in the field of Engineering. He had 02 patents to his credit and authored over 40 research publications in peer-reviewed reputed journals and conference proceedings. He had authored five books on curricula as well as cutting-edge technologies like Sensor Technology. He has served as the program committee member of various international conferences and reviewer for various international journals. His research interests include RF front-end design for 5G and beyond, IoT, and biomedical image/signal processing.

Dr. Mrs. A. M. Rajurkar is working as a Professor and Head, Department of Computer Science and Engineering at MGMs College of Engineering, Nanded, Maharashtra, India. She had received the research grants worth Rs. 32 Lakh from BARC, Mumbai, in the area of image processing. She has authored and coauthored over 70+ research publications in peer-reviewed reputed journals and conference proceedings. She is having an experience of 28+ years in the academics at various positions. She has served as the program committee member of various international conferences and reviewer for various international journals. Her research interests include content-based image processing, multimedia technology and applications, and biomedical image analysis. She received her Ph.D. degree in Computer Engineering from the Indian Institute of Technology, Roorkee, India, in 2003.

Dr. Venkat Gudivada is Chairperson and Professor in the Computer Science Department at East Carolina University. Prior to this, he was Founding Chair and Professor in the Weisberg Division of Computer Science at Marshall University and also worked as a Vice President for Wall Street companies in New York city for over six years including Merrill Lynch (now Bank of America Merrill Lynch) and Financial Technologies International (now GoldenSource). His previous academic tenure also includes work at the University of Michigan, University of Missouri, and Ohio University. His current research interests are in data management, verification and validation of SQL queries, cognitive computing, high-performance computing, information retrieval, computational analysis of natural language, and personalized learning. His research has been funded by NSF, NASA, U.S. Department of Energy, U.S. Department of Navy, U.S. Army Research Office, Marshall University Foundation, and West Virginia Division of Science and Research. He has published over 100 articles in peer-reviewed journals, book chapters, and conference proceedings. Dr. Venkat received his Ph.D. degree in Computer Science from the Center for Advanced Computer Studies (CACS), at the University of Louisiana at Lafayette.

A Deep Learning Architecture for Corpus Creation for Telugu Language

Dhana L. Rao, Venkatesh R. Pala, Nic Herndon, and Venkat N. Gudivada(✉)

Cognitive Computing Lab, East Carolina University, Greenville, NC 27855, USA
gudivadav15@ecu.edu

Abstract. Many natural languages are on the decline due to the dominance of English as the language of the World Wide Web (WWW), globalized economy, socioeconomic, and political factors. Computational Linguistics offers unprecedented opportunities for preserving and promoting natural languages. However, availability of corpora is essential for leveraging the Computational Linguistics techniques. Only a handful of languages have corpora of diverse genre while most languages are *resource-poor* from the perspective of the availability of machine-readable corpora. Telugu is one such language, which is the official language of two southern states in India. In this paper, we provide an overview of techniques for assessing language vitality/endangerment, describe existing resources for developing corpora for the Telugu language, discuss our approach to developing corpora, and present preliminary results.

Keywords: Telugu · Corpus creation · Optical character recognition · Deep learning · Convolutional neural networks · Recurrent neural networks · Long short-term memory · Computational linguistics · Natural language processing

1 Introduction

As of 2019, there are 7,111 spoken languages in the world [1]. This number is dynamic as linguists discover new spoken languages which were hitherto undocumented. Moreover, the languages themselves are in a flux given the rapid advances in computing and communications, and social dynamic and mobility. About 40% of the world languages are *endangered* and these languages have less than 1,000 speakers [1]. A language gets endangered when its speakers begin to speak/teach another (dominant) language to their children. Sadly, about 96% of the languages of the world are spoken by only 3% of the world population. About 95% of the languages might be extinct or endangered by the end of the century. Especially, the indigenous languages are disappearing at an alarming rate. To bring awareness to this issue, the United Nations has declared 2019 as The International Year of Indigenous Languages.

© Springer Nature Singapore Pte Ltd. 2020
B. Iyer et al. (eds.), *Applied Computer Vision and Image Processing*,
Advances in Intelligent Systems and Computing 1155,
https://doi.org/10.1007/978-981-15-4029-5_1

We will begin by defining some terminology. A person's *first language* is referred to as her *native language* or *mother tongue*, and this is the language the person is exposed to from birth through the first few years of life. These *critical years* encompass an extremely important window to acquire the language in a linguistically rich environment. It is believed that further language acquisition beyond the critical years becomes much more difficult and effortful [2,3]. The first language of a child enables reflection and learning of successful social patterns of acting and speaking. The first language is also an integral part of a child's personal, social, and cultural identity.

Each language has a set of *speech sounds*. Phonetics of the language is concerned with the description and classification of speech sounds including how these sounds are produced, transmitted, and received. A *phoneme* is the smallest unit in the sound system of a language. Ladefoged [4] describes all the known ways in which the sounds of the world's languages differ. Native speakers of a language learn the sounds of the language correctly. Therefore, when they speak the language, they speak natively—produce correctly accented sounds. In contrast, non-native speakers of a language speak it *non-natively*, which is evidenced by incorrectly accented sounds. We use the terms first language, $L1$, mother tongue, and arterial language synonymously.

A person is *bilingual* if she is equally proficient in two languages. However, correctly accented fluency is considered a requirement, which is difficult given that each language is associated with a set of characteristic sounds. The vocal chord development must be attuned to the distinct sounds of both the languages. Given the criticality of the mother tongue on a child's social and intellectual development, on November 17, 1999 the UNESCO designated 21 February as the *International Mother Language Day*.

There are many languages in the world which are widely spoken, but are on an accelerated path toward insignificance and eventual extinction [5]. This trend is manifesting through multiple indicators. First, the number of speakers who speak the language *natively* is declining. Second, even these native speakers lack fluency in the choice of correct words as their vocabulary is rather limited. Furthermore, they begin a sentence in their mother tongue and interject words of another language to compensate for their limited mother tongue vocabulary. Third, the dominance of English as the language of WWW is also a contributing factor. Though this is affecting most languages, there are exceptions. Languages such as Japanese, Spanish, Arabic, Turkish, German, Italian, and French have begun to flourish on the WWW. Speakers of these languages take pride in learning their mother tongue and government policies also nurture and promote these languages.

The fourth factor for the decline of languages is related to the global economy and associated mobility and migration. This is more pronounced in developing countries such as India. Fifth, the British colonialism actively promoted the use of English at the cost of native languages. Lastly, in developing countries such as India, not speaking one's mother tongue is viewed as elite social status. This is in sharp contrast with most European countries where people take pride in their mother tongue and native languages are used to study even disciplines such as medicine and engineering.

We believe that the recent and rapid advances in computing and communication technologies provide unprecedented opportunities for reversing the shift in the language use. More specifically, *computational linguistics* offers software tools and approaches to preserve and promote natural languages. However, these approaches require machine-readable corpora of diverse genre to reflect the entire language use. In this paper, we discuss the development of a machine learning-based approach for developing corpora for Telugu language. Telugu is the official language of two states in South India. There are many other spoken languages in these two states, which include Kolami, Koya, Gondi, Kuvi, Kui, Yerukala, Savara, Parji, and Kupia. This study does not address these dying languages, though the approach we develop can be used for all languages.

The remainder of the paper is organized as follows. We discuss tools and techniques for assessing language vitality and endangerment in Sect. 2. Section 3 summarizes the salient characteristics of the Telugu language. A brief overview of Convolutional Neural Networks (CNNs) and Recurrent Neural Networks is presented in Sect. 4. Existing resources and approaches for generating corpora for the Telugu language are presented in Sect. 5. Our approach to corpus generation and preliminary results are discussed in Sect. 6 and Sect. 7 concludes the paper.

2 Assessing Language Vitality and Endangerment

The first step in preserving and promoting languages is to assess their current state using a nomenclature or metric. The Fishman's Graded Intergenerational Disruption Scale (GIDS) was the first effort in this direction [6]. There are eight stages on the GIDS scale—stage 1 through stage 8, and stage 8 refers to the highest level of language disruption and endangerment. The **stage 8** indicates a state where very few native speakers of the language exist, the speakers are geographically dispersed and socially isolated, and the language needs to be reassembled from their mouths and memories. The **stage 7** reflects the language situation that most users of the language are socially integrated and ethnolinguistically active population, but passed the child-bearing age. The **stage 6** refers to language situation where intergenerational informal oralcy exists, there is demographic concentration of native speakers with institutional reinforcement.

The **stage 5** reflects a situation where the language literacy exists at home, school, and community, but there is no extra-communal reinforcement of such literacy. The **stage 4** refers to the state where the language is used in primary and secondary education, and this is mandated by education laws. The state where the language is being used in *lower work sphere* beyond the native language's geographic region is referred to as **stage 3**. The **stage 2** reflects the state that the language is used in mass media, and lower governmental services. Lastly, the **stage 1** refers to the use of the language in the higher echelons of education, occupations, government, and media, but lacks political independence. The GIDS scale is quasi-implicational meaning that the higher scores imply all or nearly all of the lesser degrees of disruption as well.

The GIDS served as a seminal evaluative framework of language endangerment for over two decades. Recently, other evaluative frameworks have been proposed. For example, UNESCO has developed a 6-level scale of endangerment [7]. Ethnologue uses yet another set of measures (five categories) to assess language vitality/endangerment [1]. Lewis et al. align the above three evaluative systems and proposed an evaluative scale of 13 levels [8], and is referred to as the E(xpanded) GIDS (EGIDS). Evaluating a language's vitality using the EGIDS involves answering five key questions about the language related to identity function, vehicularity, state of intergenerational language transmission, literacy acquisition status, and a societal profile of generational language use. To make this paper self-contained, the EGIDS scale is shown in Table 1.

Table 1. The EGIDS scale for assessing language vitality

Level number	Level name	Level description
0	International	Used in international trade and policy, and knowledge exchange
1	National	Used in education, work, mass media, and government at the national level
2	Provincial	Used in education, work, mass media, and government within major administrative subdivisions of a country
3	Wider communication	Used in work and mass media without official language status across a region
4	Educational	Vigorously used, literature sustained, and an institutionally supported education system exists
5	Developing	Vigorously used, but the literature use is not widespread and sustainable
6a	Vigorous	Used for spoken communication by all generations and the situation is sustainable
6b	Threatened	Used for spoken communication by all generations, but is losing native speakers
7	Shifting	The child-bearing generation use the language among themselves, but not transmitted to children
8a	Moribund	Only grandparent generation and older are the active users of the language
8b	Nearly extinct	Only grandparent generation and older are the only active users of the language, and they have little opportunity to use the language
9	Dormant	No one has more than symbolic proficiency of the language
10	Extinct	No one uses the language

Language Endangerment Index (LEI) is another metric for assessing the levels of language endangerment [9]. LEI is based on four factors: intergenerational transmission, absolute number of speakers, speaker number trends, and domains of use. Compared to other language endangerment assessments, LEI can be used even if limited information is available.

Dwyer [10] explores the uses and limits of various language vitality/endangerment tools through case examples of assessment, including successful language revitalization and maintenance efforts. The author also discusses the role of Non-Governmental Organizations (NGOs) in linguistic and cultural maintenance, especially in the Tibetan context.

Mihas et al. through a research monograph [11] address many complex and pressing issues of language endangerment. This volume specifically addresses language documentation, language revitalization, and training. The case studies of the volume provides detailed personal accounts of fieldworkers and language activists engaged in language documentation and revitalization work.

Lüpke [12] argues that the language vitality assessments used for African languages are rooted in Western language ideologies, and therefore, are inappropriate for the African context. The author proposes an alternative set of vitality parameters for African languages. Through a research monograph, Essegbey et al. [13] bring together a number of important perspectives on language documentation and endangerment in Africa.

A few resources are available for finding linguistic characteristics of languages and assessing their vitality/endangerment. The World Atlas of Language Structures (WALS) [14] is a free, online resource. It is a large database of the phonological, grammatical, lexical properties of the world languages. The WALS also features an interactive reference tool for exploring the database (https://www. eva.mpg.de/lingua/research/tool.php).

The Catalogue of Endangered Languages (ELCat) is another online resource for information on the endangered languages of the world [15]. The ELCat project is a partnership between Google, Alliance for Linguistic Diversity, University of Hawai'i at Mānoa Linguists, and the LINGUIST List at Eastern Michigan University. This project is sponsored by a grant from the National Science Foundation.

The Open Language Archives Community (OLAC) is an international partnership of institutions and individuals whose goal is to create a worldwide virtual library of language resources [16]. The OLAC is a free online service. Linguistic Linked Open Data (LLOD) is another free, cloud service for linguistic data [17]. It logically integrates diverse license-free, linguistic data resources and provides a unified search feature. Its linguistic resources include corpora; lexicons and dictionaries; terminologies, thesauri, and knowledge bases; linguistic resource metadata; linguistic data categories; and typological databases. Ethnologue is a commercial resource for the world language data [1].

3   The Telugu Language

Telugu is the official language of two states in southern India—Andhra Pradesh and Telangana. Telugu is also spoken in the Yanam district of Puducherry (a union territory of India). It is also spoken by a significant number of linguistic minorities in other states of India including Odisha, Karnataka, Tamil Nadu, Kerala, and Maharashtra. Telugu is a member of the Dravidian language family and there are over 215 million speakers for this language family. Among the languages of the Dravidian family, Telugu is the most widely spoken language. Per BBC news article, Telugu is the fastest growing language in the United States [18].

Contrary to the popular myth and propaganda, India has no national language. The constitution of India recognizes 22 languages as **scheduled languages** and Telugu is one of them. Telugu is also one of the six languages to have the **classical language of India** designation, which is bestowed by the Government of India.

According to the 2001 census of India, Telugu has the third largest number of native speakers in India at 74 million. Furthermore, Telugu ranks 13th in the Ethnologue list of most-spoken languages in the world. However, these ranks have fallen after a decade. Per 2011 census of India [19], Telugu slipped to the fourth position in terms of the largest number of native speakers in India and to the 15th place in the Ethnologue list of most widely spoken languages worldwide.

The Telugu script is an **Abugida**, which is derived from the Brahmi script. Abudiga is a segmental writing system where most consonants are immediately followed by a vowel, which form syllables. Each consonant-vowel sequence is written as a unit. The Telugu *Varnamala*/alphabet consists of 57 symbols, of which 18 are *achulu*/vowels, 36 are *hallulu*/consonants, and 3 are vowel modifiers. Of the 18 vowels, 2 of them are not used now. Even among the consonants, two of them have fallen out of use. The Telugu Varnamala is also called *aksharamulu*. The three vowel modifiers are considered as belonging to both vowel and consonant groups, and are referred to as *ubayaksharamulu*. The number of syllables in the language is approximately equal to the product of the number of constants and the number of vowels (i.e., $36 \times 18 \simeq 648$). From an Optical Character Recognition (OCR) point of view, the number of classes is very high.

Unlike English, there is no distinction between upper- and lower-case letters in Telugu. The script is written from left to right and the basic units of writing are syllables. Telugu words are pronounced exactly the way they are spelled. The Telugu writing system won the second place in The World Alphabet Olympics held in 2012 (https://languagelog.ldc.upenn.edu/nll/?p=4253).

Some features of *metrical poetry* are unique to the Telugu language. The Telugu poetry (called *padyalu*) employs an elaborate set of rules called *chandhas* for defining structural features. The *Chandhas* also apply to prose and it generates rhythm to the literature and poetry. The assigned unicode code-points for the Telugu language are 0C00–0C7F (3072–3199).

The Telugu language has vast literature. It has unique literary traditions, for example, *Ashtavadhanam*. The latter is a public performance of an *Avadhani* (the performer) whose goal is to demonstrate her sharpness of memory and retention, mastery of the language, literature, grammar, and linguistic knowledge. In *Ashtavadhanam*, eight peers take turns in posing questions to the *Avadhani* and also distract her with challenges. The *Avadhani* answers to peer's questions must be constructed in a way to adhere to certain grammatical constructions and other linguistic constraints. *Satavadhanam* and *Sahasravadhanam* are advanced versions of *Ashtavadhanam*, where 100 and 1000 peers, respectively, ask questions and distract the *Avadhani*. These events last over several weeks.

The main challenges that the Telugu language faces today are lack of new literature, fast declining language speakers who can speak the language natively, lack of institutional support, and societal apathy toward the language. These factors will reduce the language to a mere spoken language in the near-term and eventual extinction in the medium-term. Paradoxical it may sound, very few current generation speakers speak the language natively. Their native language vocabulary is dismal, ability to read classic texts is abysmal, and the number of qualified language teachers is astonishingly small and rapidly shrinking. Even the so-called professionals in the Telugu mass media have severe language performance issues.

The other challenges include the dominance of English on the World Wide Web (WWW), the colonial past of India, federal government policies, strong desire of people for migration to other countries, and changing cultural and societal values. The current Indian government equates nationalism with having one language (i.e., Hindi) under the motto "One Nation, One Language." The federal government aggressively enforces Hindi on non-Hindi speaking population with the eventual goal of replacing English with Hindi. The Telugu movie industry is also another endangerment for the language, where the actors and actresses lack language proficiency and set a bad example for the younger generation.

The recent language policy introduced by the government of Andhra Pradesh fast-tracks the trajectory of Telugu language extinction. In November 2019, the government issued an executive order that all instruction in elementary and primary schools be delivered only in English, which was hitherto done in Telugu. The students will still learn Telugu as a language, but physical sciences, mathematics, and social sciences will all be taught in English. In private schools, teaching Telugu is optional. Replacing instruction in mother tongue (Telugu) with a foreign language (English) is a flawed policy and is a disaster in the making. This policy runs contrary to the research that suggests the critical need for learning in native language in primary and secondary schools [20–24].

4 CNNSs and RNNs

Computational Linguistics [25] plays a central role in preserving and promoting natural languages. However, it requires corpora of different genre in machine-readable format. Almost all significant and classical works in the Telugu language are copyright-free but are not in machine readable form. Though OCR is a solved problem for languages including English, Korean, Spanish, Mandarin, Turkish,

Arabic, and Japanese, it is an unsolved problem for many Indian languages including Telugu. Machine Learning (ML) in general and Deep Learning (DL) in particular offer unparalleled opportunities for solving the OCR problem for the *resource-poor* languages. Recently, ML and DL have been actively investigated in numerous domains and seminal works which include [26–31]. In the following, we provide a brief introduction to a major deep learning architecture used for OCR—Convolutional Neural Networks (CNNs).

The use of neural networks for classification problems dates back to the early days of machine learning, with the development of perceptron, one of the first algorithms for binary classification (Fig. 1). The perceptron models a neuron, by combining multiple inputs $(x_1, x_2, \ldots, x_n)$ to generate one output, similar to how a biological neuron takes its input from multiple other neurons connected to its dendrites, and produces one output through its axon. The perceptron assigns different weights to each input $(w_1, w_2, \ldots, w_n)$, in addition to its bias (i.e., w_0). If the value of the inputs multiplied by their weights, plus the bias, is larger than a threshold then its output is one, otherwise it is zero. This is the responsibility of the *activation function*.

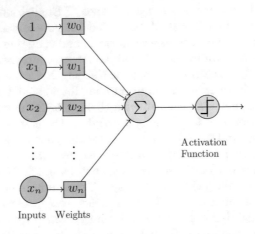

Fig. 1. A perceptron

Perceptrons can be extended to multiclass classification by using one perceptron for each class. This creates a basic neural network in which all inputs are connected to all outputs, i.e., a neural network with an input layer and an output layer. This model was improved by adding additional, hidden layers between the input and output layers, creating a multilayer, feedforward neural network, with the nodes from each layer connected to all the nodes in the subsequent layer. A feedforward neural network is shown in Fig. 2. This network has four layers—an input, an output, and two hidden layers. The first hidden layer has four neurons and the second hidden layer has three of them.

One of the first applications of neural networks was in computer vision, namely handwritten digit recognition. An image, represented by an array of integers corresponding to the pixel values in the image, would be given as an input to the neural network, and the output would indicate the numerical value of the digit in the image. One main observation with this application is that the intensity of each pixel is not as significant as the differences between that pixel and adjacent pixels. This is because, due to the lighting conditions when each image is taken, a pixel might be lighter or darker. Thus, the emphasis should be on "local" and "differences."

Convolutional Neural Networks (CNNs) take this observation into consideration, and use intermediate layers that are not fully connected. Instead, in these layers, each node takes its inputs from a limited number of adjacent pixels—a patch of an image—and apply a filter to it. The role of each filter is to identify specific features in an image, such as a horizontal line, a vertical line, and so on. The output of these layers is passed to a fully connected neural network. For example, in an OCR task, if only the convolution filters for the horizontal line and for the diagonal line are activated, it might indicate that the handwritten digit is 7.

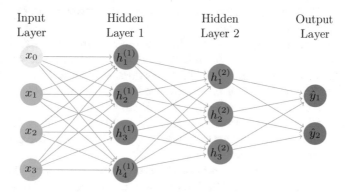

Fig. 2. A feedforward neural network

CNNs are specialized for processing grids of values. Another variation of multilayer neural networks are Recurrent Neural Networks (RNNs), which specialize in processing a sequence of values. One of the key characteristics of RNNs is that they allow cycles in the network, which enables them to create a memory of previous inputs that persists in the internal state of the network. Thus, they are appropriate for tasks in which previous inputs have an influence arbitrarily far into the future, such as in language-related tasks, where a current word influences the probability of seeing subsequent words, or a sequence of letters influence the combination of subsequent letters for creating a valid word.

One of the limitations of RNNs is that the influence of a particular input on the hidden layer, and implicitly on the output layer, either increases or decreases exponentially due to the cycle in the network. This is known as the *vanishing gradient problem*. One method that addresses this problem is Long Short-Term Memory (LSTM) architecture. The LSTM introduces memory blocks—a set of recurrently connected subnets. Each memory block has one or more self-connected memory cells as well as input, output, and *forget units* which enable write, read, or reset operations for the memory cells, respectively. This allows memory cells to save and retrieve information over long periods of time, making them suitable for applications such as learning context-free languages.

5   Existing Resources and Approaches for Generating Corpora for the Telugu Language

Google Books (https://books.google.com/) is an ambitious project of Google, Inc. Its goal is to scan books provided by publishers and authors through the Google Books Partner Program, or by Google's library partners through the Library Project. Google also partners with magazine publishers to digitize their archives. The scanned documents are converted into machine-readable text through Optical Character Recognition (OCR) technology. Google provides a full text search capability over these digitized archives. As of October 2019, the number of scanned books exceed over 40 million. However, most of these books are no longer in print or commercially available.

A criticism about Google Books is that the errors introduced into the scanned text by the OCR process remain uncorrected. Furthermore, the digitized documents are not organized in the form of corpora to enable computational linguistics research. Lastly, some critics dubbed Google Book project as *linguistic imperialism* enabler—the transfer of a dominant language to other people. Majority of the scanned books are in English and this entails disproportionate representation of natural languages. We noticed some classical Telugu works such as *Vemana Satakam* in Google Books.

The Million Books Project is a book digitization project led by Carnegie Mellon University from 2007 to 2008. This project partnered with government and research entities in India and China and scanned books in multiple languages. The project features over 1.5 million scanned books in 20 languages: 970,000 in Chinese; 360,000 in English; 50,000 in Telugu; and 40,000 in Arabic. *The Million Books Project* is now replaced by HathiTrust Digital Library (https://www.hathitrust.org/). Most of the the HathiTrust digitized collections are protected by copyright law and thus are not fully viewable.

Commercial APIs for OCR include IBM Watson Discovery API, Microsoft Azure Computer Vision API, Amazon Web Services Rekognition API, and Google Vision API. Our experimentation with all these services failed to produce results that are simply not useful (see Figs. 3, 4, 5, and 6). The Google Vision API performed slightly better than the others, still the results are not useful. We attribute the poor quality of OCR results from these systems to their grandiose goal of one system for hundreds of languages.

Tesseract (https://opensource.google/projects/tesseract) is an open-source OCR engine from Google. Google claims that Tesseract can recognize more than 100 languages out of the box. Tesseract can also be trained to recognize other languages.

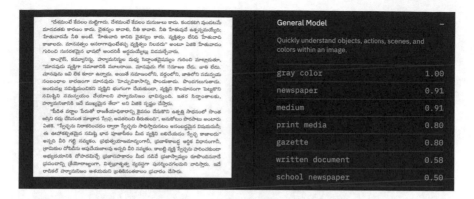

Fig. 3. OCR results from IBM Watson Discovery API

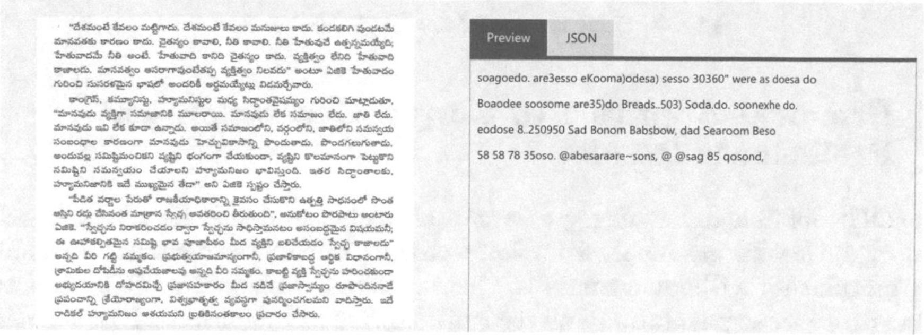

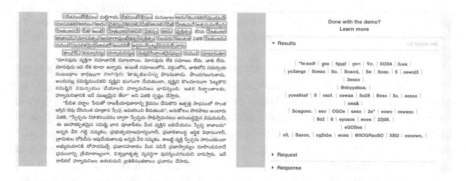

Fig. 4. OCR results from Microsoft Azure Computer Vision API

Fig. 5. OCR results from Amazon Web Services Rekognition API

https://github.com/TeluguOCR lists 8 code/corpora/font repositories for Telugu OCR. However, we were unable to make any of the code repositories compile and run. This is partly because of the code incompatibilities with the current versions of software libraries.

Fig. 6. OCR results from Google Vision API

6 Proposed Approach to Corpus Generation and Preliminary Results

The OCR for Telugu language poses several challenges due to its complex script and agglutinative grammar. Complex words in Telugu are derived by combining morphemes without changes to their spelling or phonetics. OCR is a two stage process—segmentation and recognition. The techniques used for segmenting Roman scripts are quite similar to each other. In such a script, the segmentation corresponds to identifying and demarcating units of written text. A unit corresponds to a contiguous region of text—a connected component. A robust segmentation algorithm should be able to effectively deal with noise from scanning, skew, erasure, and font variations. The recognition task is essentially a multiclass classification problem.

The proposed deep learning architecture for Telugu OCR employs a combination of CNNs and RNNs. Our model has three advantages. First, it requires no detailed annotations and learns them directly from the sequence labels. Second, the model learns informative feature representations directly from the segmented syllable images. This eliminates the handcrafting of features as well as some preprocessing steps such as the binarization component localization. Lastly, our model is lighter than a CNN model and requires less storage space.

The architecture of our model has three stacked components. The components from bottom to top layer are CNN, RNN, and a transcription layer. The CNN extracts the feature sequence, RNN predicts every frame of feature sequences passed from the CNN, and the top transcription layer translates the frame predictions into a label sequence. The entire network has a single loss-function.

We chose Amazon Web Services (AWS) infrastructure for implementing the proposed deep learning architecture. AWS infrastructure comes with pre-installed frameworks for deep learning. Amazon Web Services Rekognition API provides excellent recognition accuracy for languages including English, Arabic, Chinese, Finnish, French, German, Hebrew, Indonesian, Italian, Japanese, and a few others. Obviously, Telugu is not one of these languages. We experimented with PyTorch and CUDA deep learning frameworks before deciding on Tensor-Flow.

We experimented with a few algorithms (binary, Gaussian, and MeanC) for converting a scanned text image into a binary thresholded image. We used Hough Line Transform (HLT) for skew detection. When HLT failed to detect skew, we used Minimum Bounding Rectangle (MBR) as a substitution. Figure 7 shows the word-level segmentation of our implementation on a scanned text image.

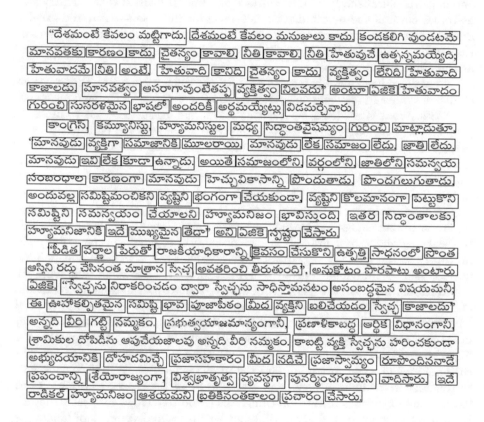

Fig. 7. Word-level segmentation performed by our system on a scanned text image

Our preliminary results indicate better performance for Telugu OCR compared to IBM Watson Discovery API, Microsoft Azure Computer Vision API, Amazon Web Services Rekognition API, and Google Vision API. This is expected given that these systems target hundreds of languages, whereas our system specifically focused on Telugu OCR. We plan to conduct additional evaluation of the system and further improve classification accuracy. Our end goal for this research is to develop and open-source a robust Telugu OCR system, which in turn will enable creation of Telugu copora of various genre.

7 Conclusions

Language and culture are intricately intertwined. Language endangerment and cultural endangerment go hand in hand. The recent advances in computing, communications, and computational linguistics offer unparalleled opportunities for preserving and promoting linguistic diversity, and ensuring cultural autonomy and cultural pluralism. In a multi-cultural and linguistically diverse country such as India, cultural self-determination is crucial for India to succeed and excel in human development. Much to the chagrin of vested interests, the fact remains that India is much like the European Union and it should have been aptly named as the Unite States of India.

Though we have highlighted the case for the Telugu language, the issues are similar for other Indian languages such as Kannada, Marathi, Odiya, and Malayalam [32]. What is needed is strategies, tactics, and execution plans to preserve, promote, and celebrate linguistic diversity and the associated cultures. Toward this goal, the International Association for the Development of Cross-Cultural Communication issued a declaration at its 22nd seminar on Human Rights and Cultural Rights held in 1987 at Recife, Brazil. This declaration is referred to as the *Universal Declaration of Linguistic Rights* or *The Recife Declaration* and requires reformulation of national, regional, and international language policies.

Among other things, the Recife Declaration states that " ... Recognizing that the learning and use, maintenance and promotion of languages contribute significantly to the intellectual, educational, sociocultural, economic and political development of individuals, groups, and states.   ... Asserting that linguistic rights should be acknowledged, promoted, and observed, nationally, regionally and internationally, so as to promote and assure the dignity and equity of all languages.   ... Aware of the need for legislation to eliminate linguistic prejudice and discrimination and all forms of linguistic domination, injustice and oppression, in such contexts as services to the public, the place of work, the educational system, the courtroom, and the mass media.   ... Stressing the need to sensitize individuals, groups, and states to linguistic rights, to promote positive societal attitudes toward plurilingualism and to change societal structures toward equality between users of different languages and varieties of languages. ...   "

It is often said that the threatened languages are frequently surrounded by indifferent and unsympathetic insiders rather than the hostile outsiders (e.g., imposition of Hindi by the Government of India on non-Hindi-speaking populations). The need of the hour is not to wait for the governments to implement policies and provide resources to reverse the language shift and trajectory toward extinction. What is needed is grassroots movements to bring widespread awareness of the language and cultural endangerments, followed by strategy, tactics, and clean execution. In the next stage, this awareness should be steered toward influencing electoral outcomes and effecting government policies and resource allocations.

The confluence of handheld computing devices, machine learning, computational linguistics, social media, and native language enthusiasts is an unstoppable force for language revitalization. Telugu youth with computing knowledge and skills can play an extraordinary role by creating open-sourced language learning apps including cross-word puzzles, grade-appropriate vocabulary lists, modern dictionaries, and short stories [33]. If the native speakers of the Telugu language do not lead this effort, who else will? The time is now and any further delay will cause irreversible damage.

This work is supported in part by the National Science Foundation IUSE/PFE:RED award #1730568.

References

1. Eberhard, D.M., Simons, G.F., Fennig, C.D. (eds.): Ethnologue: Languages of the World, 22nd edn, SIL International, Dallas, Texas (2019). http://www.ethnologue.com
2. Lenneberg, E.H.: Biological Foundations of Language. Wiley, New York, NY (1967)
3. Miozzo, M., Rapp, B. (eds.): Biological Foundations of Language Production. Psychology Press, Special Issues of Language and Cognitive Processes (2011)
4. Ladefoged, P., Maddieson, I.: The Sounds of the World's languages. Wiley-Blackwell, New York, NY (2009)
5. Hale, K.: Endangered languages **68**, 1–42 (1992). https://doi.org/10.2307/416368
6. Fishma, J.A.: Reversing Language Shift: Theoretical and Empirical Foundations of Assistance to Threatened Languages. Multilingual Matters (1991)
7. United Nations Educational: Scientific and Cultural Organization: UNESCO's language vitality and endangerment methodological guideline: Review of application and feedback since **2003**, (2011)
8. Lewis, M.P., Simons, G.F.: Assessing endangerment: expanding fishman's GIDS. Revue roumaine de linguistique **LV** (2), 103–110 (2010). https://www.lingv.ro/RRL-2010.html
9. Lee, N.H., Way, J.V.: Assessing levels of endangerment in the Catalogue of Endangered Languages (ELCat) using the Language Endangerment Index (LEI). Lang. Soc. **5**, 271–292 (2016). https://doi.org/10.1017/S0047404515000962
10. Dwyer, A.M.: Tools and techniques for endangered-language assessment and revitalization. In: Vitality and Viability of Minority Languages, Trace Foundation Lecture Series Proceedings. New York, NY (2011). http://www.trace.org/about
11. Mihas, E., Perley, B., Rei-Doval, G., Wheatley, K. (eds.): Responses to Language Endangerment: In Honor of Mickey Noonan. Studies in Language Companion, John Benjamins (2013)
12. Lüpke, F.: Ideologies and typologies of language endangerment in Africa. In: Essegbey, J.A., Henderson, B., Mc Laughlin, F. (eds.) Language Documentation and Endangerment in Africa, Handbook of Statistics, Chap. 3, pp. 59–105. John Benjamins (2015). https://doi.org/10.1075/clu.17.03lup
13. Essegbey, J., Henderson, B., Laughlin, F.M. (eds.): Language Documentation and Endangerment in Africa. Culture and Language Use, John Benjamins (2015)

14. The world atlas of language structures (wals) (2013). http://wals.info/
15. Catalogue of Endangered Languages (ELCat) (2019). http://www. endangeredlanguages.com/
16. OLAC: The open language archives community (2019). http://olac.ldc.upenn.edu/
17. LLOD: linguistic linked open data (2019). http://linguistic-lod.org/
18. BBC News: Do you speak telugu? welcome to America (2018). https://www.bbc. com/news/world-45902204
19. of India, G.: Data on language and mother tongue (2011). http://censusindia.gov. in/2011Census/Language_MTs.html
20. What is mother tongue education? (2019). https://www.rutufoundation.org/what-is-mother-tongue-education/
21. Jain, T.: Common tongue: the impact of language on educational outcomes. J. Econ. Hist. **77**(2), 473–510 (2017). https://doi.org/10.1017/S0022050717000481
22. Noormohamadi, R.: Mother tongue, a necessary step to intellectual development. Pan-Pacific Association of Applied Linguistics **12**(2), 25–36 (2008). https://files. eric.ed.gov/fulltext/EJ921016.pdf
23. Savage, C.: The importance of mother tongue in education (2019). https://ie-today. co.uk/Blog/the-importance-of-mother-tongue-in-education/
24. Seid, Y.: Does learning in mother tongue matter? evidence from a natural experiment in Ethiopia. Econ. Educ. Rev. **55**, 21–38 (2016). https://doi.org/10.1016/j. econedurev.2016.08.006
25. Clark, A., Fox, C., Lappin, S.: The Handbook of Computational Linguistics and Natural Language Processing. Wiley-Blackwell, New York, NY (2012)
26. Barber, D.: Bayesian Reasoning and Machine Learning. Cambridge University Press, New York, NY (2012)
27. Faul, A.: A Concise Introduction to Machine Learning. Chapman and Hall/CRC, Boca Raton, Florida (2019)
28. Goldberg, Y.: Neural Network Methods in Natural Language Processing. Synthesis Lectures on Human Language Technologies. Morgan & Claypool, San Rafael, California (2017)
29. Ji, Q.: Probabilistic Graphical Models for Computer Vision. Academic Press, Cambridge, Massachusetts (2019)
30. Koller, D., Friedman, N.: Probabilistic Graphical Models: Principles and Techniques. The MIT Press, Cambridge, Massachusetts (2009)
31. Murphy, K.: Machine Learning: A Probabilistic Perspective. Adaptive Computation and Machine Learning. The MIT Press, Cambridge, Massachusetts (2012)
32. Economist, T.: Language identity in India: one state, many worlds, now what? (2013). https://www.economist.com/johnson/2013/06/25/one-state-many-worlds-now-what
33. Open-source software can revitalize indigenous languages (2019). https://en. unesco.org/news/open-source-software-can-revitalize-indigenous-languages-0

Entropy-Based Facial Movements Recognition Using CPVM

Vikram Kehri[1]([✉]), Digambar Puri[2], and R. N. Awale[1]

[1] Department of Electrical Engineering, VJTI, Mumbai, India
vakehri@el.vjt.ac.in, rnawale@vjti.el.ac.in
[2] Department of Electronics and Telecommunication, Dr. B.A.T.U, Lonere, India
digspuri@gmail.com

Abstract. This work presents a novel classifier called cascaded principle vector machine (CPVM) to fulfill the goal of high-accuracy facial expression recognition. First, the facial electromyogram (FEMG) signals are acquired by the two-channel wireless data acquisition device. Second, the adaptable signal processing algorithm is proposed for FEMG signals. In this algorithm, the discrete wavelet transform (DWT) is applied to FEMG signals for decomposing it into its frequency sub-bands. Statistical analysis is applied to these sub-bands to extract three unique features, namely Shannon entropy, Tsallis entropy, and Renyi entropy. Thirdly, the CPVM is implemented to classify these extracted features. The proposed CPVM is a combination of principal component analysis (PCA) and least-square support vector machine (LSSVM). By using PCA, both the goals of the dimensionality reduction of input features vector and the selection of discriminating features can be reached. Then, LSSVM combined with a one-against-one strategy is executed to classify features. The results show that the best 95% classification accuracy was achieved by the proposed CPVM classifier.

Keywords: Facial electromyogram (FEMG) · Wavelet transform (WT) · Cascaded principle vector machine (CDVM) · Linear discriminant analysis (LDA) · Least square support vector machine (LSSVM)

1 Introduction

Facial expression recognition plays a crucial role in designing mechatronic systems. FEMG signal classification has been widely accepted in many applications such as muscle–computer interaction (MuCI) [1]. For the patients with critical disabilities as a result of strokes, neuro-diseases, and muscular dystrophy, MuCI has been proposed as a promising way to improve the quality of their lives. Controlling assistive devices, such as hands-free wheelchairs, is an instance in this area. For designing such a system, strong human–computer interfaces are needed [2]. Recognizing the facial expression through bioelectrical action and transforming into control commands for the system have been the focus of this study. The main focus of this presented work in this area is on how to achieve the best facial expression recognition accuracy. Like other pattern recognition

© Springer Nature Singapore Pte Ltd. 2020
B. Iyer et al. (eds.), *Applied Computer Vision and Image Processing*,
Advances in Intelligent Systems and Computing 1155,
https://doi.org/10.1007/978-981-15-4029-5_2

problems, facial expression recognition challenges researchers with important issues such as feature selection and classifier design.

FEMGs are random in nature and are generated by a facial muscle action potential. FEMG is responsible for measuring facial muscle activity [3]. The amplitude range for FEMG signals differs from 0 to 12 mV and the frequency ranges from 0 to 480 Hz [4].

In the literature, various research works related to EMG classification have been proposed (e.g., [4–10]). Mostly in all the previous works, various EMG features were extracted, including different time-domain features such as IEMG and wavelength (integral EMG), VAR (variance), ZC (zero-crossing), and WAMP (Willison amplitude) [4–6]. The frequency-domain features are histogram, statistical features, and most powerful wavelet features [7–9]. Among all these features, sign change (SSC) feature was found to be insignificant.

Regarding the classifier, there are several classifiers implemented in different systems [4, 6–10]. Among them, Bayes classifier (BA), nonlinear discriminant analysis (NDA), and principal component analysis (PCA) belong to the statistical analysis, whereas KNN is based on instance-based learning. Recently, the advance classifier mostly belongs to intelligent approach, including fuzzy systems and artificial neural networks (ANN) [7, 11, 12].

Literature review has shown good work on the FEMG classification, but still, there remain few problems that need to be investigated. In most of the previous works, few FEMG patterns are available for classifier training and testing purposes, which directly affect the classification results [13]. The best classifier will produce good results even for the more number of subjects. In this paper, the FEMG signals are acquired from 30 different subjects. Therefore, the complexity of the feature transformation mechanism and the classification process was more difficult.

2   Materials and Methods

2.1   FEMG Signal Acquisition Protocol

The signal acquisition protocol for experimental work is approved by the ethical committee formed by the Department of Electrical, VJTI, Mumbai. The FEMG signals were recorded by two-channel myon-made aktos-mini wireless data acquisition device, as shown in Fig. 1. The FEMG signals were recorded from 30 participants (20 males and 10 females) in the age group 18–40. The participants were requested to perform three unique facial expressions. The three expressions considered are a voluntary smile, lip pucker, and frown movement tasks. The correct position for surface electrode placement plays an important work, in order to acquire the meaningful FEMG signals [14]. The muscles are assessed in this work, including zygomaticus major, corrugator supercilii, and lateral frontalis. Each expression was recorded for 5 s of duration and performed twice by all the participants. Hence for each expression, ten (5 × 2)-second informative data is recorded. For three facial movements of each subject, 20,000 datasets (2 [No. of channels] × 10 s [informative signal] × 1000 [sampling frequency]) were recorded.

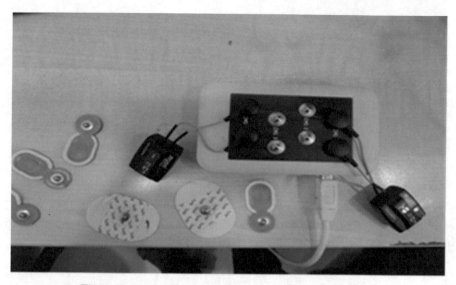

Fig. 1 The myon-aktos-mini wireless data acquisition system

2.2 Features Extraction

DWT is applied to FEMG datasets for decomposing them into their frequency sub-bands. Depending on the choice of a mother wavelet function, the value of the wavelet coefficients is generated [15]. In this work, we have applied a widely used symlet (sym3) wavelet corresponding to the actual FEMG. The decomposition level is restricted to four in order to have a reasonable computational complexity. Statistical analysis is applied to these sub-bands to extract the FEMG features, which characterize the distribution of the wavelet coefficient. Three different features, namely Shannon entropy (SE), Tsallis entropy (TE), and Renyi entropy (RE), are to be considered for FEMG classification.

2.2.1 Shannon Entropy Shannon entropy is basically used to estimate the entropy of probability density distributions around some limits. The entropy can provide additional information about the importance of specific datasets [16]. Shannon entropy is defined as the measures of the randomness in the information being processed and determined by:

$$E = -\sum_{i=1}^{n} w_i \log_2 w_i \tag{1}$$

Here w_i represents the wavelet coefficients at the ith level of decomposition and E represents entropy of FEMG datasets.

2.2.2 Tsallis Entropy Tsallis entropy contains additional parameter β which used to make it more or less sensitive to the shape of probability distributions. If the parameter is

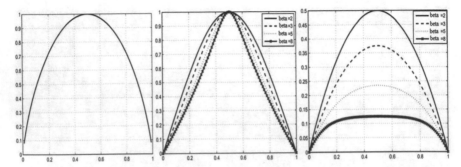

Fig. 2 A plot of Shannon, Renyi and Tsallis entropies for several positive values of β

set to one ($\beta = 1$), then Tsallis entropy represents a similar shape as of Shannon entropy shown in Fig. 2. The Tsallis entropy is defined as [16]:

$$E_T = \frac{1}{1 - \beta}\left(1 - \sum_{i=1}^{n} w_i^{\beta}\right) \tag{2}$$

In Eq. 2, w_i represents the wavelet coefficients at the ith level of decomposition and E_T represents Tsallis entropy of FEMG datasets.

2.2.3 Renyi Entropy The Renyi entropy has similar properties as the Shannon entropy but contains additional parameter β. The Renyi entropy is defined as:

$$E_R = \frac{1}{1 - \beta} \log\left(\sum_{i=1}^{n} w_i^{\beta}\right) \tag{3}$$

Comparison of Shannon, Renyi, and Tsallis entropies for two probabilities $p1$ and $p2$ where $p1 = 1 - p2$ is depicted in Fig. 2. Figure 2 presents different plots of Renyi and Tsallis entropies for several positive values of β [16].

3 Proposed CPVM Classifier

CPVM is a cascaded architecture of PCA, also called the principal component analysis and the least square support vector machine (LSSVM). The proposed architecture of the CPVM classifier is a combination of two layers, first is PCA and the second is LSSVM. CPVM receives extracted features from FEMG datasets and classifies it to correct expression class.

3.1 Principle Component Analysis

PCA is a technique of extracting important components from a large set of FEMG data. PCA is more useful when dealing with random data [17]. It extracts the set of new variables from large sets of FEMG datasets, and these new variables are defined as principal components (PC). The maximum variance in the FEMG dataset represents the first PC. The second PC extracted from the remaining variance in the dataset is uncorrelated to the first PC. The third PC tries to define the variance which is not explained by the first two PC, and so on [18]. All succeeding PC analysis follows the same methodology, that is, they extract the remaining variation without being correlated with the previous component. In general, for $(q \times n)$ dimensional data, min $(q - 1, n)$ principle component can be extracted.

3.2 LSSVM

Suykens and Vandewalle have proposed the concept of LSSVM classifiers [11]. The methodology of LSSVM classifier is a kernel-based learning machine very similar to SVM classifier. The advantage of LSSVM over SVM is that it improves the process during testing and training stage. LSSVM algorithms mainly depend on an alternate mechanism of SVM models suggested in [19]. Assume a given training dataset of M data points $\{w_l, z_l\}_{l=1}^{M}$ with input data, $k_l \in R^M$ and output data $z_l \in r$ where R^M is the M-dimensional vector space and r is the single-dimensional vector space [20]. Three input variables are used for an LSSVM model in this study. Hence, in this work, we take $k = [a, b, c]$ and $z = f$. In the feature space

$$z(k) = w^T \beta(k) + p \tag{4}$$

where β $(\cdot)$ maps the input data into a feature space $w \in R^M$, $b \in r$, w is the weight vector, and p is the threshold in Eq. 6. The given optimization problem is:

$$\text{Minimize } \frac{1}{2w} \cdot w^T w + \Upsilon \frac{1}{2} \sum_{c=1}^{M} e_c^2 \tag{5}$$

subject to

$$z(k) = w^T \beta(k_c) + p + e_c; \ c = 1, 2, 3 \ldots M \tag{6}$$

Here f is obtained by solving Eq. 7.

$$f = z(k) = \sum_{c=1}^{M} \lambda_c C(k, k_c) + p \tag{7}$$

In Eq. 7, $C(k, k_c)$ represents the kernel function of the LSSVM classifier [20]. Table 1 depicts the types of kernel function with their mathematical equations [21].

4 Results and Discussion

The most important parameter to estimate the CPVM performance is classification accuracy. The proposed CPVM classifier classifies this expression into appropriate class. The redundancy in FEMG features sets is reduced by implementing the PCA method [22]. After that, the LSSVM classifier was classified according to these extracted features. The different kernel functions depicted in Table 1 are implemented in order to determine the best one.

The steps for the proposed work are as follows:

Table 1 Types of kernel functions with their mathematical equation

S. no.	Types of kernel functions	Equations
1	Polynomial kernels	$C(k, k_c) = (k * k_c + 1)^d$ where d is the degree of the polynomial
2	Gaussian kernel	$C(k, k_c) = e^{(-\frac{\|k-k_c\|^2}{2\sigma^2})}$
3	Gaussian radial basis function (RBF)	$C(k, k_c) = e^{(-\gamma\|k-k_c\|^2)}$
4	Laplace RBF kernel	$C(k, k_c) = e^{(-\frac{\|k-k_c\|}{\sigma})}$

- Wavelet coefficient is calculated from FEMG datasets using DWT method.
- FEMG features such as SN, TN, and RN are estimated using a decomposed wavelet coefficient.
- These extracted features are classified by the proposed CPVM classifier.
- First, PCA reduces most of the redundant data from FEMG features sets.
- The classification process was carried out using LSSVM-based classifier by using different kernel functions.

Table 2 depicts the value of Renyi, Tsallis (for a different value of β), and Shannon entropies, calculated through decomposed wavelet coefficients. In Table 2, three different classes represent three different facial expressions. An extensive range of kernel function parameter values is examined in order to find the best results and selected after several trials. Finally, the CPVM classifier was designed.

Table 2 Renyi, Tsallis, and Shannon entropies extracted from decomposed wavelet coefficients

	Class	Beta (β)						
		0.9	1.3	1.5	2.0	3.0	4.0	5.0
Renyi entropy	1	0.5643	0.5433	0.5211	0.5022	0.4895	0.4233	0.4022
	2	0.6604	0.6321	0.6123	0.6022	0.5632	0.5164	0.4966
	3	0.7062	0.7033	0.6987	0.6921	0.6521	0.6322	0.5965
Tsallis entropy	1	0.7206	0.6139	0.5974	0.4233	0.3211	0.3165	0.3022
	2	0.5260	0.4744	0.4023	0.3498	0.3037	0.2983	0.2893
	3	0.6589	0.5641	0.5038	0.4269	0.3611	0.3254	0.3022
Shannon entropy	1	0.931						
	2	0.765						
	3	0.877						

Overall, it is obvious that the maximum (95%) and minimum (75%) classification accuracy is obtained by CPVM through RBF and polynomial kernel function, respectively. Laplace RBF and Gaussian kernel function also exceed 90% accuracy, in order to classify the extracted features. According to the results depicted in Table 3, it is concluded that the best kernel function for the CPVM classifier in terms of classification accuracy is RBF. In all the experiments, cross-validation Type-Hold out "80-20" type has been implemented. Figure 3 shows the confusion matrix representing 95% classification accuracy with the CPVM RBF kernel function. Figure 4 depicts the confusion matrix representing 93.3% classification accuracy with the CPVM Gaussian RBF kernel function. Figure 5 shows the confusion matrix representing 91.7% classification accuracy with the CPVM Laplace RBF kernel function. Figure 6 shows the confusion matrix representing 75% classification accuracy with the CPVM polynomial kernel function.

Table 3 Comparison of classification accuracy to recognize three different facial expressions through CPVM with a different kernel function

S. no.	CPVM kernel functions	Classification accuracy (%)
1	Polynomial kernels	75
2	Gaussian kernel	93.3
3	Gaussian RBF kernel	95
4	Laplace RBF kernel	91.7

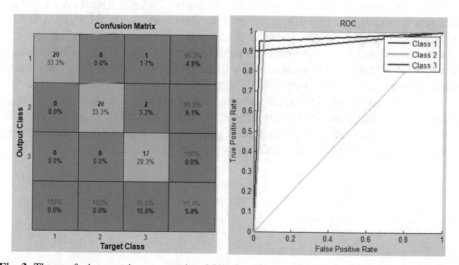

Fig. 3 The confusion matrix representing 95% classification accuracy of CPVM with the RBF kernel function and its ROC plot

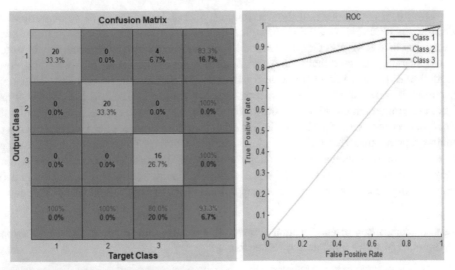

Fig. 4 The confusion matrix representing 93.3% classification accuracy of CPVM with the Gaussian kernel function and its ROC plot

5 Conclusion

This paper presents a novel approach for recognizing three different facial expressions. The signals were acquired by the two-channel wireless device from 30 different subjects. For recognition of different facial expressions, three different features were extracted from decomposed wavelet coefficients: Shannon entropy, Tsallis entropy, and Renyi entropy. As discussed, the CPVM classifier with four different kernel function algorithms was executed for expression recognition. The performance of CPVM with RBF is reasonably better than other combinations. Based on training and testing accuracy results, the proposed technique provides better accuracy (95%) in less training time. It was concluded that FEMG patterns are distributed as Gaussian forms where parameters can be optimally considered by the RBF algorithm.

The proposed study discussed in this paper used as an interface in the muscles—computer system to design artificial devices like an intelligent wheelchair [23–25]. This study can also be implemented to propose a voice signal recognition device for communicating with voiceless people based on facial expression [26].

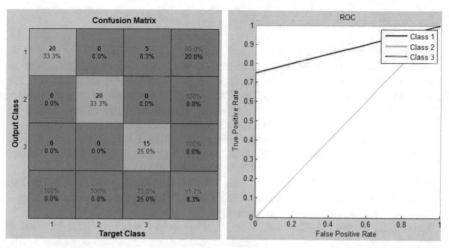

Fig. 5 The confusion matrix representing 91.7% classification accuracy of CPVM with the Laplace RBF kernel function and its ROC plot

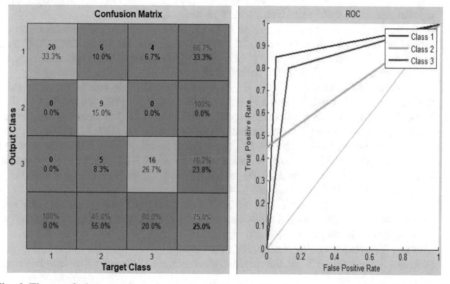

Fig. 6 The confusion matrix representing 75% classification accuracy of CPVM with the polynomial kernel function and its ROC plot

References

1. Ferreira, A., Silva, R.L., Celeste, W.C., Bastos Filho, T.F., Sarcinelli Filho, M.: Human–machine interface based on muscular and brain signals applied to a robotic wheelchair. J. Phys. Conf. Ser. **90**(1) (2007). https://doi.org/10.1088/1742-6596/90/1/012094
2. Hamedi, M., Salleh, S., Ting, C., Astaraki, M., Noor, A.M.: Robust facial expression recognition for MuCI: a comprehensive neuromuscular signal analysis. IEEE Trans. Affect. Comput. **9**(1), 102–115 (2018). https://doi.org/10.1109/taffc.2016.2569098
3. Kehri, V., Ingle, R., Awale, R.N.: Analysis of facial EMG signal for emotion recognition using wavelet packet transform and SVM. In: International Conference on Machine Intelligence and Signal Processing (MISP 2017). IIT-Indore, India (2017)
4. Hamedi, M., Salleh, S.-H., Noor, A.M., Tan, T.S., Kamarul, A.: Comparison of different time-domain feature extraction methods on facial gestures' EMGs. In: Progress in Electromagnetics Research Symposium Proceedings, March, pp. 1897–1900 (2012)
5. Tamura, H., Manabe, T., Tanno, K., Fuse, Y.: The electric wheelchair control system using surface-electromyogram of facial muscles. In: World Automation Congress (WAC) (2010)
6. Wei, L., Hu, H.: EMG and visual based MUCI for hands-free control of an intelligent wheelchair. In: 8th World Congress on Intelligent Control and Automation, pp. 1027–1032 (2010)
7. Yang, G., Yang, S.: Emotion recognition of electromyography based on support vector computer. In: 3rd International Symposium on Intelligent Information Technology and Security Informatics, vol. 2, pp. 298–301 (2010)
8. Kehri, V., Awale, R.N.: EMG signal analysis for diagnosis of muscular dystrophy using wavelet transform, SVM and ANN. Biomed. Pharmacol. J. (2018)
9. Arjunan, S.P., Kumar, D.K.: Recognition of facial movements and hand gestures using surface electromyogram (SEMG) for MuCI based applications. In: 9th Biennial Conference of the Australian Pattern Recognition Society on Digital Image Computing Techniques and Applications, pp. 1–6. Glenelg, Australia (2008)
10. Firoozabadi, M., Oskoei, M.A., Hu, H.: A human-computer interface based on forehead multichannel biosignals to control a virtual wheelchair. In: ICBME08. Tehran, Iran (2008)
11. Yang, Y.G., Yang, S.: Study of emotion recognition based on surface electromyography and improved least squares support vector machine. J. Comput. **6**(8), 1707–1714 (2011)
12. Hamedi, M., Salleh, S.-H., Astaraki, M., Noor, A.M., Harris, A.R.A.: Comparison of multilayer perceptron and radial basis function neural networks for EMG-based facial gesture recognition. In: 8th International Conference on Robotic, Vision, Signal Processing & Power Applications, pp. 285–294. Singapore (2014)
13. Vikram, K., Ingle, R., Thulkar, D., Awale, R.N.: A machine learning approach for FEMG pattern recognition system to control hands-free wheelchair (TRIBOINDIA). In: An International Conference on Tribology (2018)
14. Thulkar, D., Bhaskarwar, T., Hamde, S.T.: Facial electromyography for characterization of emotions using LabVIEW. In: International Conference on Industrial Instrumentation and Control (ICIC), Pune (2015)
15. Englehart, K., Hudgin, B., Parker, P.A.: A wavelet-based continuous classification scheme for multifunction myoelectric control. IEEE Trans. Biomed. Eng. **48**(3), 302–311 (2001). https://doi.org/10.1109/10.914793
16. Maszczyk, T., Duch, W.: Comparison of Shannon, Renyi and Tsallis entropy used in decision trees. ICAISC (2006)
17. Ingle, R., Oimbe, S., et al.: Classification of EEG signals during meditation and controlled state using PCA, ICA, LDA and support vector machines. Int. J. Pure Appl. Math. (2018)

18. Chen, Y., Yang, Z., Wang, J.: Eyebrow emotional recognition using surface EMG signals. Neurocomputing **168**, 871–879 (2015)
19. Hamedi, M., Salleh, S.-H., Noor, A.M., Harris, A.R.A., Majid, N.A.: Multiclass least-square support vector computer for myoelcctric-based facial gesture recognition. In: 15th International Conference on Biomedical Engineering, Jan. 2014, pp. 180–183
20. Suykens, J.A.K., Vandewalle, J.: Least squares support vector machine classifiers. Neural Process. Lett. **9**(3), 293–300 (1999)
21. Hamedi, M., Salleh, S.-H., Noor, A.M.: Facial neuromuscular signal classification by means of least square support vector machine for MuCI. J. Soft Comput. **30**, 83–93 (2015)
22. Gruebler, A., Suzuki, K.: A wearable interface for reading facial expressions based on bioelectrical signals. In: International Conference on Kansei Engineering and Emotional Research, March 2010
23. Wei, L., Hu, H., Zhang, Y.: Fusing EMG and visual data for hands-free control of an intelligent wheelchair. Int. J. Humanoid Rob. **8**(4), 707–724 (2011)
24. Tsui, C., Jia, P., Gan, J.Q., Hu, H., Yuan, K.: EMG-based hands-free wheelchair control with EOG attention shift detection. In: IEEE International Conference on Robotics and Biomimetics (ROBIO 2007), pp. 1266–1271
25. Gibert, G., Pruzinec, M., Schultz, T., Stevens, K.: Enhancement of human-computer interaction with facial electromyographic sensors. In: Proceedings of the 21st Annual Conference of the Australian Computer-Human Interaction Special Interest Group on Design Open, Melborn, Australia, 2009, ACM Press, pp. 1–4
26. Gruebler, A., Suzuki, K.: Design of a wearable device for reading positive expressions from facial EMG signals. IEEE Trans Affect. Comp. **5**(3), 227–237 (2014)

A Novel Approach of Image Processing to Identify Type and Name of Snake

Mohini Niraj Sheth$^{(\boxtimes)}$ and Sanjay L. Nalbalwar

Dr. Babasaheb, Ambedkar Technological University, Lonere-Raigad 402103, MS, India
mohiniraj@dbatu.ac.in, nalbalwar_sanjayn@yahoo.com

Abstract. Snake is an important part of our flora and fauna system. Most of snake species are found in nearby forest area. As we know that majority of people of Maharashtra are living in villages, snakes might enter into residential area; people brutally kill them out of fear as there is lack of awareness about snake species. Saving snakes is important as most of the snake species are under extinction. Though different anti-venoms are available for different snake bites, one must be aware of different snake species (venomous/semi-venomous/non-venomous) and their names so that a person can get proper first aid. Many people are curious about snake species; we are living in a digital world, so by simply taking picture of that snake one must get the basic information about that specific snake. In this project we have developed a system which uses PCA and different distance algorithms in MATLAB to identify the type and present the basic information about snake on liquid crystal display using Arduino Uno. This system will be helpful for study of snake, saving them for extinction and getting the basic information about snakes.

Keywords: Principal component analysis · Manhattan distance · MATLAB

1 Introduction

As we know that the main source of income of people in Maharashtra is farming and day-by-day population is increasing at a high rate, so providing shelter for all is a major challenge. As a result, people live near forest area to fulfill their basic requirements. Natural habitats are reducing due to which they find their own way in nearby residential areas which may be risky. In rural areas there are many misconceptions about snakes. Therefore, it is essential to create awareness about snakes. If someone is interested in knowing the basic information about snakes and as we are living in modern era to provide information in less time with more accuracy, we are implementing a system which requires image of that specific snake. By simply giving the image of a snake to our system, he/she would get the name and type of that specific snake.

For this system various snake images are required. Collecting those images is a challenge as it requires well-trained snake catchers. The images required for our system are obtained from different parts of Raigad District of Maharashtra state with the help of snake catchers and internet (indiansnakeorg.com). Nearly 14 various snake species have been found in Raigad District. Out of them two species are semi-venomous, namely

© Springer Nature Singapore Pte Ltd. 2020
B. Iyer et al. (eds.), *Applied Computer Vision and Image Processing*,
Advances in Intelligent Systems and Computing 1155,
https://doi.org/10.1007/978-981-15-4029-5_3

common vine snake and common cat snake. Number of images used for this category is 25. Around eight species are non-venomous, namely beaked worm snake, checked keelback, Elliot shield tail, common kukri, common sand boa, common trinket and Indian rock python. Number of images used for this category is 50. Nearly four species are venomous, namely Russel's Viper, Common Krait, Najanaja and Saw Scaled Viper. Number of images used for this category is 25.

2 Related Studies

Many researchers published their research on classification, detection and identification of animal category and bird category from the images. In 2012, Alexander Loos and Andreas Ernst published research paper entitled, 'Detection and identification of Chimpanzee faces in the wild' [1] and they used space representation classification method. Andreia Marini and Jacques Facon published paper in 2013 based on color histogram using SIFT algorithm entitled, 'Bird species classification based on color feature' [2]. Principal component analysis (PCA) is one of the promising methods for feature extraction [3]. Lian Li and Jinqi Hong used PCA algorithm for feature extraction and Fisher discriminate and Mahalanobis distance for identification of species. They published their work 'Identification of fish species based on image processing and statistical analysis research' in 2014 [4]. In 2014, Santosh Kumar and Sanjay Kumar Singh used PCA, LDA, ICA, local binary pattern methods in their work 'Biometric recognition for pet animal' [5] for snake classification from image. Alex James extracted taxonomical relevant features and used nearest neighbor classifier for classification in 2017 [6].

3 Methodology

We are proposing a system, which displays the name and type of the snake. This system has two main parts, that is, image processing module and another is embedded module.

Image processing module consists of image pre-possessing, feature extraction and finding measure of similarities; all these processes are carried out in MATLAB. Embedded module consists of Arduino Uno board and LCD. Input for image processing module is test snake image, and the desired output will display on LCD (Fig. 1).

3.1 Image Pre-processing

For enhancing image quality, reducing size of image and converting image into gray-scale image, we perform some image pre-processing techniques in MATLAB.

3.2 Feature Extraction

For extraction of features, simple principal component analysis (PCA) algorithm is used. PCA is a technique that can be used to simplify the dataset. Principal component analysis requires centralized data, that is, mean subtracted data. To obtain centralized data, we have subtracted row mean from image data. Subtracting the mean is equivalent to

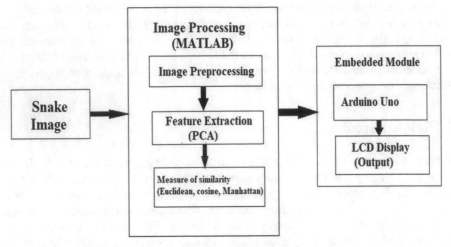

Fig. 1 Block diagram of the system

translating coordinate system to location of mean. To get eigenvalues of mean subtracted data, scattering matrix is calculated. For principal component, that is, feature vector of higher eigenvalues, we have sorted the eigenvalues in descending order. All these feature vectors of training images are stored for finding minimum distance with feature vector of test image.

3.3   Measure of Similarities

To find the measure of similarities between feature vector of training and test image, we are using cosine distance, Euclidean distance and Manhattan distance algorithm. The image which has less distance sends the information to Arduino Uno and displays name and type of that snake on liquid crystal display. Cosine distance gives a metric of similarity between two vectors. It measures cosine of angle between two vectors. The Euclidean distance is the straight-line distance between two pixels [7]. It is square root of sum of difference between corresponding elements of two vectors. Manhattan distance is the measure of distance between two vertices. It is sum of difference of corresponding element of two vectors.

3.4   Arduino Uno and LCD

Arduino Uno is programmed in such way that it accepts input from MATLAB and gives information to liquid crystal display for display name and type of snake.

4   Algorithm

1. Read the image and perform image pre-processing technique
2. Compute row mean and subtract it from image data

3. To find the eigenvalue, first calculate scattering matrix of centralized data.
4. Sort the eigenvalue in descending order
5. Compute feature vector of image.
6. For getting the name and the type of snake we find out measure of similarities between feature vector of training image and test image.
7. The image which has minimum distance is sent to Arduino to display name and type of snake on LCD.

5 Discussion about Result

Having calculated the feature vectors and distance between the two feature vectors, the training image closest to the given test image is returned as a result of query. So, we decide to use PCA for feature extraction, and cosine, Euclidean and Manhattan distance to measure similarity between images. If the subject of the test image and the subject of the training image closest to the given test image are the same, then a correct match is said to have occurred or else it is taken as an incorrect match.

The above approach is tested on the whole database and we got different values. To decide threshold value for different distance algorithms, we used trial-and-error method. For getting more accurate results we decided threshold value as:

For Euclidean distance the threshold value is **25**.
For Manhattan distance the threshold value is **210**.
For cosine distance the threshold value is **150** (Table 1).

Table 1 Comparison between distance measuring techniques

Technique used for identification	No. of total samples	No. of accurate identification	Accuracy rate (%)
PCA + cosine distance	86	15	17.44
PCA + Euclidean distance	86	20	23.26
PCA + Manhattan distance	86	60	69.77

From Table 1, it is clear that we get more accurate results using PCA + Manhattan distance.

The results below mentioned are obtained using PCA algorithm for feature extraction and to measure the similarity we calculated using Manhattan distance. For displaying the type of snake, we use 'N' for non-venomous type of snake. 'S' for semi-venomous type of snakes. 'V' for venomous type of snake (Table 2).

Table 2 Output of the system

Input image	Output display on LCD	Description of output
	N Common Kukri.	The name of snake is Common Kukri and the type of snake is non-venomous
	S cat.	The name of snake is Common Cat and the type of snake is semi –venomous
	V naja7.	The name of snake is Naja Naja and the type of snake is venomous

6   Conclusion

A simple PCA algorithm is used in this system for feature extraction. After finding minimum Manhattan distance we got the exact name and the type of snake. From this, it can be concluded that this system can be used for better understanding of snake species. Unnecessary killing of snake can be stopped. Identification of type of snake will also help to decide first-aid in case of snake bite. Apart from this, the present work is useful for snake catchers and people who study snakes.

Alex James gathered data from five types of venomous snakes for his work. From around 207 snake images he extracted 31 taxonomical relevant features and for classification KNN classifier is used. However, this study used data of 15 snakes among which 8 were non-venomous, 2 were semi-venomous and remaining 4 were venomous type of snakes. Most of them are found in Raigad district of Maharashtra state. For feature extraction of 86 images, the author used PCA algorithm, and for identifying snake from images Manhattan distance is used.

In future for extraction of feature and classification, we can use SVM and neural network. We can add more snake images in our database. We can display basic information which includes anti-venom. Furthermore, we can make an Android application of this system. For real time we can use different sensors for detection of snake.

References

1. Loos, A., Ernst, A.: Detection and identification of chimpanzee faces in the wild. In: 2012 IEEE International Symposium on Multimedia. 978-0-7695-4875-3/12, IEEE. https://doi.org/10.1109/ism.2012.30 (2012)
2. Marini, A., Facon, J.: Bird species classification based on color feature. In: IEEE International Conference on Systems, Man, and Cybernetics (2013)
3. Patil, M.N., Iyer, B., Arya, R.: Performance evaluation of PCA and ICA algorithm for facial expression recognition application. In: Proceedings of Fifth International Conference on Soft Computing for Problem Solving. Advances in Intelligent Systems and Computing, vol. 436, pp. 965–976. Springer, Singapore (2016)
4. Li, L., Hong, J.: Identification of fish species based on image processing and statistical analysis research. In: IEEE International Conference on Mechatronics and Automation Tianjin, China (2014)
5. Kumar, S., Singh, S.K.: Biometric recognition for pet animal. J. Soft. Eng. Appl. 7, 470–482 (2014)
6. James, A.: Snake classification from images. PeerJ. Preprints https://doi.org/10.7287/peerj.preprints.2867v1. CC BY 4.0 Open Access (2017)
7. Kumar, T.N.R.: A real time approach for indian road analysis using image processing and computer vision. IOSR J. Comput. Eng. (IOSR-JCE), 17(4). e-ISSN: 2278-0661, p-ISSN: 2278-8727, Ver. III (2015)

IMEXT Text Summarizer Using Deep Learning

R. Menaka$^{(\boxtimes)}$, Jeet Ketan Thaker, Rounak Bhushan, and R. Karthik

School of Electronics Engineering, Vellore Institute of Technology, Chennai, India
menaka.r@vit.ac.in

Abstract. This research focuses on the problem of creating a short summary for every page in a physical book. This will enable students, professors, lawyers and others to get the most important points of every page in the book in a summarized form. Motivated by the fact that users do not have the time to read lengthy materials, the research aims to make reading a less tedious activity by summarizing long paragraphs into short sentences. In this research, we model the summarizer using deep learning algorithms using bi-RNNs and attention mechanisms. Furthermore, a hardware model is made to turn and capture images of pages from a book automatically using raspberry pi.

Keywords: Abstractive text summarization · Bi-RNN · Attention · News dataset · Raspberry pi · Page turner · Image to summary · Seq2seq · Word embeddings

1 Introduction

In the modern digital age, there is an abundance of information available to each and every one of us. The irony of the situation is that in this age people don't have enough time to read the amount of information available. The data over the internet is unstructured and in order to get the information required, one has to go through numerous searches and skim the results manually. This problem is even more cumbersome when it comes to the case of a physical book. In order for people to navigate through huge amount of data easily, there is a great need for a focused summary that encapsulates the prominent details of the text document. Furthermore, summaries are viable because of the following features [1]: the reading time is effectively reduced by summaries; during documents researches, selection process is made easier; effectiveness of indexing is improved by automatic summarization; the algorithms of automatic summarizers are less biased than human summarizers; personalized summaries are beneficial in query answering systems as personalized information is provided.

Automatic text summarization is classified broadly as either extractive or abstractive. In extractive summarization, the algorithm picks out the most important lines from the input paragraph and calls it a summary. These are the systems which generate summaries by copying parts from the document by deploying various measures of importance and then combine those parts together to present a summary. The important rank of the sentence is based on the linguistic and statistical features. In abstractive summarization, the

© Springer Nature Singapore Pte Ltd. 2020
B. Iyer et al. (eds.), *Applied Computer Vision and Image Processing*,
Advances in Intelligent Systems and Computing 1155,
https://doi.org/10.1007/978-981-15-4029-5_4

algorithm understands the paragraph and tries to make a novel summary using phrases and making up of sentences not found in the input passage. These are the systems which generate new phrases, either by rephrasing or using words that were not in the original text. Abstractive approaches for summarization are harder and complex. To obtain a perfect abstractive summary, the model needs to first completely understand the document and then try to express that understood context in short, by using new words and phrases. This method has complex capabilities of generalization, paraphrasing and to suitably incorporate real-world knowledge. Abstractive methods are usually more concise and human-like, whereas extractive methods do not fully summarize but rather shorten the given paragraph. Genism [2] is one such open-source library for unsupervised modeling of topics and natural language processing, which uses present-day statistical machine learning. This popular library uses extractive techniques to summarize documents. In this research, though, abstractive summarization technique will be used as it provides a much more generalized summary which is easier to read and digest.

There are different researches that talk about abstractive summarization but the primary motive of these researches has been on the algorithm for summary generation. But their work doesn't equip people to summarize material from a physical book. The primary users who are in need of summarization are students, professors, medical practitioners and so on, who mainly rely on physical books and the traditional way of reading. Hence there is a need to bridge this gap and equip this section of people with a summarization tool by integrating it with a hardware model. In this research a new algorithm for abstractive summarization is discussed; also a hardware part is added that will make the process for the end user easier. The hardware comprises a page turning and image-capturing mechanism which will convert physical books into digital ones and each image will be processed and turned to a summary, so people even have the gist of a physical book. An effort to improve over the existing sequence to sequence model [3] is made with the introduction of a bi-directional recurrent neural network in the encoder part of the network. Furthermore, an attention mechanism [4] is used at the top of the encoder layer to form the best representation of context vectors.

The structure of the further contents of this research is as follows. In Sect. 2, related works in the field of text summarization are named. In Sect. 3, the methodology for IMEXT summarizer is discussed along with the hardware implementation of the project. Section 4 comprises the experimental results obtained using this summarization algorithm and a comparison is drawn between summaries generated from the short and long articles. Finally, in Sect. 5, the total work done in the research is concluded, and a discussion about future works of this project is made.

2 Related Works

Owing to the fact that literary data as computerized archives rapidly collect to enormous measures of information and the recent boom in deep learning technologies, the research topics on automatic text summarization have attracted a lot of attention recently. In addition to the efforts of this research, previous works have made lots of efforts deploying various algorithms to summarize texts. The different approaches researchers have used can be broadly classified into five categories as: (1) *rule-based approaches*, (2) *iterative-based approaches*, (3) *deep reinforcement approaches*, (4) *graph-based approaches* and (5) *deep supervised learning approaches.*

3   IMEXT Summarizer Methodology

In this work we propose a system that lets users summarize physical books. The methodology is broken down into two different sections: one is the hardware workflow and the other section explains the algorithm used to summarize input passages.

3.1   Hardware Methodology

A page turning device and an image-capturing mechanism is introduced in this section. The idea is to make a model that will automatically turn pages of a book capturing images of these pages. Once an image is captured it needs to be converted into machine-readable text. One popular method is to use optical character recognition (OCR) to convert images into texts. This research uses the python library PyTesseract to handle the OCR. This library expects the input images to be crisp and have a clear distinction between words and background. To achieve this, the image is first converted to grayscale to remove color from it. This gray-scaled image is then binary thresholded to give out an output image that is suited for PyTesseract. The image processing part is done with OpenCV 2.4.13. Figure 1 shows the block diagram of the proposed hardware setup. The project uses a wheel connected to a 9 V DC motor positioned at the top corner of the book.

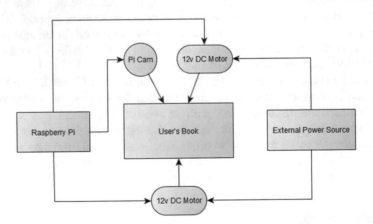

Fig. 1 Proposed model for a page turner

When the motor gets power the wheel turns, raising up the page with it. The amount of time the motor runs is set by the raspberry pi. There is another 9 V DC motor positioned at the middle-bottom of the book with a rigid object (e.g. a pencil) attached to it. Once Motor 1 raises a page of the book, it stops and waits for Motor 2. This motor now starts functioning and turns the page. There is a pi-camera attached to this setup which captures images of each page and is in sync with the two motors. The timing is set such that no component clashes with another.

Motor 1 is run for 0.4 s which raises the page of the book by the right amount. After a wait of 1 s, Motor 2 is run for 3.63 s ensuring the page gets turned properly. After another wait of 1 s, pi-camera takes an image of the new page and stores it to memory. Figures 2 and 3 depict the hardware implementation of the proposed model.

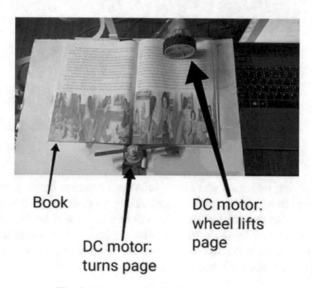

Book   **DC motor:
              wheel lifts**

DC motor: page

turns page

Fig. 2 Components to assist page turning

3.2 Algorithm for IMEXT Summarizer

The approach used for summarizing in this research is the deep supervised learning approach. The dataset used for learning is, "The Signal Media One-Million News Articles Dataset [5]." The dataset contains a collection of a variety of news articles and headlines approximately equaling one million that are mainly in English. The articles in this dataset are treated as passages and their corresponding headings are regarded as their summaries. The aim of the algorithm is to create an appropriate headline to a random article given from this dataset which captures the main gist of the article. The structure used for achieving this aim is a sequence-to-sequence architecture containing an encoder–decoder network as inspired by [3] with an attention layer stacked on the top of the encoder network which has shown to increase performance and decrease computational computations [4]. The main difference between our algorithm and the other algorithms is the way the encoder network is set up. Instead of using a one-directional RNN as used in [6], this research uses a bi-directional RNN network with LSTM cells. The motivation for using a bi-RNN is due to the reasoning that when humans try to summarize an article they do not look only at the current and previous words to make a decision on which words are to be used in the summary. Humans read the entire passage

Fig. 3 Pi-camera to scan pages

first, thereby gaining both past and future information which helps in making a better decision as to which words to choose in generating a summary. Bi-RNNs will give the same opportunities to the network, although the trade-off will be the added computational complexity gained by the network. With the increase in computing power seen in recent times, this trade-off will diminish. The following sections will explain the different components of the algorithm in detail.

3.2.1 Pre-processing of the dataset The first step is to obtain and pre-process the dataset. The dataset contains articles and headings in a text format, for the machine to understand individual words and different sentences. These articles and headlines have to be converted into numbers. The process of converting words into relevant vectors that encapsulate some meaning of the word is called word embedding. But before creating word embeddings, the dataset needs to be cleaned up. The algorithm used for summarizing is computationally expensive and the project is run on *Google Colab*, so we first filter out smaller articles and headings. The max length of the article is kept at 116 words and the max length of headings is kept at 13 words. So, all articles with length greater than 116 and headings with length greater than 13 are discarded. The same process is later repeated to obtain larger article sizes and see the performance of the network based on the length of articles and headings. The dataset now reduces to 5190 articles and headlines where the max lengths of these articles and headings are 116 and 13, respectively. The texts are then lowercased and fullstops, commas, exclamation marks, question marks, quotes, semicolons, colons and so on are removed. The dataset contained a lot of "\n", "\xa0", "\r" and "\t" characters which are also removed and replaced with a space. To convert these articles into word vectors, a pre-trained GloVe model [7] is used. GloVe [8] stands for global vectors, which is a count-based approach toward learning word embeddings. The dimensions of these word vectors are fixed as 100. GloVe gives us vector representation of various words. We find all the words used in our dataset by using *TreeBankWordTokenizer* on the dataset. This technique tokenizes articles and headings into words, and the process takes into consideration English grammar rules while tokenizing. Some words with the same meaning like "cat" and "cats" appear in the dataset. To normalize this redundancy, *Lemmatization* process is used on every word. This normalizes every word by bringing words to their root forms. All these words are given a code (rank) according to their frequency in the dataset with the most common word given a code of 0, the second most common word given a code of 1 and so on. Words in GloVe are parsed according to this coding scheme, and for each word in the dataset, the value of that word is added to an embedding matrix at the corresponding code position of the word. For out of vocabulary words we pick an irregular 100-dimensional vector from a normal distribution with mean 0 and standard deviation 0.1. The size of the matrix is [total_number_of_words, 100].

All the words in the articles and headings are changed into their corresponding codes. Now, a machine can only operate on a fix length of data. The max length of articles is 116 but not all articles are that long. To get around this problem we zero pad the articles and headings which are shorter than 116 and 13, respectively, to get a fixed length.

3.2.2 The Encoder Network For the encoder network, a three-layered bi-RNN [9] architecture is used. The cells of all three layers are kept as LSTM cells [10] to overcome the problem of short-term memory of a recurrent neural network. The input of the encoder network is a one-hot encoded representation of the articles and headings according to the word-codes. Each word is represented as a sparse vector with all spaces filled with zeros other than the space at the code value of the given word which is filled with one. The initial weight matrix of the first layer from the input one-hot encoded vectors to the bi-RNN cells is initialized as the embedding matrix. This will allow the already pre-trained weights to be trained on further and refine performance. The number of neurons used in each LSTM cell is 50 and no initial states for the LSTM cells are given. The outputs from the three-layered bi-RNN network are 116 forward cells containing 50 neurons and 116 backward cells containing 50 neurons. We concatenate these to get 116 cells containing 100 neurons. An *attention layer* [4] is stacked on top of these cells to get the alpha values for each word in the article.

$$\alpha_{ts} = \frac{\exp(score(h_t, \bar{h}))}{\sum_{s'}^{S} \exp(score(h_t, \bar{h}_s))} \begin{bmatrix} attention \\ weights \end{bmatrix} \tag{1}$$

$$a_t = f(c_t, h_t) = \tanh(W_c[c_t; h_t]) \begin{bmatrix} attention \\ vector \end{bmatrix} \tag{2}$$

These values represent how important each word is to generate a particular word in the summary. To obtain these alpha values a small neural network is trained on top of each of the 116 cells with input from the final hidden states of the decoder network too. The number of hidden neurons is experimentally set to 30 with the first and second layers' activation functions set as tanh and *ReLu*, respectively. The weights are *Xavier initialized* [11]. The number of output neurons for this small network will be the same as number of words in the headings that is 13. These output values are softmaxed to get the alpha (importance) values. The weighted average according to these alpha values is taken with the combined bi-RNN output. These form the *context vectors* [12] used for decoding.

$$c_t = \sum_s \alpha_{ts} \bar{h}_s \ [context \ vector] \tag{3}$$

3.2.3   The Decoder Network For the decoder network a simple RNN network of length 13 is used. The cells are LSTM cells with number of neurons set as 50 each. The context vectors produced go into the input of this network. To actually obtain words from the output of the RNN network, a densely connected neural network is used on top of it. This neural network maps these 50 neurons to 492,476 neurons; 492476 being the number of unique words in our dataset. The weights for this network are *Xavier initialized* [11] and activation function used is *ReLu*. Each neuron represents a single word from our dataset and the neuron corresponding to a particular word is got by looking at the code of that word found in the pre-processing step. The outputs from this network are compared to the one-hot encoded headings and the measure of loss is quantified by *cross entropy loss*. The error correction factors are backpropagated using *Adam Optimizer* [13] with the learning rate set to 0.002. The complete network architecture is illustrated in Fig. 6.

4   Experimental Results

The final results are headings that are automatically generated from the fed input articles. The network was first trained on short articles and the results were noted. The end sequences, "the" in results is due to zero padding the input articles and headings during training. The code 0 corresponding to the most common word represents the word "the", which makes the network think that zero-padded sequences are actually articles and headings ending with "the" sequences. The final cross entropy loss for the short articles test set was 0.9894. Figure 4 shows the actual headings and the predicted headings by this network.

The same network was also trained on longer articles with max number of words of 213 and max number of words in headings of 33. The training time for these articles on Google Colab was much longer than the previous training time. The final cross entropy loss for long articles was 1.0683. Figure 5 shows the predicted and actual headings for these longer articles. For our evaluation metric ROUGE scores are used. ROUGE represents recall-oriented understudy for gisting evaluation. It is basically a set of metrics [14] for assessing automatic summarization of texts and machine translation. It works by contrasting an automatically delivered summary or translation against a set of reference outlines (normally human-created). To get a good quantitative value, computation of precision and recall is required using overlaps.

Recall in the context of ROUGE shows how much of the reference summary is the system summary recovering. Precision is measured as the amount of system summary that is relevant. The precision aspect becomes crucial for summaries that are concise in nature. F-measure is calculated using these precision and recall values. The different types of ROUGE scores are ROUGE-N, ROUGE-S and ROUGE-L. ROUGE-N measures

```
predicted_heading[0:10]
```

```
['the look ahead the the the the the the the the the the',
 'sudan james strengthen dar relations the the the the the the the the',
 'station by pass on 09 21 2015 1110 pm the the the the the',
 'template # 55551 the the templates the the the the the the the the',
 'emmys the the the the the the the the the the the the',
 'instagram giveaway the the the the the the the the the the the the',
 'video the tec up the the the the the the the the the the',
 'gilbert glenn the the the the the the the the the the the the',
 'nci downgraded by thestreet lower hold ncit the the the the the the',
 'mr grumpy isis fetch the the the the the the the the the the']
```

```
actual_heading[0:10]
```

```
['a look ahead the the the the the the the the the the the',
 'sudan to strengthen dar relations the the the the the the the the',
 'station 2 pass on 09 21 2015 1110 pm co 2 the the',
 'template # 55551 from newest templates the the the the the the the the',
 'emmys time the the the the the the the the the the the the',
 'instagram giveaway the the the the the the the the the the the the',
 'st paul tec the the the the the the the the the the the',
 'gilbert glenn the the the the the the the the the the the the',
 'nci downgraded by thestreet to hold ncit the the the the the the',
 'mr grumpy isis fetch the the the the the the the the the the']
```

Fig. 4 The predicted and actual headings for short articles

unigram, bigram and trigram overlap. In this research ROUGE-N: unigram and bigrams overlapping and F1 scores are used as metrics. The scores are generated with help of 30 test documents (Fig. 6).

Table 1 illustrates these scores. The ROUGE 1 scores show the similarity between unigrams of the predicted headings and the actual headings. The recall for short articles is higher, which means it remembers more of the heading than the long articles. Precision shows how much relevant information is available in the heading. For shorter articles there is more relevant information and for longer articles there tends to be more words that are not as relevant. The ROUGE 2 scores show fluidity and continuity in sentences as they compare bigrams in predictive headings and actual headings. The more bigrams are matched, the more is the structure to the sentence. For shorter articles there is more structure as compared to longer articles. The ROUGE 1 scores for longer articles are higher than ROUGE 2 scores, and by approximately 24%. This means even for longer articles the algorithm with less training time picks up keywords but is not able to give a structure to them. With more data and training time these scores will improve. Figure 7 shows the above comparison visually to give a better idea.

```
predicted_heading[0:10]

['update update update medtronic medtronic medtronic evp the the the the the the the the the the the
 'flowers foods foods president ceo exercises options the the the the the the the the the the the th
 'four four private derby his his stripes the the the the the the the the the the the the the the th
 '9 9 9 15 15 $ worth worth the the the the 8 the the the the the the the the the the the the the the
 'pacific pacific could default on on million the the the the the the the the the the the the the th
 'bassmasters bassmasters auction off a pheasant the the the the the the the the the the the the the
 'intelligence intelligence intelligence intelligence sacked surprise shake up the the the the the t
 'invasion invasion invasion akwa ibom govt house by the the the the the the the the the the the the
 'when when when t20 t20 world cup home in the the the the the the the the the the the the the the t
 'swedish swedish electric electric d109 the the the the the the the the the the the the the the the
```

```
actual_heading[0:10]

['company update medtronic inc nysemdt - medtronic evp & cfo gary ellis to speak at wells fargo
 'flowers foods president and ceo exercises options the the the the the the the the the the the
 'four legged private derby earns his stripes the the the the the the the the the the the the th
 'tuesday 9 15 15 prize puzzle is worth $ 5000 from channel 8 and kxoj the the the the the the t
 'pacific lutheran could default on $ 54 million in bonds the the the the the the the the the th
 'bassmasters to auction off a pheasant hunt the the the the the the the the the the the the the
 'algeria intelligence chief sacked in surprise shake up the the the the the the the the the the
 '" invasion of akwa ibom govt house by dss barbaric " the the the the the the the the the the t
 'when dhoni brought the t20 world cup home in 2007 the the the the the the the the the the the
 'swedish wood bodied electric d109 the the the the the the the the the the the the the the the
```

Fig. 5 The predicted and actual headings for longer articles

5 Conclusions and Future Work

In this research, an idea to summarize physical books is proposed. A page turning and image-capturing device is outlined using motors and pi-camera. Furthermore, an algorithm to summarize passages is made and implemented. Evaluation of the efficiency of the proposed model is done using ROUGE scores. The performance of the model is being around 70% on ROUGE 1 criteria and around 60% on ROUGE 2 criteria for short articles. While the performance is decent for long articles, that being around 65% on ROUGE 1 criteria and 40% on ROUGE 2 criteria. The model works well for short passages but works decent for longer passages. This is due to the increase in computational complexity as longer articles imply longer RNN sequences which hinder training. With better hardware the algorithm promises to show better results on long articles as well. The model can be improved by using a dropout layer to avoid over-fitting. The training corpus can be made larger so that the network generalizes better. The pre-processing part can be improved by identifying and omitting stop words and taking care of syntactic and grammatical connections between words. A system is to be made to automatically transfer images of pages from raspberry pi to the OCR and summarization code. This will enable users to keep the page turner device on overnight to get a summarized digital version of their book. In all, this research shows the first steps in creating a summarizer which is capable of summarizing a physical book and aims to make reading a less tedious activity by summarizing long paragraphs into short sentences. This will enable students, professors, lawyers and others to get the most important points of every page in the book in a summarized form and aid the users who do not have the time to read lengthy materials.

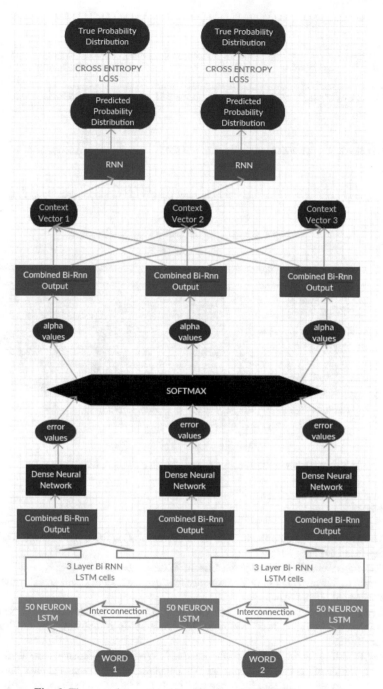

Fig. 6 The complete network architecture for IMEXT summarizer

Table 1 Evaluation of the model using ROUGE scores

Dataset	Rouge 1 Recall	Rouge 1 Precision	Rouge 1 F1-Score	Rouge 2 Recall	Rouge 2 Precision	Rouge 2 F1-Score	Rouge N F1-Score (avg)
Short articles (30 articles)	0.701171817	0.727927054	0.7142989	0.5960591	0.601231527	0.598634147	0.65646656
Long articles (30 articles)	0.643567832	0.692654798	0.6672096	0.3889788	0.412354662	0.400325820	0.53376775

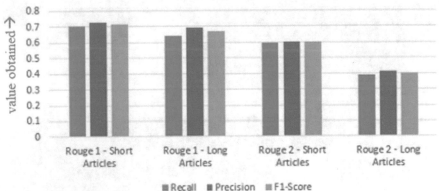

Fig. 7 Performance comparison between short and long articles

References

1. Torres-Moreno, J.-M.: Automatic Text Summarization. Cognitive Science and Knowledge Management, 1st edn (2014)
2. Genism. https://en.wikipedia.org/wiki/Gensim?oldformat=true
3. Sutskever, I., Vinyals, O., Le, Q.V.: Sequence to sequence learning with neural networks. In: Proceedings of NIPS (2014)
4. Vaswani, A., Shazeer, N., Parmar, N., Uszkoreit, J., Jones, L., Gomez, A.N., Kaiser, Ł., Polosukhin, I.: Attention is all you need. In: Advances in Neural Information Processing Systems (2017)
5. Corney, D., Albakour, D., Martinez, M., Moussa, S.: What do a million news articles look like? In: Proceedings of the First International Workshop on Recent Trends in News Information Retrieval co-located with 38th European Conference on Information Retrieval (2016)
6. Nallapati, R., Zhou, B., dos Santos, C.N., Gülçehre, Ç., Xiang, B.: Abstractive text summarization using sequence-to-sequence RNNs and beyond. In: The SIGNLL Conference on Computational Natural Language Learning (2016)
7. Glove. https://nlp.stanford.edu/projects/glove/

8. Pennington, J., Socher, R., Manning, C.D.: GloVe: global vectors for word representation. In: Proceedings of EMNLP (2014)
9. Schuster, M., Paliwal, K.K.: Bidirectional recurrent neural networks. IEEE Trans. Signal Process. **45**(11), 2673–2681 (1996)
10. Hochreiter, S., Schmidhuber: Long short-term memory. In: Neural Computer. MIT Press (1997)
11. Glorot, X., Bengio, Y.: Understanding the difficulty of training deep feedforward neural networks. In: Proceedings of the Thirteenth International Conference on Artificial Intelligence and Statistics (2010)
12. Caid, W., et al.: Context vector-based text retrieval. Fair IsaacCorporation (2003)
13. Kingma, D., Ba, J.: Adam: a method for stochastic optimization. arXiv:1412.6980 (2014)
14. Lin, C.-Y.: Rouge: a package for automatic evaluation of summaries. In: Text Summarization Branches Out: Proceedings of the ACL-04 Workshop, Association for Computational Linguistics (2004)

Lane Detection and Collision Prevention System for Automated Vehicles

Rajat Agrawal[1($\boxtimes$)] and Namrata Singh[2]

[1] Department of Computer Science and Engineering,
Naraina College of Engineering and Technology, Kanpur, India
rajat.visitme@gmail.com
[2] Department of Computer Science and Engineering, Naraina Vidya Peeth Engineering and
Management Institute, Kanpur, India
nam2817120@gmail.com

Abstract. This study gives the efficient model of the working process of lane detection and collision prevention system using computer vision and machine learning (ML) for decision-making. Lane detection technique is widely used in autonomous vehicle to keep tracking the road lanes, while collision detection is an advance feature used in intelligent vehicle to detect and prevent any type of collision. Here, we capture live video and process it to extract 3D information from the feed to detect road lanes and to detect objects for analysing the distance of it from the vehicle to prevent collisions. Collision detection and prevention is based on the object detection and recognition using ML model which is capable of detecting any objects, like vehicle, pedestrians, animals and so on.

Keywords: Lane detection · Object detection · Collision prevention · Machine learning · R-CNN · Tensor flow

1 Introduction

According to statistics more than 140,000 persons are killed in road accident in India every year. And almost all the accidents that occurred are due to human error. It became more challenging for the country with huge population to run awareness campaigns for all. So regular technical advancement is made in vehicles to reduce the accident rate, but still there is no such technology present which is intelligent enough like humans but with no or negligible errors which can completely eliminate the accidents.

So, we are providing an artificial intelligence system to vehicles which is based on machine learning. The system detects the chances of collision from objects on the road using machine learning object detection algorithm, and alerts if any collision chances detected. It can also be applied with adaptive cruise control system to control the vehicle speed accordingly. It detects vehicles even faster than human eyes and takes required actions faster than human reaction time. To keep the vehicle within lanes, lane detection is also performed by detecting lanes edges. This system can be applied to both semi-automated and fully automated vehicles.

© Springer Nature Singapore Pte Ltd. 2020
B. Iyer et al. (eds.), *Applied Computer Vision and Image Processing*,
Advances in Intelligent Systems and Computing 1155,
https://doi.org/10.1007/978-981-15-4029-5_5

Advantages of our proposed system are reduced collisions/accidents, reduced traffic congestion, reduced accidents due to tailgating and due to human errors like tiredness, stress and frustration during long-distance driving.

Our system is cost-effective as reduction in medical treatment cost of accident injuries and reduced automobile repairing cost. In addition, development cost is less due to single hardware device, that is, ordinary camera and almost all the software tools used are open source.

2 Literature Survey

As defined in [1–3], Canny edge detector is used to smooth the image and remove any noise in the captured video frame and detect the edges of the lanes. Hough transform algorithm is used for detecting the straight road lanes [1, 4], and by determining the bending direction of the curve lines, curved road lanes are detected [2].

EDLines algorithm proposed by Nguyen et al. [5] is used for line segment detection which works faster if applied over specific region (lines between 85° and 90° with horizontal axis in anti-clockwise, i.e. lanes). And the horizontal detected lines between the lanes, that is, vehicles' edges are used to detect the vehicles.

In [6] the pixel difference in the captured image/video frame is analysed by using global threshold edge detector algorithm to detect the lanes.

The concept of adaptive cruise control system defined in [7] by Saravanan and Anbuelvi is the modified form of cruise control system that controls the vehicle speed automatically according to the detected front vehicle in the same lane and also makes decisions on overtaking the front vehicle if adjacent lane is vacant. The technique is helpful in collision prevention.

The idea of using Bluetooth device for collision prevention system is proposed by Das and Sengupta [8]. When any car is present within the range of Bluetooth then it will give indication of that car and further attached sensors will activate which send interrupt to the lane departure alerting system or anti-lock braking system. The work presented in [9] by Jain introduced the concept of machine learning for lane detection. A CNN model is trained over road lanes images. The model performs image processing to extract features and recognise patterns of lanes in the images/frames.

The idea of using Adaboost algorithm along with other algorithms to improve the overall performance is proposed by Ju et al. [10]. It used to detect pedestrians and vehicles. And the combination of Haar-like features and Edgelet-Shapelet features with Adaboost is used for advanced pedestrian detection system by G. R. Rakate, S. R. Borhade, P. S. Jadhav and M. S. Shah mentioned in [11].

3 Methodology

3.1 Flowchart

See Fig. 1.

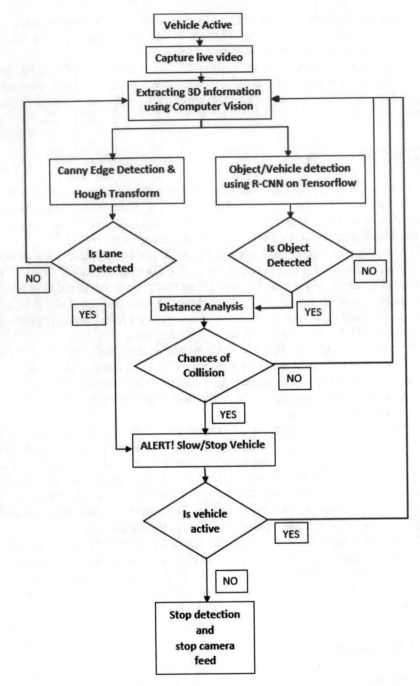

Fig. 1 The flowchart of the entire system

3.2 Algorithm

Steps:

1. Vehicle start/active
2. Camera activate—capturing live video
3. Image processing using computer vision—extracting 3D information from live video frames
4. Edge detection using Canny edge detector
5. Lane detection using Hough transform

    a. Lane detected; go to next step
    b. Lane not detected; go to step 3

6. Object/vehicle detection using region-convolutional neural network (R-CNN) on tensor flow

    a. Object detected; go to next step
    b. Object not detected; go to step 3

7. Distance analysis of detected object

    a. Chances of collision; go to next step
    b. NO collision chances; go to step 3

8. Alert warning! Slow/Stop vehicle

    a. Vehicle stopped; go to next step
    b. Vehicle active; go to step 3

9. Stop object and lane detection and stop camera feed.

Live Video Capture: We are using an ordinary camera to capture live video using computer vision. OpenCV an open-source real-time computer vision library is used in our approach. The captured video frame is shown in Fig. 2a.

Lane Detection: In this approach, frames are processed first, where coloured frames are converted into greyscale frames. Then we apply Canny edge detector [12] for edge detection, as shown in Fig. 2b. In the processed frames a lot of useless data/noises are present which should be filtered before applying the lane detection algorithm. Thus, we define a trapezoid-shaped region of interest (ROI) as shown in Fig. 2c, which filters all the useless data available in the frame. After getting the required edges of lanes further we apply Hough transform to detect lanes. To increase the detection accuracy, we remove all the horizontal, vertical and very small edges and extract only those edges whose slope is greater than 0.5 and less than −0.5 for right and left lane, respectively. The results are demonstrated in Fig. 2d.

$$slope = (y2-y1)/(x2-x1) \tag{1}$$

(a)   (b)

(c)   (d)

Fig. 2 **a** Image of road on a rainy day. **b** Edge detected by Canny edge detector. **c** Applied trapezoid shape region of interest to reduce noise. **d** Detected lanes represented by purple lines

Vehicle/Object Detection: In our object detection algorithm we use a machine learning model for object detection. Machine learning is an advance technique which has high accuracy in object detection. Here we feed the captured frames as an input to the pretrained object detection model, that is, faster_rcnn_resnet101_coco, which extracts the patterns and features present in the frame and detect objects (vehicles, pedestrians, animals etc.) if present and then enclosed them in rectangular frame with the object label and accuracy, as shown in Fig. 3.

Collision Detection and Prevention: The frames marked with detected objects are further utilised for analysing the collision chances and to prevent it if any. The detected objects have scores based on their presence and these scores are used to analyse the collisions. If there is any chance of collision (with any objects like vehicle, pedestrian, animals etc.) then our algorithm will detect it efficiently.

For collision prevention in semi-automated vehicle, our system will alert the vehicle operator/driver with-in time to take any/all required action(s). And in fully automated vehicle, our system is attached with the cruise control system which can control the vehicle speed, to slow down or to stop the vehicle, whatever is required as per situation to prevent any collision with-in time.

(a)   (b)

(c)   (d)

Fig. 3 **a** Detected trucks, car and person. **b** Detected animal (cow), trucks, motorcycle and persons. **c** Detected cars, stop sign and persons. **d** Detected animal (dog), trucks, motorcycle and person

4 Observation Table

See Tables 1 and 2.

5 Results

See Fig. 4.

Table 1 Observations based on literature survey

Author	Publication year	Edge detection algorithm	Lane detection	Other algorithm/technique	Vehicle detection	Pedestrian or objects detection (other than vehicle)	Methodology	Result
Jia He, Hui Rong, Jinfeng Gong, Wei Huang	2010	Canny edge detection algorithm	Hough transform	n/a	n/a	n/a	Removing horizontal edges/lines from detected edges	Visual display of detected lanes on displayer (screen)
Van-Quang Nguyen, Changjun Seo, Heungseob Kim, and Kwangsuck Boo	2017	EDLines algorithm	Lines almost parallel to vertical axis and have angle between 85° and 95° with horizontal axis in anticlockwise	Kalman filter for position tuning	Detecting horizontal and vertical edges of the vehicle	n/a	Removing horizontal, vertical and short edges, and creating sliding windows to verify the vehicle	

(continued)

Table 1 (*continued*)

Author	Publication year	Edge detection algorithm	Lane detection	Other algorithm/technique	Vehicle detection	Pedestrian or objects detection (other than vehicle)	Methodology	Result
Ming-Jer Jeng, Chung-Yen Guo, Bo-Cheng Shiau and Liann-Be Chang, Pei-Yung Hsiao	2009	Global threshold edge detector	The pixel difference in the captured frame is analysed by the algorithm to detect the lanes	n/a	n/a	n/a	Difference between each pixel is detected using a 3 × 3 mask	Warning/Alert
Ganlu Deng, Yefu Wu	2018	Canny edge detection algorithm	Hough transform	n/a	n/a	n/a	For straight lanes: polar angle range – 90° to 90° and for curved lanes: analysing the deviation point	

(continued)

Table 1 (*continued*)

Author	Publication year	Edge detection algorithm	Lane detection	Other algorithm/technique	Vehicle detection	Pedestrian or objects detection (other than vehicle)	Methodology	Result
Yue Dong, Jintao Xiong, Liangchao Li, Jianyu Yang	2012	Canny edge detection algorithm	Using endpoints of the detected edges in frame(s)	n/a	n/a	n/a	Lane's edge is detected by Canny detector and its endpoint is used to detect lanes	
Jyun-Min Dai, Lu-Ting Wu, Huei-Yung Liny, Wen-Lung Tai	2016	Sobel operator	Hough transform	Lucas-Kanade optical flow	Cumulative density function (CDF) to detect front vehicle	n/a	Lines at angle range 15° to 85° and from 95° to 165° are considered for lanes and CDF for vehicle detection	Warning/Alert
P. Saravanan, M. Anbuelvi	2009	n/a	n/a	Adaptive cruise control (ACC) for collision prevention	n/a	n/a	Adaptive cruise control system for speed control and overtaking vehicle	Collision Prevention from front vehicle and overtaking vehicles

(*continued*)

Table 1 *(continued)*

Author	Publication year	Edge detection algorithm	Lane detection	Other algorithm/technique	Vehicle detection	Pedestrian or objects detection (other than vehicle)	Methodology	Result
Prabal Deep Das, Sharmila Sengupta	2017	n/a	n/a	Drunken driving alerting system using alcohol sensor	Vehicle(car) detection using Bluetooth device	n/a	Nearby cars in Bluetooth range will be detected and send warning to anti-lock braking system	Theft and accidental warning
Aditya Kumar Jain	2018	n/a	CNN	n/a	n/a	n/a	Tracks/lanes are detected using a convolutional neural network	Left, right and straight lane detection. Stop vehicle in case of no lane
Ting-Fung Ju, Wei-Min Lu, Kuan-Hung Chen, Jiun-In Guo	2014	n/a	n/a	n/a	Image classifier and Adaboost algorithm	Image classifier and Adaboost algorithm	Training a ML image classifier on positive samples for object detection	Vision-based moving objects detection

Table 2 Observations based on algorithm implementation and testing

Tests	Edge detection	Lane detection	Vehicle detection	Pedestrian/object detection	Methodology	Result
	Canny edge detection algorithm on region of interest (ROI)	Hough transform (considering edges almost parallel to vertical axis)	Region-convolutional neural network (R-CNN) with tensor flow	Region-convolutional neural network (R-CNN) with tensor flow	Canny edge detector is applied on trapezoid-shaped ROI over which Hough transform detects the road lanes. A machine learning algorithm, i.e. R-CNN detects objects and then collision chances are determined	Lane detection and advanced collision prevention
Sunny day	100% accuracy	98% accuracy	99% accuracy	98% accuracy		
Rainy day	100% accuracy	97% accuracy	99% accuracy	98% accuracy		

6 Conclusion

In this paper, we propose a vision-based advance lane detection and collision prevention system. An ordinary camera is used for live video capture. Lane detection is performed by Canny detector and Hough transform on the processed data. Extracted 3D information helped in vehicle/object detection and analysis of collision chances using machine learning.

Performance of our system is impressive with very high accuracy. Ultimately it is capable enough to prevent collision and save lives which are more important than anything else in this world.

(a)   (b)

(c)   (d)

Fig. 4 **a** Original captured frame on rainy day. **b** Lane, vehicles and pedestrian detected with alert warning. **c** Original captured frame on rainy day. **d** Lane, vehicles, pedestrians and animal (cow) detected with alert warning. **e** Original captured frame on rainy day. **f** Lane, vehicles and pedestrian detected with alert warning. **g** Original captured frame on sunny day. **h** Lane, vehicles and pedestrian detected with alert warning

Fig. 4 (*continued*)

References

1. He, J., Rong, H., Gong, J., Huang, W.: A lane detection method for lane departure warning system. In: 2010 International Conference on Optoelectronics and Image Processing. IEEE, Haikou, China (11–12 Nov. 2010)
2. Deng, G., Wu, Y.: Double lane line edge detection method based on constraint conditions hough transform. In: 2018 17th International Symposium on Distributed Computing and Applications for Business Engineering and Science (DCABES). IEEE, Wuxi, China (19–23 Oct. 2018)
3. Dong, Y., Xiong, J., Li, L., Yang, J.: Robust lane detection and tracking for lane departure warning. In: 2012 International Conference on Computational Problem-Solving (ICCP). IEEE, Leshan, China (19–21 Oct. 2012)
4. Dai, J.-M., Wu, L.-T., Liny, H.-Y., Tai, W.-L.: A driving assistance system with vision-based vehicle detection techniques. In: 2016 Asia-Pacific Signal and Information Processing Association Annual Summit and Conference (APSIPA). IEEE, Jeju, South Korea (13–16 Dec. 2016)
5. Nguyen, V.-Q., Seo, C., Kim, H., Boo, K.: A study on detection method of vehicle based on lane detection for a driver assistance system using a camera on highway. In: 2017 11th Asian Control Conference (ASCC). IEEE, Gold Coast, QLD, Australia (17–20 Dec. 2017)

6. Jeng, M.-J., Guo, C.-Y., Shiau, B.-C., Chang, B.-C., Hsiao, P.-Y.: Lane detection system based on software and hardware codesign. In: 2009 4th International Conference on Autonomous Robots and Agents. IEEE, Wellington, New Zealand (10–12 Feb. 2009)
7. Saravanan, P., Anbuelvi, M.: Design of an enhanced ACC for collision detection and prevention using RTOS. In: 2009 International Conference on Advances in Computing, Control, and Telecommunication Technologies. IEEE, Trivandrum, Kerala, India (28–29 Dec. 2009)
8. Das, P.D., Sengupta, S.: Implementing a next generation system to provide protection to vehicles from thefts and accidents. In: 2017 International Conference on Innovations in Green Energy and Healthcare Technologies (IGEHT). IEEE, Coimbatore, India (16–18 March 2017)
9. Jain, A.K.: Working model of self-driving car using convolutional neural network, Raspberry Pi and Arduino. In: 2018 Second International Conference on Electronics, Communication and Aerospace Technology (ICECA). IEEE, Coimbatore, India (29–31 March 2018)
10. Ju, T.-F., Lu, W.-M., Chen, K.-H., Guo, J.-I.: Vision-based moving objects detection for intelligent automobiles and a robustness enhancing method. In: 2014 IEEE International Conference on Consumer Electronics—Taiwan. IEEE, Taipei, Taiwan (26–28 May 2014)
11. Rakate, G.R., Borhade, S.R., Jadhav, P.S., Shah, M.S.: Advanced pedestrian detection system using combination of haar-like features, adaboost algorithm and edgelet-shapelet. In: International Conference Computational Intelligence & Computing Research, pp. 1–5. Coimbatore (18–20 Dec. 2012)
12. Gonzalez, R.C., Woods, R.E.: Digital image Processing, 2nd edn. Prentice-Hall, New Jersey (2002)

Hardware Implementation of Histogram-Based Algorithm for Image Enhancement

Renuka M. Chinchwadkar[1](✉), Vaishali V. Ingale[1], and Ashlesha Gokhale[2]

[1] Department of Electronics & Telecommunication,
College of Engineering Pune, Pune, India
chinchwadkarrenuka@gmail.com, vvi.extc@coep.ac.in
[2] EsciComp-India Pvt. Ltd, Pune, India
ashleshaag@gmail.com

Abstract. Image enhancement plays an important role in all fields of applications, like biomedical, satellite images, graphical applications and many others. In today's world everything needs to work in real time, hence this study talks about the implementation of image enhancement technique. This technique uses hardware approach for analysing the quality of an image which gives less time and can be worked for parallel processing applications. As histogram of an image contains details such as pixel intensity, distribution of grey levels helps to understand and analyse the quality of an image. Here, we have proposed an efficient architecture for image enhancement technique and was simulated using Verilog HDL. This study outlines brightness enhancement and contrast stretching method which permits better visual appearance of an image. Also here we have proposed design for finding region of interest in an image by developing eight different grey levels which can be used in detecting object in an image as it is one of the major areas of research in biometric application.

Keywords: Histogram · Brightness enhancement · Contrast stretching · Region of interest · FPGA

1 Introduction

Today's world is the world of social media where everyone shares information mostly in the form of images; hence image enhancement is a most important operation in image processing. Image enhancement primarily focused on improving brightness, contrast and better visual appearance of an image for human perception. Due to poor resolution of camera, bad weather conditions or lack of light at the time of capturing an image, quality of an image degrades; hence by applying pre-processing techniques we can achieve better quality of an image. Image carries information in the form of features, like brightness, contrast and colour space and it has become easier to decide what exact operations need to be performed on image data. The histogram of an image is much suitable for predicting the quality of an image and taking operations required to perform on an image. These image-processing algorithms give accurate result in short time without losing important

© Springer Nature Singapore Pte Ltd. 2020
B. Iyer et al. (eds.), *Applied Computer Vision and Image Processing*,
Advances in Intelligent Systems and Computing 1155,
https://doi.org/10.1007/978-981-15-4029-5_6

features of an image. When they are implemented on hardware they can solve all these problems.

There are many features to analyse an image such as colour, shape, texture but here we have adopted method based on histogram of an image. Histogram of an image tells us about the number of intensity values present in an image, distribution of grey levels, and also maximum and minimum pixel intensity of an image. The grey values must be equally distributed for good quality image.

Most of the digital devices like camera use RGB colour space for capturing an image, which depends more on lighting conditions. While capturing an image if there is not much light present in the surrounding, it can directly affect the quality of an image. Hence almost all digital appliances have built-in image enhancement technique.

The aim of this paper is to develop a digital system that highlights all details of an image. One of the applications is face detection where we need to detect faces from complete image by ignoring other background details in that image; hence by implementing object detection and extraction technique using hardware we get more accurate result in less time. Here we have developed an algorithm for finding the region of interest with the help of histogram which is easy to implement on FPGA. Hardware approach provides parallel processing, due to which it requires least time to complete a process as all operations are performed with minimum one or two clock cycles. Due to this, speed of execution of task increases.

In the paper [1], the author presents spatially adaptive enhancement algorithm to overcome the inferior quality of images from CMOS image sensor. The complete pre-processing algorithms implemented on hardware reduce the number of logic gates on chip, so the cost gets reduced. This paper motivates for understanding and analysing few other applications related to our work.

In the paper [2], the author presents the FPGA implementation of image enhancement method for automatic vehicles number plate (AVPN) detection. When an image is captured at night, the image obtained will be blurry and noisy due to absence of light, in such cases pre-processing of an image is highly recommended before optical character recognition (OCR); here FPGA plays an important role.

In the paper [3], the author suggested a method for image enhancement of infrared images. Using histogram equalisation technique normal image will give good result but while getting good brightness contrast effect of infrared images traditional algorithm fails. Hence, here the author suggested some changes in histogram equalisation by rearranging grey-scale values. So, by using grey-transforming functions one can target the object from background.

Here we have developed a digital system for images which are not dedicated to particular set of images, which improves the quality of image and at the same time detect object from it. This technique is used in biomedical application to enhance images of CAT, MRI, and so on. Hardware platform provides high speed, less memory utilisation and better accuracy as they work on parallelism and hence they are used for all real-time applications rather than software which take many cycles, and all operations are run sequentially which in turn increases the response time and the resource utilisation also increases.

2 Methodology

Histogram improves the visual appearance of an image. Histogram of an image depicts the problems that originate during image acquisition such as dynamic range of pixels, contrast and many others.

In this paper we have proposed method for obtaining eight different grey levels of input image from RGB colour space into grey-scale image. This process converts each pixel to equivalent grey level. The methods involved in this work are histogram, brightness enhancement, contrast stretching and region of interest of an image which are implemented on FPGA using Verilog. Apart from this we have also mentioned file handling in Verilog.

For all the below-mentioned algorithms we have used the grey-scale image with eight grey levels.

2.1 Hardware Implementation of Histogram of an Image

Here we have considered a 256×256 image with 8 bits per pixel, hence memory required to represent an image is

$$\text{Image size} = \text{No. of Rows} * \text{No. of Columns} * \text{Bits per pixel}$$
$$= 256 * 256 * 8 = 512 \, \text{Kb} \tag{1}$$

For the said image we have considered RTL module with 8×256 decoder and 256 counters, each of 16 bits for histogram implementation. While selecting bit depths for counter we need to think an extreme condition like dark image as it consists of only 0 pixel intensity for complete image. Histogram of such image will have maximum value of 65,536. For 0 pixel to store 16-bit counter is used.

The minimum time required to implement this histogram on our device is 1.89 ns.

2.2 Brightness Enhancement

Human eye doesn't recognise all colours with same brightness (luminosity). Hence [4], NTSC (National Television System Committee) proposed some weights as green 0.59, red 0.39 and blue 0.11 for RGB image. Mathematically, it can be represented as

$$\text{RGB Luminance} = 0.3 * R + 0.59 * G + 0.11 * B \tag{2}$$

Luminance is nothing but equivalence brightness present in grey-scale. Histogram can be used for checking the brightness in an image. Brightness is the maximum light present in an image. If in the image most of zero-intensity or low-intensity pixels are present, then there is a need to apply amplitude scaling for an image. In amplitude scaling brightness can be applied in an image by analysing histogram. If the histogram of an image is towards lower pixels then the image is at the darker side, and by adding offset to all pixels brightness can be increased. The offset can be decided after studying the details of histogram. For better appearance of an image we can use two offsets: one for background and another for object.

Mathematically [5], it can be stated as

$$F(x, \ y) = \ I(x, \ y) + k1; \quad if0 <= I(x, y) <= 50$$
$$F(x, \ y) = \ I(x, \ y) + k2; \quad Otherwise \tag{3}$$

where

k1, k2 are two offset values for background and object in an image. I(x, y) is the pixel intensity of original image in x, y location. F(x, y) is the pixel intensity of an enhanced image.

Figure 1 shows the DUT for brightness enhancement,

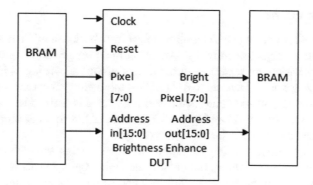

Fig. 1 DUT for brightness enhancement

2.3 Contrast Stretching

Contrast is a significant statistical textural feature which is defined as the measure of difference of intensity between a pixel and its neighbouring pixel of the image. Lower value shows low contrast and higher value shows high contrast. An image with very bright and very dark parts (i.e. many distinct values) has a dynamic range.

Low-contrast image contains less intensity values, while high-contrast image has many distinct intensity values. Histogram can be used to adjust the contrast of an image. If the histogram of an original image has grey values only in low-intensity region, then we say that the object is with low contrast. Such object cannot be differentiated from background as the difference between the intensity of object and background is very small. To increase the contrast of an object in an image, we have used contrast stretching method which adds most of high grey values to make object lighter and sharper. This method can be used in all biomedical applications to identify malignant tissue which is too hard for the medical professional to detect.

Mathematically [5], it can be stated as

$$Po = \left[(Pi - c) \left(\frac{b - a}{d - c} \right) \right] + a \tag{4}$$

where

Po—new pixel intensity after contrast stretching; Pi—old pixel intensity; b, a—upper and lower limit for 8-bit image which is 0 and 255; d, c—maximum and minimum intensity of pixels present in an image.

Figure 5 shows the comparison between histogram of the original image and after contrast stretching process performed on image. Through this we are able to understand how contrast stretching works.

In original image as pixel intensities are not uniformly distributed, but after application of contrast stretching all pixel values are uniformly present in an image due to which quality of image improved.

2.4 Region of Interest

Region of interest is nothing but separating object from background in order to highlight all properties of an object in an image. An object in an image can be one or more than one in numbers which are separated into different groups according to intensity of the pixel. From histogram we have studied about how these grey values are spread from 0 to 255. After examining 50 such images the threshold value is derived which can separate an object from background. Here T1 and T2 are used as threshold to separate an object from background as shown in the flowchart of Fig. 3.

After applying these two thresholds on an image we can separate an object from background but the image shows only black and white regions. This is because all the low-intensity values in an image get brighter due to which some minor details in an image are missed. Hence, to avoid this issue we have to use eight grey levels in order to keep all properties of an object without losing other minor details from them.

The 8 bits directly describe red, green and blue grey-scale values; typically with three bits for red, three bits for green and two bits for blue.

$$\underline{001}\ \underline{001}\ \underline{01}$$
Red Green Blue

By changing the position of bits places of red, green and blue, we get eight grey-scale levels, as shown in Fig. 2.

1. 0–000 000 00
2. 37–001 001 01
3. 73–010 010 01
4. 108–011 011 00
5. 145–100 100 01
6. 180–101 101 00
7. 216–110 110 00
8. 255–111 111 11

Fig. 2 Eight different grey-scale levels

We have modified the flowchart proposed in [6] for detection of region of interest, as given in Fig. 3. We applied two thresholds: one to find background and the other for applying grey levels to an image.

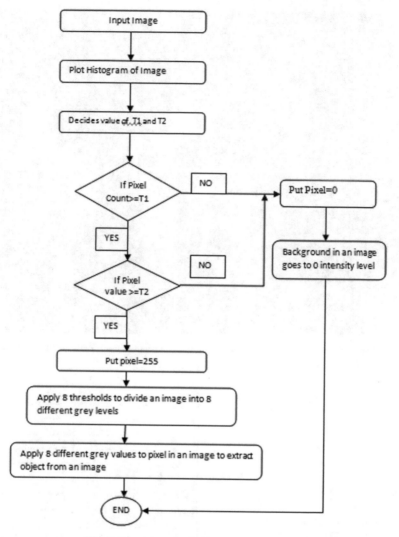

Fig. 3 Flowchart for finding region-of-interest

Here, we have used file handling in BRAM using Verilog, which involves $read-memh, $fopen and $fwrite commands. Similar to as we use in C language, these are used for reading, opening and writing contents from text file to BRAM.

3 Results

See Figs. 4 and 5.

Fig. 4 (i) Original image (top left), (ii) Brightness enhancement (top right), (iii) Contrast enhancement (bottom left), (iv) Region-of-interest (bottom right)

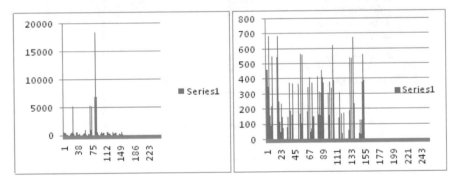

Fig. 5 Histogram of original image and histogram after contrast enhancement

4   Conclusion

Thus histogram-based image enhancement algorithm implemented on hardware provides better result as compared to software. Hardware provides high speed, less memory utilisation, better accuracy than software and they also work on parallelism, hence can be used for all real-time applications. The software code runs sequentially and takes many clock cycles which increases delay and resource utilisation. In the implemented design area utilisation on FPGA is very small, thus it can be used for images with higher dimensions. The design we have implemented is for 256×256 size image. Based on this, Table 1 shows utilisation of all units on FPGA board.

Image enhancement algorithms are easy to understand and can be useful in every field of applications and one of them can be for segmentation of satellite images such as forest, water and land. All this needs the basic information of pixel analysis which we can get through histogram and image enhancement algorithms. Hence histogram unit plays an important role and is used in all these image enhancement techniques. Thus, by designing only this unit all the image enhancement methods can be easily implemented. Principal component analysis (PCA)-based methodology is the best candidate to reduce the noise in images [7]. In future, the proposed methodology may be tested and verified using PCA and ICA techniques (Table 1).

Table 1 Design utilisation targeted device virtex-5 FPGA

DUT	Slice register	LUTs	I/O blocks
Histogram unit	4096 (32%)	5673 (45%)	26 (15%)
Brightness enhancement	–	11 (0%)	17 (2%)
Contrast stretching	4105 (32%)	5688 (45%)	36 (20%)
Region of interest	–	5 (0%)	18 (10%)

References

1. Jung, Y.H., Kim, J.S., Hur, B.S., Kang, M.K.: Design real-time image enhancement preprocessor for CMOS image sensor. IEEE Trans. Consumer Electron. **46**(1) (2000)
2. Bai, L., Li, X., Chen, Q., Zhang, B.: The hardware design of real-time infrared image enhancement system. In: IEEE International Conference on Neural Networks & Signal Processing Nanjing, China, December 14–17, 2003
3. Zhang, L., Yang, K.: Region-of-interest extraction based on frequency domain analysis and salient region detection for remote sensing image. IEEE Geosci. Remote Sens. Lett. **11**(5) (2014)
4. Greenberg, A.D., Greenberg, S.: Digital Images a Practical Guide. McGraw Hill Edition 20 (1995)
5. Shandhilya, R., Sharma, R.K.: FPGA implementation of image enhancement technique for automatic vehicles number plate detection. In: International Conference on Trends in Electronics and Informatics ICEI 2017
6. Kiadtikornthaweeyot, W., Tatnall, A.R.L.: Region of interest detection based on histogram segmentation for satellite image. In: The International Archives of the Photogrammetry, Remote Sensing and Spatial Information Sciences, vol. XLI-B7, 2016 XXII ISPRS Congress, 12–19 July 2016, Prague, Czech Republic

7. Patil, M.N., Iyer, B., Arya, R.: Performance evaluation of PCA and ICA algorithm for facial expression recognition application. In: Proceedings of Fifth International Conference on Soft Computing for Problem Solving, vol. 436, pp. 965–976 (2016)
8. Liu, G., Li, J., Ma, H.: Study of infrared image enhancement algorithm based on FPGA. In: Proceedings of the 2009 IEEE International Conferences of Mechatronics and Automation, August 9–12, Changchun, China
9. Li, L., Yu, F.: Block region of interest method for real-time implementation of large and scalable image reconstruction. IEEE Sig. Process. Lett. **22**(11) (2015)
10. Hore, A., Yadid-Pecht, O.: On the design of optimal 2D filters for efficient hardware implementations of image processing algorithms by using power-of-two terms. J. Real-Time Image Proc. **16**, 429–457 (2019). https://doi.org/10.1007/c11554-015-0550-2
11. Tsai, C.-Y., Huang, C.-H.: Real-time implementation of an adaptive simultaneous dynamic range compression and local contrast enhancement algorithm on A GPU. J. Real-Time Image Proc. **16**, 321–337 (2019). https://doi.org/10.1007/s11554-015-0532-4

Comparative Analysis of PCA and LDA Early Fusion for ANN-Based Diabetes Diagnostic System

Sandeep Sangle$^{(\boxtimes)}$ and Pramod Kachare

Department of Electronics and Telecommunication Engineering, Ramrao Adik
Institute of Technology, Nerul, Navi Mumbai, India
sandeepsangle12@gmail.com, kachare.pramod1991@gmail.com

Abstract. The paper shows the effect of PCA and LDA, for dimension reduction and early fusion, in the framework of the Diabetes disease diagnostic system. Several ANN architectures were evaluated to obtain the optimum classification model. Confusion matrix-based analysis was performed to analyze the effect of dimension reduction. PIMA Indians Diabetes dataset was used for evaluation. The diabetes detection accuracy of 87.8% was obtained using original patient records fused with the first six PCA dimensions. Similar detection accuracy was achieved using the first six LDA dimensions. Relative detection accuracy of fused features increased by 15 and 3% compared to detection accuracy with the original patient record and PCA features without fusion, respectively.

Keywords: Principal component analysis · Linear discriminant analysis · Early feature fusion

1 Introduction

Diabetes, introduced by Egyptians about thirty centuries ago, is a metabolic disease characterized by elevated blood glucose levels. International standards predict growth in the diabetes population from 366 million as per today to 522 million by 2030. Primary causes for diabetes are an unhealthy diet, physical inactivity, smoking, age factor, and hereditary reasons [1]. Clinical experts frequently utilize a 75-g oral glucose tolerance test and criteria such as 2-h plasma glucose (2-h PG) and fasting plasma glucose (FPG) value [1]. Regular exercise and a healthy diet help in controlling diabetes.

Automatic diabetes diagnosis models have been an active research area over a decade. Kayaer et al. utilized brief statistics of the PIMA Indian dataset to train five different classifiers: Gradient descent, RBF, Levenberg–Marquardt, GRNN, and BFGS quasi-Newton. Comparative performance of all five classifiers showed GRNN to be best suited for the task with a tenfold accuracy of 80.21% [2]. Polat et al. analyzed PIMA Indian dataset records using Gradient

© Springer Nature Singapore Pte Ltd. 2020
B. Iyer et al. (eds.), *Applied Computer Vision and Image Processing*,
Advances in Intelligent Systems and Computing 1155,
https://doi.org/10.1007/978-981-15-4029-5_7

Descent-Least Square SVM classifier to get tenfold accuracy of 79.16% [3]. In [4], Artificial Neural and Fuzzy Interference System was used to classify PCA analyzed features. Feature dimensionality reduction and complex classifier boosted accuracy to 89.74%. Considering the improved model accuracy due to PCA feature encoding, Esin et al. proposed LDA for feature dimensionality reduction. LDA compressed features classified using ANFIS provided maximum accuracy of 84.61% [5], less than PCA. LDA compressed features with the Morlet wavelet SVM classifier provided the best accuracy of 89.74% [6]. Thus, the literature shows the importance of feature–classifier interaction for improving diagnostic accuracy.

Most of the diabetes diagnosis models reported in the literature employ either original patient's records or transformed features. The proposed model uses a fusion of original patient records and transformed features. Two feature transformation techniques are studied: Principal Component Analysis and Linear Discriminant Analysis. The optimized feature set is selected using forward feature selection approach. Comparative analysis using MLP-ANN classifier is presented.

The paper is organized as follows: Sect. 2 presents algorithm for fusion of PCA or LDA transformed features and original patient record features. Section 3 presents comparative analysis of classifiers. Section 4 concludes the work.

2   System Development

Early fusion concatenates different features into a single representation before they are used in a classifier. In this paper, early fusion combines original features of diabetes patient and PCA or LDA transformed features into single vector. Early fusion selects the most informative features based on correlation measure using the feedforward technique. This approach reduces a large set of variables to a smaller, conceptually more coherent set of variables which retain the original information.

In this study, we have used eight clinical attributes from PIMA Indian dataset with binary classification (explained in Sect. 3). The proposed diabetes diagnosis model is implemented in two steps; features set optimization using forward feature selection approach and classification using MLP-ANN classifier.

2.1   PCA Early Fusion

Principal component analysis (PCA) is a linear dimension reduction technique used to transform the information in high-dimension correlated variables into a low-dimension uncorrelated variables [7].

PCA employs most significant k eigenvectors, corresponding to k largest eigenvalues, to transform correlated clinical features to uncorrelated features [7]. Selected eigenvalues are concatenated to original patient records, known as PCA early fusion. Figure 1 shows the process diagram for early PCA fusion. The fused features represent an additional nonlinear dimension at the input of the classifier. Number of PCA features to be fused are calculated using a forward feature selection algorithm.

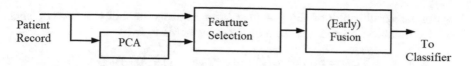

Fig. 1. The flow diagram of PCA features and process for PCA fused features

2.2 LDA Early Fusion

Linear discriminant analysis (LDA) achieves dimension reduction by increasing global variation (intra-class) and decreasing local (inter-class) variation, simultaneously. LDA finds best class-discriminating vectors within the underlying space. LDA and PCA compressed features are distinct in the respect that later is class agnostic. Selected eigenvalues are concatenated to original patient records, known as LDA early fusion. Maximum LDA features to be fused are determined using a forward feature selection algorithm (Fig. 2).

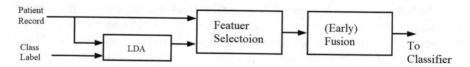

Fig. 2. The flow diagram of LDA features and process for LDA fused features

Both LDA and PCA calculate direction for dimension reduction through a linear transformation. LDA is for direction finding the base on maximizing class discrimination while PCA is agnostic to individual class labels. The direction in PCA is determined based on a maximum variance in data. In PCA, a linearly mapped dataset is the dataset to a smaller orthogonal feature space, while a linear map in LDA exploits orthogonal patterns belonging to individual classes; PCA "ignores" class labels [8].

2.3 Classifier System

MLP-ANN is a nonlinear classifier consisting of three fundamental layers: input, hidden, and output. The nonlinearity of the hidden layer maps data from the input to the output layer. It is a parallel processing system [8]. Equation 4 represents the mathematical form of layer,

$$y = g(W^T x + b) \tag{1}$$

where g is a nonlinear activation, W is weights matrix, x is the input vector, and b is a bias vector. The output of the inner layer also works as the input of the outer layer with bias value b. The fully connected layer of MLP-ANN has connected all units of the previous layer to every unit of the current layer. Equation number 5 gives example of three layer MLP-ANN

$$f(x) = g_3(W_3^T g_2(W_2^T g_1(W_1^T x + b_1) + b_2) + b_3) \tag{2}$$

The parameters of each layer are independent of the other layers, which means each unit of layer possesses a unique set of weights. The output is a result of a nonlinear activation function to a weighted sum of inputs and bias.

3   Results and Discussion

In this investigation, the PCA and LDA features are classified using MLP-ANN for the diagnosis of diabetes disease. The Pima Indians Diabetes dataset, maintained by National Institute of Diabetes and Digestive and Kidney Diseases, is used for building model and performance evaluation. All patients in this database are Pima-Indian women aged less than 21 and living in proximity to Phoenix, Arizona, USA. The dataset has two classes identified using eight clinical attributes, Table One shows brief statistics about each attribute. Diabetic and healthy subjects are represented using class labels "1" and "0", respectively. It consists of clinical records, 268 diabetics, and 500 healthy subjects [9]. To optimize the 70–30 holdout, we used a cross-validation technique. This observation corresponds to 185 (healthy)—353 (diseased) samples which are used for training samples, and testing samples 83 (healthy)—147 (diseased) which are used for modal validation (Table 1).

Table 1. Statistical summary of Pima-Indian dataset [9]

Sr. No.	Name of attributes	Avg.	Std. Dev.	Min/Max
1	Number of time pregnancies	3.8	3.4	0/17
2	Glucose level	120.9	32.0	0/199
3	Blood pressure	69.1	19.4	0/122
4	Skin thickness	20.5	16.0	0/99
5	Insulin in body	79.8	115.2	0/846
6	BMI weight in kg	32.0	7.9	0/67.1
7	Diabetes pedigree function	0.5	0.3	0.078/2.42
8	Age (year)	33.2	11.8	21/81

3.1   Performance Evaluation of Proposed Expert System

Given an input record, a diabetes diagnosis output is a class value or label, such as Yes/No and 1/0. For the predictive hypothesis that a diabetes disease is present, it is positive class, while for others it is a negative class. The confusion matrix is the most intuitive performance evolution adopted by the researcher's worldwide for Binary classifier. Each element of confusion matrix as shown in the figure is defined as

- True Diabetic (TD): Correct Diabetes Disease prediction
- False Diabetic (FD): Incorrect Diabetes Disease prediction
- True Healthy (TH): Correct Healthy prediction
- False Healthy (FH): Incorrect Healthy prediction

Three different quantitative measures namely, accuracy, specificity, and sensitivity are derived using the above elements.

The dimensionally compressed features, along with original patient record features, are applied at the input of MLP-ANN. To evaluate the optimum configuration for the current problem, we have analyzed several MLP-ANN architectures, in this investigation. Two MLP-ANN architectures, model-1: XL-7N-7N-5N-2L, model-2: XL-10N-8N-5N-2L, showed consistent performance. Here, N refers to nonlinear activation, L refers to linear activation, and X refers to the number of inputs. The output is calculated using the softmax technique. Percentage accuracy for various numbers of PCA and LDA components using MLP-ANN is shown in Figs. 3 and 4.

Figure 3 shows the comparative performance of PCA and LDA for varying numbers of components. For PCA analysis, model-1 provided the highest accuracy of 86.1% using 4 and 7 PCA components. Model-2 gives the highest accuracy of 84.3% using 3 PCA components, same as obtained using model-1. For LDA analysis, model-1 achieved the overall highest accuracy of 87.8% using 6 LDA components. The highest accuracy obtained for model-2 using 7 LDA components is 86.1%. Model-1 and model-2 provide the same accuracy 83.5%, 84.3%, 86.1%, and 83.5% using 3, 5, 7, and 8 LDA components, respectively.

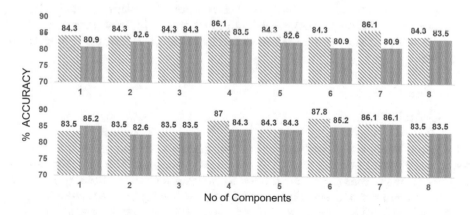

Fig. 3. Comparative analysis of PCA (top) and LDA (bottom) compressed features for various number of components

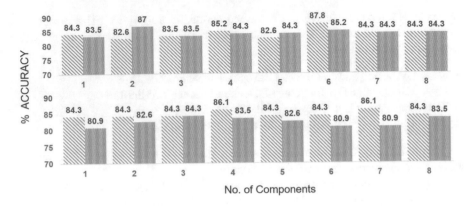

Fig. 4. Comparative analysis of early fusion using PCA (top) and LDA (bottom) features for various number of components

Figure 4 utilizes the early fusion of original patient records and with PCA or LDA compressed features. The horizontal axis denotes the number of PCA or LDA components used for early fusion. In PCA analysis, model-1 achieved the overall highest accuracy of 87.8% by fusing 6 PCA components. Model-2 provides the highest accuracy for 2 PCA components that is 87%. PCA components 3,7, and 8 provide the same accuracy 83.5%, 84.3%, and 84.3% for both models, respectively.

The LDA analysis shows the highest accuracy of 86.1% for 4 and 7 LDA components for model-1. Model-2 provides the highest accuracy of 84.3% for 3 LDA components. Similar accuracy was obtained using 3 LDA components for both model-1 and model-2.

In summary, early fusion of 6 PCA components provide the highest accuracy of 87.8%. On the other hand, only 6 LDA compressed features provide similar accuracy. Early fusion of LDA with original records reduces accuracy by 1.7%. Model-1 achieved the overall highest accuracy using LDA features compared to model-2.

4   Conclusion

The paper studies the performance of the diabetes disease diagnostic system using early fusion of PCA and LDA compressed features. The literature shows the performance comparison for different dimensions of PCA. PCA compressed features provides diagnosis accuracy of 85.2% for five layers (8I-10N-8N-5L-2O) bottleneck MLP-ANN classifier. In this paper, original patient record features are fused with PCA and LDA compressed features and the model is trained using MLP-ANN and SVM. A fusion of original features and first six PCA compressed features provide accuracy as high as 87.8%, a relative increase of 3%.

On the other hand, LDA alone gives an accuracy of 87.8% while fusion with original patient record features reduces accuracy to 86.1%. In summary, PCA features are suitable for early fusion with source data features, while LDA performs better without early fusion. Its performance can be further analyzed for task-specific parameters and increased training data.

References

1. Sangle, S., Kachare, P., Sonawane, J.: PCA fusion for ANN-based diabetes diagnostic. In: Computing, Communication and Signal Processing, pp. 583-590. Springer, Singapore, (2019). https://doi.org/10.1007/978-981-13-1513-8_59
2. Kayaer, K., Yildirim, T.: Medical diagnosis on Pima Indian diabetes using general regression neural networks. In: Proceedings of the International Conference on Artificial Neural Networks and Neural Information Processing (ICANN/ICONIP), vol. 181, p. 184 (2003). https://pdfs.semanticscholar.org/ef31/2e378325707b371c4727f6b1f9225fc03a9f.pdf
3. Polat, K., Gne, S., Arslan, A.: A cascade learning system for classification of diabetes disease: generalized discriminant analysis and least square support vector machine. Expert. Syst. Appl. **34**(1), 482–487 (2008). https://doi.org/10.1016/j.eswa.2006.09.012
4. Polat, K., Gne, S.: An expert system approach based on principal component analysis and adaptive neuro-fuzzy inference system to diagnosis of diabetes disease. Digit. Signal Process. **17**(4), 702–710 (2007). https://doi.org/10.1016/j.dsp.2006.09.005
5. Calisir, D., Dogantekin, E.: An automatic diabetes diagnosis system based on LDA-wavelet support vector machine classifier. Expert. Syst. Appl. **38**(7), 8311–8315 (2011). https://doi.org/10.1016/j.eswa.2011.01.017
6. Dogantekin, E., et al.: An intelligent diagnosis system for diabetes on linear discriminant analysis and adaptive network based fuzzy inference system: LDA-ANFIS. Digit. Signal Process. **20**(4), 1248–1255 (2010). https://doi.org/10.1016/j.dsp.2009.10.021
7. Lindsay, L.S.: A tutorial on principal components analysis lindsay I Smith 26 February 2002. http://www.cs.otago.ac.nz/cosc453/student_tutorials/principal_components.pdf
8. Bishop, C.M.: Pattern recognition: machine learning **128** (2006). https://www.isip.piconepress.com/publications/reports/1998/isip/lda/lda_theory.pdf/
9. Pima Indians Diabetes.: https://archive.ics.uci.edu/ml/machine-learning-databases/pima-indians-diabetes/

Copy-Move Image Forgery Detection Using Shannon Entropy

Dayanand G. Savakar and Raju Hiremath[(⊠)]

Department of Computer Science, Dr. P.G. Halakatti P G Centre, Rani Channamma University, Toravi-Vijayapur, Karnataka, India
dgsavakar@gmail.com, hiremathrm@gmail.com

Abstract. Forgery detection plays a very important role in today's fast-moving world of technology. Videos and images are having a wide range of applications in authentication, communication, military, banking, and so on. There are many tools available in the market to tamper an original image. In this technique of copy and paste forgery, the required part of digital image is copied and then pasted on the same image in order to hide the targeted image. And the tampered image looks similar to the original image. The operator does many pre- and post-operation on the original image such as rotation, flipping, resizing, compressing, and many more. Thus, it is very difficult to detect such kind of unauthorized alterations. In this article, the authors will be discussing the forgery detection of an image using Shannon's entropy method and similarity and dissimilarity measurements. The proposed method gives higher accuracy than the methods which are mentioned in the comparative study.

Keywords: Forgery · Tampered · Shannon · MATLAB · Entropy · Compressing · Authentication · Unauthorized · Measurements · Copy-move

1 Introduction

The present scenario is being noticed within the developing digital technologies and also traditional concepts which are recognized as "seeing is believing" and it is no more valid. The information nowadays is preserved within the digital form and also in digital images which can be further manipulated easily, and therefore forgery of these became a serious topic that has to be considered. The images containing the secret information are being easily doctored with the usage of Adobe Photoshop which is easily available. This process has mainly led to serious consequences which further reduces the trustworthiness and also creates the false belief within the real-world applications. Thus trustfulness of these images has to be taken seriously by the process of verification of the reliability and the wholeness of the digital images which has become a major issue.

© Springer Nature Singapore Pte Ltd. 2020
B. Iyer et al. (eds.), *Applied Computer Vision and Image Processing*,
Advances in Intelligent Systems and Computing 1155,
https://doi.org/10.1007/978-981-15-4029-5_8

The forgery detection techniques are being categorized mainly in a number of domains such as intrusive and non-intrusive, where intrusive domain is also referred to as the non-blind technique and non-intrusive is referred to as the blind domain of forgery detection technique. The intrusive method is referred to as the non-blind because the method requires some of the digital information that has to be embedded within the original image when it is being generated and also it is notified that this domain has a very limited scope. The examples that are associated with this method are mainly digital watermarking and also using digital signature concerned techniques which are of the camera, and also that not most of the digital device provides this feature. Further the non-intrusive method which is also considered as the blind method does not require any of the embedded information. A digital image is recognized to be forged when the originality of the image is tampered by the application of various transformations, such as the rotation, scaling, and resizing. It is also being noted that an image is tampered with the addition of the noise or by the removal or by the addition of some of the objects mainly to hide the real information. The commonly used image forgery method is copy-move forgery where a part of the original image is copied and pasted on the other part either once or many times to hide some of the existing information. This ease and the effectiveness of the copy-move forgery make it possible to be most of the common forgery which is used mainly to alter the content of the image.

So it becomes necessary to find out forgery in an image. There are two methods to determine digital image forgery: active and passive methods (Fig. 1).

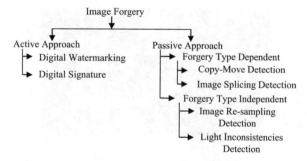

Fig. 1 Digital image forgery methods

1.1 Active Method

This method needs watermark, signature, or key for processing.

1.2 Passive Method

The passive method is different from active method, as it does not need a watermark, signature, or key. Passive approach mainly includes pixel-based, camera-based, physical-based, geometric-based, and format-based images. Passive method includes splicing, retouching, and cloning of an image.

Splicing approach. This is the most aggressive method. Splicing is a method used in many forgery algorithms wherein a part of an image is taken and pasted on different images. This method is composed of two images to produce a new image (Fig. 2).

(a)   (b)   (c)

Fig. 2 **a** First image, **b** second image, **c** splicing forgery image

Retouching approach. Retouching is a method which is applied to enhance image features. Retouching is mainly used in magazines and Photoshop's. It can be considered as less harmful. Here no image is being altered but only it is enhanced with quality or compression, and so on (Fig. 3).

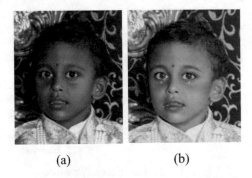

(a)   (b)

Fig. 3 **a** Original image, **b** retouched image

Cloning approach/copy-move. The final one is cloning or copy-move which is almost the same as cloning where a duplicate image is produced using an original image. From the figure, it is visible that the original image is having two pens but in the forgery image, the red pen is missing. Henceforth a part from an original image is copied and moved (Fig. 4).

(a)   (b)

Fig. 4 **a** Original image, **b** copy-move forged image

2 Literature Survey

Harpreet Kaur and Jyoti Saxena proposed a system [Published in IEEE and Springer] where key-point-based forgery is detected instead of block matching. In this paper the authors discuss different key-point-based methods, like SIFT, SURF, ORB, and BRISK [1]. Jiming Zheng and Liping Chang proposed a system [Published in IEEE and Springer] based on a tactic of extracting a Harris corner from an image as a key-point and extracting binary feature descriptors whereby finding for similar feature descriptors [2]. Mohamadian proposed a theory [Published in IEEE and research gate] to detect copy-paste tampering forgery. He used a simple method based on scale-invariant feature transformation algorithm. This algorithm is used to perform detections but cannot detect flat copied region [3].

Salam A. Thajeel and Ghazali Bin Sulong proposed a system [Published in IEEE and research gate] that is built using a technique to find likeness and connection between the altered part and real image [4]. L. Li, S. Li, and H. Zhu proposed a system [Published in IEEE and Springer] where forgery is detected by first filtering an image and divided into overlapping circular blocks. Then rotational invariant uniform local binary pattern is extracted as feature vector, and the feature vectors are compared to determine whether the block is altered [5]. G. Muhammad proposed a theory [Published in research gate] on blind copy-move image forgery detection using undecimated dyadic wavelet transformation algorithm [6]. M. Zimba and S. Xingming propose a system [Published in research gate] that involves discrete wavelet transform (DWT) and principal component analysis-eigenvalue decomposition (PCA-EVD). This system works efficiently but does not work on scaled or rotated images [7]. Y. Huang, W. Lu, W. Sun, and D. Long proposed a system [Published in research gate] where improved DCT is used to detect a forgery in an image. This principle involves applying improved DCT over an image to extract features, image is sorted lexicographically sorted, and forgery blocks will be matched [8].

Q. Wu proposed a method theory [Published in IEEE and research gate] using a log-polar fast Fourier transforms to detect the duplicated image region that may be rotated or re-scaled. This method has lower complexity for feature extraction [9]. Bayram, S., Sencar, H. T., and Memon, N. proposed theory [Published in IEEE and Springer] where altered part of an image can be detected even on lossy compressed image, and which has been severely modified using counting bloom filters [10]. G. Li, Q. Wu, D. Tu, and S. Sun. proposed a theory [Published in IEEE and research gate] of blind forensic approach to detect edited regions of an image using DWT and singular value decomposition (SVD) [11]. A. N. Myna proposed a theory [Published in IEEE and Springer] based on log-polar coordinates and by using phase correlation as the similarity criterion. In this technique, they have first applied wavelet transformation to reduce the dimension of the image. The time complexity is lower in this method but is not robust against geometric operations [12].

3   Copy-Move Forgery Detection Techniques

This section describes about different forms of techniques that are concerned to copy-move forgery detection.

3.1   Exhaustive Search

This method is a very basic approach where image and its circular shifted version are overlaid on each other to check for closely matched segments of the image. Although this method is effective, it is very complex and resource-demanding in terms of computational power, so implementing an exhaustive search for bigger image.

3.2   Autocorrelation

Fridrich et al. proposed autocorrelation in [13] where autocorrelation is implemented using Fourier transform. Here original and dilated image segments will generate higher peaks in an autocorrelation for shifts in segments for the copy-moved image. In the natural image, segments contain the power in lower frequency. But if we autocorrelate the original image we get higher peaks in corners and its neighboring pixels, so the original image is passed through before performing autocorrelation.

3.3   Block-Matching

We are going to discuss this approach extensively here as we are going to use this in our research here. The best method to approach is to divide image to overlapping or non-overlapping blocks of m × n. and the features of the block are calculated and matched with every block. But here analyzing block size is very difficult because if the block size is more than forged area then exact blocks do not result. If block size is less than forged area then that particular does not fit in one block, so it cannot be compared properly.

4   Proposed System

The proposed system detects if there was copy-move done on an image using Shannon entropy and the intensity matrix as features and using Euclidean distance to calculate similarity score between the blocks. The system works in three different phases: entropy and intensity matrix are used because images tend to keep the same values for the mean value of entropy and intensity of pixel in blocks even if the images are flipped or rotated.

4.1 Phase 1

Here the RGB image is divided into 4×4 overlapping blocks, and Shannon entropy for red, green, and blue columns are calculated separately using Eq. 1 and average of each column is taken and stored in the feature matrix.

Then the intensity of each block is calculated using Eq. 3. In addition, they are stored in feature vector for matching later (Fig. 5).

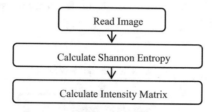

Fig. 5 Phase 1

Shannon Entropy. It is being noted that the Shannon entropy equation is recognized to estimate an average minimum number of bits that are required to encode a string of symbols which are based on the frequency of the symbols. The Shannon entropy of A is

$$H(A) = -\sum_{i=1}^{n} pi \log_2 pi \tag{1}$$

where p_i is the probability of a given symbol. The minimum average number of bits per symbol is

$$\text{numBits} = [H(A)] \tag{2}$$

The entropy is not at all responsible to make out any of the statement that is considered for the compression efficiency which can be achieved by prediction. It provides a lower bound for compression of data. The complexity of the algorithm theory is related to this area. Consider an infinite set of data and the data set can be examined for randomness. And if the data set is not random then there will be some program that will generate it and data set can be compressed.

Intensity Matrix. In the study of intensity matrix, the intensity of a window I is equal to the average intensity of pixels as in

$$m = \frac{1}{a^2} \sum_{i=1}^{a} \sum_{j=1}^{a} I_{ij} \tag{3}$$

4.2 Phase 2

In this phase, the RGB image is converted into grayscale image. Again the image is divided into 4 × 4 overlapping blocks, and Shannon entropy for grayscale image is calculated separately using Eq. 1 and the average entropy is taken and stored in feature matrix. Then the Intensity of each block is calculated using Eq. 3 and stored in feature vector for matching later.

Here same RGB image is converted to grayscale image and divides the image into 4 × 4 overlapping blocks then Shannon entropy and intensity matrix calculated and stored in feature vectors (Fig. 6).

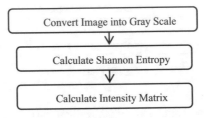

Fig. 6 Phase 2

4.3 Phase 3

In this phase similarity score of each element of the feature vector for the individual block is calculated with all the remaining blocks. To check similarity score Euclidean distance is used. Euclidean distance is used to calculate the distance between two points on a plane. It is calculated using the following equation:

$$d(a, b) = \sqrt{(b1 - a1)^2 + (b2 - a2)^2} \tag{4}$$

Euclidean distance for more than two points is

$$d(a, b) = \sqrt{\sum_{i=0}^{n}(bi - ai)^2} \tag{5}$$

Similarity score is calculated using the following equation:

$$SC = 1 - d(a, b) \tag{6}$$

Similarity score will be in the limit of 0 and 1, where zero means it is not similar original block and 1 means it is exactly as the original block. It can be converted to a percentage to get a similarity score between 0 and 100%. If the measured similarity score is above the threshold, mark that matching block as red and blue. Then "block-matching" is performed between red- and blue-marked blocks, and pixel sensitivity is calculated for the same. If 95% of block-matching is met with pixel sensitivity of more than or equal to 0.0033 then that image is classified as "copy-move detected image" (Fig. 7).

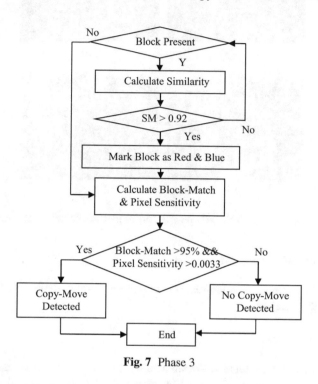

Fig. 7 Phase 3

5 Experimental Results

In order to test the proposed model, we have downloaded some images from the internet and edited using image editing software to add, remove, and retouch the image so that we can create forged images for analysis. Here are some of the results of the proposed model (Fig. 8).

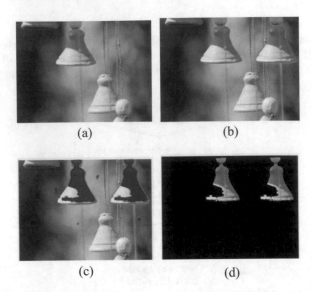

Fig. 8 **a** Original image, **b** forged image, **c** forged part marked, **d** forged part cropped

6 Performance Analysis

Here we adopt a technique called receiver operating characteristic (ROC) analysis for our performance analysis.

6.1 ROC

ROC is used as a tool to evaluate discriminate effects among different systems or methods [14]. ROC is used as follows: An image is either forged or not forged. To find out if the image is forged the user has to perform some test on image. Just because images are tested that does not mean they are done many times. It is based on the test made that does not mean all tests are correct. To avoid this kind of challenges one or more tests are conducted on the same image. So here positive test means image is forged and test also found to be forged. A negative test means the image is not forged but tests are showing as forged.

To analyze forged image, sample groups are created with all the possibilities. Let p_i be considered as the probability for the image to obtain positive test were the prevalence (p) concerned to the positive test within the data set is theoretical (Table 1).

$$P = \text{mean}(pi) \tag{7}$$

The level of test (Q) is

$$Q = \text{mean}(qi) \tag{8}$$

We also define

$$P' = 1 - P \text{ and } Q' = 1 - Q \tag{9}$$

Table 1 Relation between measurement probabilities, prevalence, and level of test

Tests	Test results		
Positive	TP	FN	P
Negative	FP	TN	P'
	Q	Q'	1

There are four possible outcomes of the test conducted. They are true positive (TP), true negative (TN), false positive (FP), and false negative (FN).

Values described by the above-given data can be used to calculate different measurements of the quality of the tests.

Sensitivity (**SE**) is the probability of having positive forgery among the positively tested images.

$$SE = \frac{TP}{TP + FN} = \frac{TP}{P} \tag{10}$$

Specificity (**SP**) is the probability of having negative forgery among negatively tested images.

$$SP = \frac{TN}{FP + TN} = \frac{TN}{P} \tag{11}$$

Precision (**p**) is the probability of positive predictive value.

$$p = \frac{TP}{TP + FP} \tag{12}$$

Fmeasure (**r**) is measure of true positive rate.

$$r = TP/(TP + FN) \tag{13}$$

Accuracy is measurement of total accuracy of system.

$$\text{Accuracy} = \frac{(Tp + Tn)}{(Tp + Tn + Fp + Fn)} * 100 \tag{14}$$

6.2 The ROC Curve

The ROC system gives the results in the [0, 1], and it is recognized that "0" always represents negative and "1" represents positive test. By introducing threshold in between we can test against the known forged image set with tested image. The ROC curve is obtained by plotting the vertical axis representing the sensitivity along with the horizontal axis representing the reversed scale specificity, which is also called a false-positive rate (Fig. 9).

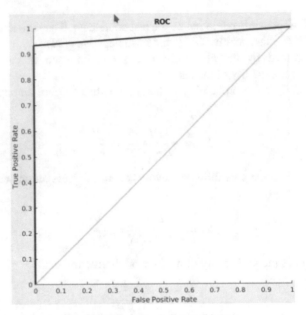

Fig. 9 ROC chart of all tested images

6.3 Overall Image Analysis

For our test purpose, we considered 35 sample images including 30 tampered images. Table 2 shows the results we found with experiment (Fig. 10; Tables 2, 3 and 4).

From Eq. 14 Accuracy $= \frac{28+5}{28+5+0+2} * 100 = 94.29\%$

From Eq. 10 Sensitivity $= SE = 93.33\%$

From Eq. 11 S Pecificity $= SP = 100\%$

From Eq. 12 Precision $= p = 100\%$

From Eq. 13 F measure $= r = 96.55\%$.

Table 2 Experimental results

Image file name	Actual type	Detected as	Result	Image file name	Actual type	Detected as	Result
Org_Img1	Original	Original	TN	Tamp_Img14	Forged	Forged	TP
Org_Img2	Original	Original	TN	Tamp_Img15	Forged	Forged	TP
Org_Img3	Original	Original	TN	Tamp_Img16	Forged	Forged	TP
Org_Img4	Original	Original	TN	Tamp_Img17	Forged	Forged	TP
Org_Img5	Original	Original	TN	Tamp_Img18	Forged	Forged	TP
Tamp_Img1	Forged	Forged	TP	Tamp_Img19	Forged	Forged	TP
Tamp_Img2	Forged	Forged	TP	Tamp_Img20	Forged	Forged	TP
Tamp_Img3	Forged	Forged	TP	Tamp_Img21	Forged	Forged	TP
Tamp_Img4	Forged	Forged	TP	Tamp_Img22	Forged	Forged	TP
Tamp_Img5	Forged	Forged	TP	Tamp_Img23	Forged	Forged	TP
Tamp_Img6	Forged	Original	FN	Tamp_Img24	Forged	Forged	TP
Tamp_Img7	Forged	Forged	TP	Tamp_Img25	Forged	Forged	TP
Tamp_Img8	Forged	Forged	TP	Tamp_Img26	Forged	Forged	TP
Tamp_Img9	Forged	Forged	TP	Tamp_Img27	Forged	Forged	TP
Tamp_Img10	Forged	Forged	TP	Tamp_Img28	Forged	Forged	TP
Tamp_Img11	Forged	Forged	TP	Tamp_Img29	Forged	Forged	TP
Tamp_Img12	Forged	Forged	TP	Tamp_Img30	Forged	Forged	TP
Tamp_Img13	Forged	Original	FN				

$TP = 28, TN = 5, FP = 0, FN = 2$

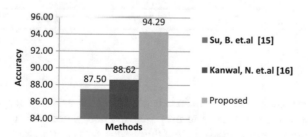

Fig. 10 Accuracy comparison chart

Table 3 Comparison with other methods

Method used	Merits/demerits
DCT	Not compatible with AWGN
PCA	Shows the behavior of robustness against AWGN and JPEG compression
DWT-SVD	Lower time complexity
DWT-log polar coordinates	Has a lower rate of time complexity and also the concerned geometric operations are not discussed
SIFT	Shows the robustness behavior with respect to geometric transformation
FMT	It is observed that it is robust to scaling bussing along with noise addition and SEEG compression
Zernike	It is observed to have a high detection rate mainly not for the seating
PCA-EVD	Will not work in rotation and scaling
Improved DCT	Unable to work within an image that the area distorted by JPEG compression, blurring or AWGN
PCA on DCT	It is robust against noise and JPEG compression and also to achieve invariance to illumination
Proposed	Very efficient against rotated or flipped images. Will not work on compressed

Table 4 Accuracy comparison

	Su, B. et al. [15]	Kanwal, N. et al. [16]	Proposed
Accuracy	87.50	88.62	94.29

7   Conclusion

Shannon entropy method discussed in this paper can be used for copy-move forgery detection in the digital images. The proposed method works well even with flipped images which are not easily possible by other methods. The proposed method can also be used in blurred and brightness adjusted images. The accuracy of the proposed model with various test images is observed to be 95%. By comparing with other methods, the tested proposed method is very efficient in terms of accuracy. Check Table 4 for accuracy compared with other methods. However, this method does not work with small copy-moved images. This method only works with uncompressed image formats. Table 3 mentions different methods that various authors used and lists its merits and demerits and tries to compare with the proposed method.

As the digital image forgery is increasing, the need for forgery detection algorithms is also increasing day-by-day. In this article, the authors have discussed different forgery methods and algorithms to detect the forgery of a digital image. Copy-move forgery is detected for pixels size >0.03% of image size. The currently used methodology is showing 94.29% accuracy performance for reasonable copy-move forgery. The big issues are copied and move forgery where a part of the image is copied and moved to some other place or another image. In this article, it has been cleared of the concept regarding copy-move forgery detection.

8 Future Work

Future work for the proposed model can be worked on improvising methods to work with compressed images like JPEG. Also, artificial neural networks can be used for classification instead of Euclidean distance as it gives more accurate results but increases computation complexity and programming complexity. The proposed model can be modified to work with small-sized copy-moved images.

References

1. Kaur, H., Saxena, J.: Key-point based copy-move forgery detection and their hybrid methods: a review. J. Int. Assoc. Adv. Technol. Sci. **6** (2015). ISSN-4265-0578
2. Zheng, J., Chang, L.: Detection of Region-duplication Forgery in image-based on key points' binary descriptors. J. Inf. Comput. Sci. **11**(11), 3959–3966 (2014)
3. Mohamadian, Z., Pouyan, A. A.: Detection of duplication forgery in digital images in uniform and non-uniform regions. In: Paper Presented at the UKSim 15th International (2013)
4. Thajeel, S.A., Sulong, G.B.: State of the art of copy-move forgery detection techniques: a review. IJCSI **10**(6), 2 (2013)
5. Li, L., Li, S., Zhu, H.: An efficient scheme for detecting copy-move forged images by local binary patterns. J. Inf. Hiding Multimed. Signal Process. **4**, 46–56 (2013)
6. Muhammad, G., Hossain, M.S.: Robust copy-move image forgery detection using undecimated wavelets and Zernike moments. In: Proceedings of the Third International Conference on Internet Multimedia Computing and Service, pp. 95–98 (2011)
7. Zimba, M., Xingming, S.: DWT-PCA (EVD) based copy-move image forgery detection. Int. J. Digital Content Technol. Appl. **5**(1), 251–258 (2011)
8. Huang, Y., Lu, W., Sun, W., Long, D.: Improved DCT-based detection of copy-move forgery in images. Forensic Sci. Int. **206**, 178–184 (2011)
9. Wu, Q., Wang, S., Zhang, X.: Log-polar based scheme for revealing duplicated regions in digital images (2011)
10. Bayram, S., Sencar, H. T., Memon, N.: An efficient and robust method for detecting copy-move forgery. In: Paper presented at Acoustics, Speech and Signal Processing. IEEE International Conference on ICASSP (2009)
11. Li, G., Wu, Q., Tu, D., Sun, S.: A sorted neighborhood approach for detecting duplicated regions in image forgeries based on DWT and SVD. In: IEEE International Conference on Multimedia and Expo, pp. 1750–1753 (2007)
12. Myna, A., Venkateshmurthy, M., Patil, C.: Detection of region duplication forgery in digital images using wavelets and log-polar mapping. In: International Conference on Computational Intelligence and Multimedia Applications, pp. 371–377 (2007)

13. Fridrich, A.J, Soukal, B.D, Lukáš, A.J.: Detection of copy-move forgery in digital images. In: Proceedings of Digital Forensic Research Workshop (2003)
14. Kalin, L.W.: Recceiver Operating Characteristic (ROC) Analysis. UMINIF (2018)
15. Su, B., Yuan, Q., Wang, S., Zhao, C., Li, S.: Enhanced state selection markov model for image splicing detection. EURASIP J. Wirel. Commun. Netw. **2014**(1), p. 7 (2014)
16. Kanwal, N., Girdhar, A., Kaur, L.; Bhullar, J.S.: Detection of digital image forgery using fast fourier transform and local features. IEEE Trans. Autom. Comput. Technol. Manage. (2019)

Application of GIS and Remote Sensing for Land Cover Mapping: A Case Study of F-North Ward, Mumbai

Darshan Sansare[✉] and Sumedh Mhaske

Department of Civil & Environmental Engineering, VJTI,
Matunga, Mumbai 400019, India
darshansansare@gmail.com

Abstract. Socio-economic survey plays an important role for development of society. Land cover assessment is a vital part of these surveys. The surface of land covered by water, bare soil, vegetation, urban infrastructure, and so on is referred as 'Land Cover'. In the city like Mumbai the land cover is haphazardly raised in past few years. The maximum amount of land cover is found increased in the form of concrete surfaces. It is responsible for water-logging issues in many parts of the city. Out of which, it is found that F/North ward consists of maximum numbers of water-logging spots. Therefore, F/North ward is particularly selected as a present study area. Since losses and damages due to flooding and water-logging are huge, decision-making governing bodies require accurate information of the affected area for developing appropriate flood protection and drainage system for minimizing the severity. For establishing sustainable development of any city, it is mandatory to monitor the ongoing process of land cover pattern considering a certain time period. The detailed micro-level information about land cover of a particular region will be very beneficial for the policy makers to make regulatory policies and social environmental programs to save the environment. The purpose of this study is to comprehensively investigate factors which are responsible for frequent water-logging in F-North ward of Mumbai in terms of detail database of land cover and creating thematic map for the study area using GIS tools and techniques.

Keywords: Land cover · Water-logging · GIS database · F-North ward · Mumbai

1 Introduction

Mumbai is extremely limited by its geography and occupies a small area of land about 458.28 km^2. There is an increase in population in the urban areas as people have migrated from the villages to the cities in search of livelihood. Urbanization is the prime cause of the alterations in hydrologic and hydraulic processes, loss of current drainage capacity and occurrence of flooding situations in urban areas. The migration of the rural population to the city combined with an increase in population, urbanization and limited land availability has resulted in an increase in concrete cover. As concrete is impervious by

© Springer Nature Singapore Pte Ltd. 2020
B. Iyer et al. (eds.), *Applied Computer Vision and Image Processing*,
Advances in Intelligent Systems and Computing 1155,
https://doi.org/10.1007/978-981-15-4029-5_9

nature, there is a limited scope for water infiltration and hence the rain water is directly converted to runoff. As most of the drains get clogged due to waste materials, there are increased chances of occurrence of water-logging.

Land cover change and its severe impacts on surrounding environment have become a leading research area in the recent years [1–3]. Significant change in LULC has impacted on hydrology, and increase in peak discharge and the clogged drains has resulted in frequent occurrence of water-logging in Mumbai. Study has suggested solution in the form of porous concrete for storm water management of Mumbai [3–5]. Land cover maps play significant role in planning and monitoring of small-scale programs at regional level and large scale at national level. This information not only provides a clear understanding of land utilization aspects but also it is useful for study of the changes that are happening in our ecosystem. Environmental assessment includes identification of undeveloped and vacant land, planning of green space for recreational purposes and other urban planning applications. This survey consists of both nonspatial and spatial data sets.

For evaluation of LULC, toposheets and satellite images provide proper spatially distributed data [6–8].

Geographic information system (GIS) is a computer-based information system which capture, store, analyze and display geographically referenced information and also generate database in the form of attribute table. Mhaske and Choudhury [9] used geographic information system (GIS) and global positioning system (GPS) to obtain the soil index property maps for Mumbai city. About 450 soil testing reports inclusive of soil properties, like specific gravity, ground water level, liquid limit, moisture content and so on, were collected from various sites to create a database useful for geotechnical professionals.

In coastal cities such as Mumbai, area of land is very limited and most valuable. The aim of the present study is to analyze the land cover of the year 2018 by using manual digitizing process for F-North ward catchment area in Mumbai, India. This paper is projected to generate detail database of land use cover and generating 2D and 3D maps for F-North ward of Mumbai city using open-source software.

2   Study Area

Mumbai is considered as financial nerve center of India and it is among the top-ten largest urban agglomerations in the world. The area is covered between latitude 18° 53′ N–19° 15′ N and longitude 72° 48′ E–73° 00′ E. Mumbai is inherently vulnerable to floods (Government of Maharashtra 2006). Mumbai and its suburbs are distributed in total of 24 wards. F-North is the ward consisting of more flood spots compared to the other wards of Mumbai. Hence this ward was selected for the study. The area of F-North ward is 12.9 km^2. As per Sansare et al. 2019, the study has carried out risk analysis, hazard analysis and its thematic mapping for F-North ward and entire Mumbai city [10, 11] (Fig. 1).

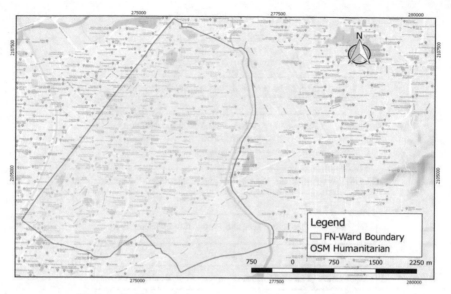

Fig. 1 Thematic map for study area F-North ward boundary

3 Methodology

Data collected include: satellite images from USGS, toposheet and the field surveyed data. The data related to F-North ward, Mumbai is collected from MCGM (Municipal Corporation of Greater Mumbai) website and is georeferenced in Q-GIS to locate the F-North ward boundary. Similarly, polygon layers are created for inner boundaries, outer boundaries of city and for creating parcels, for example, buildings, parks, industrial zones, and so on. Line layer is created for internal roads of F-North ward.

The map is digitized using the above layer classification which is needed to generate the attribute table which contains details on the dimensions which aid in the calculation of land cover area. The study analyzes the land cover of the year 2018 by using manual digitizing process for F-North ward catchment area in Mumbai, India. Land cover of 3116 buildings (including slums), areas under infrastructures, road, open spaces (like barren land, farmland, grass, park, playground, railway, recreation ground, scrub, sports ground, wetland and water), parcel (commercial, residential, slum residential, educational, industrial, parking, railway site, sports center, substation) and so on are digitized in QGIS software by using polygon, segment layer to analyze total area. About 342 roads are digitized by using line (segment) layer, and the actual width of the road has put in terms of buffer, and an attribute table is generated for finding total area.

Flow chart of proposed methodology

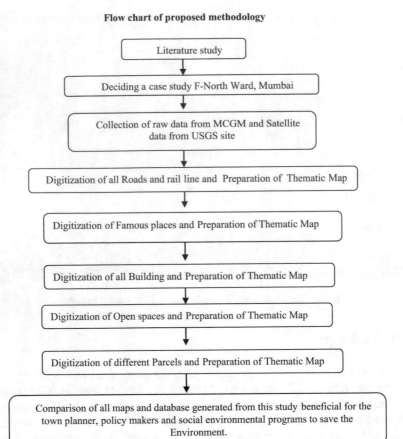

Literature study

Deciding a case study F-North Ward, Mumbai

Collection of raw data from MCGM and Satellite data from USGS site

Digitization of all Roads and rail line and Preparation of Thematic Map

Digitization of Famous places and Preparation of Thematic Map

Digitization of all Building and Preparation of Thematic Map

Digitization of Open spaces and Preparation of Thematic Map

Digitization of different Parcels and Preparation of Thematic Map

Comparison of all maps and database generated from this study beneficial for the town planner, policy makers and social environmental programs to save the Environment.

4   Results and Discussions

See Figs. 2, 3, 4, 5, 6, 7 and Table 1.

5   Conclusion

The database generated from this study gives the clear and detail information regarding 12.9 km^2 land cover distribution of F-North ward, Mumbai.

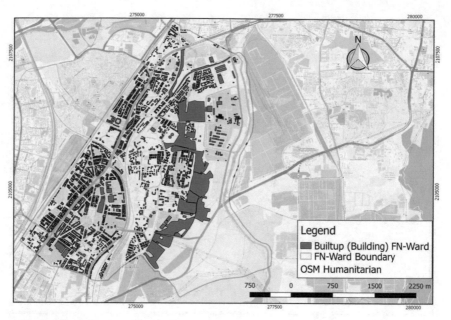

Fig. 2 Thematic map of F-North ward which shows total constructed area (building)

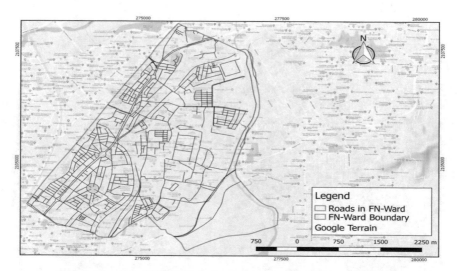

Fig. 3 Thematic map of F-North ward which shows roads distribution

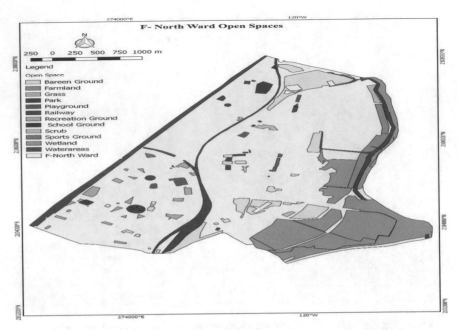

Fig. 4 Thematic map of F-North ward which shows open spaces distribution

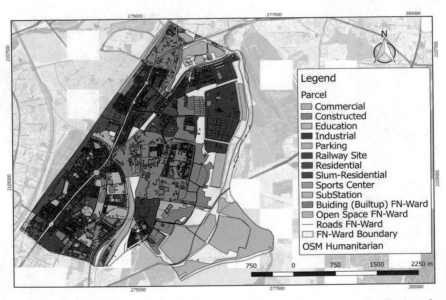

Fig. 5 Thematic map of F-North ward which shows all layers of land covers distribution of year 2018

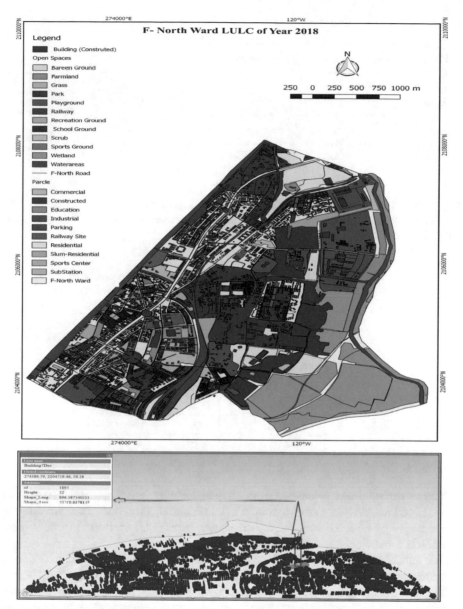

Fig. 6 Thematic map of F-North ward which shows land cover distribution of year 2018 in 2D and in 3D

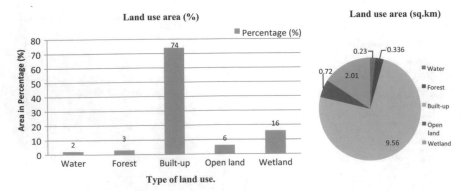

Fig. 7 The above column chart indicates the land cover in percentages w.r.t. total catchment (FN-ward) for different geographical features, and pie chart indicates land cover area in (km^2) over the period considered

Table 1 Land cover assessment of the study area for the year 2018

S. no.	Types of land use	Year—2018	
		Area (km^2)	% Total
1	Water	0.23	2
2	Forest	0.336	3
3	Built-up	9.56	74
4	Open land	0.72	6
5	Wetland	2.01	16
	Total	12.856	100

- The sum of total area calculated from attribute table for building (constructed, slum) = 2.348 km^2, and for parcel (commercial, residential, slum-residential educational, industrial, parking, railway site, sports center, substation) calculated from attribute table is 8.4 km^2.
- The sum of total area under road calculated from attribute table is 0.72 km^2 and area under footpath after considering width of 1.5 m on both sides of road = 0.40 km^2.
- The sum of total open spaces (barren land, farmland, grass, park, playground, railway, recreation ground, scrub, sports ground, wet land, water) area calculated from attribute table is 3.41 km^2 from that of wet land = 2.01 km^2, water = 0.23 km^2.

The study proves that GIS and remote sensing provide an efficient way to land cover identification and area calculation.

To achieve sustainable urban development and to control the haphazard development of cities, it is necessary to associate the governing civic bodies with the urban development planning. The planning models shall be generated so that inch by inch of available land can be utilized efficiently in most optimal and rational way.

Changes in land cover are rapid, pervasive and can have serious impacts on people, economy and environment of any country. Applications of land cover maps are useful for natural resource management and resource extraction activities. Also, it is beneficial for the governing bodies to make regulatory polices to control urban expansion by encroachments, and thereby protecting wildlife habitat. In addition to this, land cover maps are useful for deciding legal boundaries for property and tax evaluation target detection.

Acknowledgments. The authors are grateful to the Municipal Corporation of Greater Mumbai and VJTI, Mumbai for cooperation for this study. Also thankful to Dr. Babasaheb Ambedkar Research and Training Institute (BARTI) for awarding fellowship for Ph.D. Research purpose.

References

1. Amini, A., Ali, T., Ghazali, A., Aziz, A., Aikb, S.: Impacts of land-use change on stream flows in the Damansara Watershed Malaysia. Arab J. Sci. Eng. **36**(5), 713–720 (2011)
2. Chen, Y., Xu, Y., Yin, Y.: Impact of land use change scenarios on storm-runoff generation in Xitiaoxi basin. China. Ouat. Int. **208**, 121–128 (2009)
3. Fox, D.M., Witz, E., Blance, V., Soulie, C., Penalver-Navarro, M., Dervirux, A.: A case study of land-cover change (1950-2003) and runoff in a Mediterranean Catchment. Appl. Geogr. **32**(2), 810–821 (2012)
4. Mhaske, S.Y., Choudhury, D.: GIS-based soil liquefaction susceptibility map of Mumbai city for earthquake events. J. Appl. Geophys. **70**(3), 216–225 (2010)
5. Sansare, D.A., Mhaske, S.Y.: Analysis of land use land cover change and its impact on peak discharge of storm water using GIS and remote sensing: a case study of Mumbai City, India. Int. J. Civ. Eng. Technol. (IJCIET) **9**(11), 1753–1762 (2018)
6. Kulkarni, A.T., Bodke, S.S., Rao, E.P., Eldho, T.I.: Hydrological impact on change in land use/land cover in an urbanizing catchment of Mumbai: a case study. ISH J. Hydraul. Eng. **20**(3), 314–323 (2014)
7. Sansare, D.A., Mhaske, S.Y.: Storm water management of Mumbai City by using pervious concrete on GIS based platform. IJRTE **8**(2), 4400–4006 (2019)
8. Sansare, D.A., Mhaske, S.Y.: Risk analysis and mapping of natural disaster using QGIS tools for Mumbai City. Disaster Adv. **12**(10), 14–25 (2019)
9. FFC (Fact Finding Committee), Maharashtra State Govt. Committee Report, pp. 31–130 (2006)
10. Miller, S.N., Kepner, W.G., Mehaffey, M.H., Hernandez, M., Miller, R.C., Goodrich, D.C., Kim Devonald, K., Heggem, D.T., Miller, W.P.: Integrating landscape assessment and hydrologic modeling for land cover change analysis. J. Am. Water Resour. Assoc. **38**, 915–929 (2002)
11. Sansare, D.A., Mhaske, S.Y.: Natural Disaster Analysis and Mapping using Remote Sensing and QGIS Tools for FNorth ward, Mumbai City. India. Disaster Adv. **12**(1), 40–50 (2019)

A Novel Approach for Detection and Classification of Rice Leaf Diseases

Yash Kumar Arora, Santosh Kumar, Harsh Vardhan$^{(\boxtimes)}$, and Pooja Chaudhary

Department of Computer Science and Engineering, Graphic Era deemed to be University, Dehradun 248001, Uttarakhand, India
arorayash1998@gmail.com, amu.santosh@gmail.com, harshsam80@gmail.com, poojachaudharypcm@gmail.com

Abstract. India stands tall as one of the world's largest rice-producing countries. A major part of Indian agriculture consists of rice as the principal food crop. Rice farming in India is challenged by diseases that can infest and destroy the crops causing detrimental losses to the farmers. Thus, the detection of diseases like "leaf smut", "brown spot", and "bacterial leaf blight" becomes a need of the hour. In this paper, we have proposed a way that can efficiently detect and classify these three diseases through image processing. The research can help in knowing if the rice crop is infested with the diseases or not. Images of the infected crop can be used in a real-life scenario and one can know if it is infested with any of the three diseases mentioned. The detection and classification of these diseases have been made possible using various state-of-the-art classification models, like support vector machine (SVM), random forest, KNN, naïve Bayes, and neural network.

Keywords: Rice leaf disease · Classification · Detection · Image processing

1 Introduction

The most exquisite fact about India is the dependency of this country on agriculture. Agriculture and its allied activities make for the country's 80% sources of livelihood while its contribution to the GDP is around 15%. Rice tops the list of important and unique crops in India [1]. Unlike so many other crops, rice can flourish in wet and humid conditions. Rice serves as both, a cash crop as well as a food crop. Farmers in different parts of the country face umpteen challenges in the preservation and nourishment of this crop against diseases. Therefore, early identification of rice diseases and taking remedial measures to control them has become utmost important [2]. Rice diseases cause more than 10% loss in rice production annually, which is catastrophic for the farmers. Diseases like bacterial leaf blight, brown spot, and leaf smut can ransack the rice fields causing a lot of damage.

Bacterial leaf blight forms lesions on the leaves that start from the tip and spread to the leaf base [3]. It is caused by bacteria named Xanthomonas oryzae. The disease can spread easily through water and wind, infecting the whole rice paddy within a few weeks. Brown spot is a fungus infecting the coleoptile, panicle branches, leaves, and

© Springer Nature Singapore Pte Ltd. 2020
B. Iyer et al. (eds.), *Applied Computer Vision and Image Processing*,
Advances in Intelligent Systems and Computing 1155,
https://doi.org/10.1007/978-981-15-4029-5_10

spikelets. This disease strikes most severely at the maturity stage of the crop, causing a reduction in yield. The work in one of the researches by Santanu Phadikar et al. [4] helped us understand the disease better and allowed us to find ways to detect it. It forms numerous big spots on the leaves which can prove fatal for the crop. The fungus can travel through air infecting majority of the crops in the field. The brown spot fungus has been known to thrive in the seeds for more than four years. Leaf smut is also a fungus that can form small raised black spots on the rice leaves. This noxious disease makes the crop vulnerable to other diseases which can ultimately kill the crop. It spreads through contact of a healthy leaf with infected pieces of leaves in soil.

Humans can detect the unusual formation of lesions or spots on the leaves but identifying the disease that has infected the rice leaf can be cumbersome. Moreover, the local conditions can cause the symptoms to look similar to other diseases. This causes implementation of gratuitous and unnecessary remedial measures that prove to be futile in the treatment of the crop. Through image processing, the classification and detection of these diseases prove to be more efficient and accurate. In our research, image processing allows for the timely detection and precise classification of "bacterial leaf blight", "brown spot", and "leaf smut" so that the right remedial measures can be used in the treatment of the crops.

2   Related Works

An overwhelming increase in the field of image processing for plants-related diseases has been seen in the past few years. Rice disease identification using pattern recognition techniques [4] can be made possible by classifying the diseased leaf using SOM neural network. Through extraction of the infected parts of the leaves, the train images were obtained. The research includes identification of diseases such as leaf blast and brown spot. The research gave an accuracy of 92% in the usage of RGB of the spots for classification [5]. Diseases like brown spot and blast diseases were detected using pattern recognition. Our research enables the classification and identification of three different rice leaf diseases, namely "brown spot", "leaf smut", and "bacterial leaf blight" through image processing.

A survey on the classification and detection of rice diseases includes techniques like machine learning and image processing for the classification and detection of diseases that provide an insight into the various aspects of these techniques for the detection of rice diseases [6]. Another important research found in this field focuses on the classification and detection of diseases in leaves of plants by using image processing [7]. The major techniques used in the research are K-means clustering, GLCM, and BPNN. Our approach introduces use of classification models like SVM for the classification and detection of rice diseases.

Our research has optimized the use of image processing techniques for detecting the specific diseases occurring in rice. In another research application of SVM is discussed for the detection of rice diseases by focusing on features like color and shape that allow the detection of three diseases, namely rice sheath blight, bacterial leaf blight, and rice blast. The research enabled us to study the use of SVM in classification of other diseases like "leaf smut", "brown spot", and "bacterial leaf blight". The use of SVM in this research gave us an insight into the implementation of different effective classification models for the classification of rice leaf diseases. Another study based on Matlab application shows the diagnosis of diseases like narrow brown spot disease and brown spot disease in crops [8].

3 Methodology

In this paper, our approach is shown in Fig. 1 which includes processes like image acquisition, image processing, feature extraction, detection, and classification. The images of healthy as well as diseased leaves were captured and edited for uniformity. Several feature extractors like Inception v3, Visual Geometric Group 16, and Visual Geometric Group 19 were used to find inherent properties in the images captured. The images were then processed for detection and classification using state-of-the-art classification models like SVM, random forest, naïve Bayes, and neural network. Our research showed that more than 90% of the images were classified accurately using SVM. The detailed description of these processes is provided in the sub-sections.

Fig. 1 Our proposed approach

3.1 Image Acquisition

Diseased rice leaves infected with three diseases, namely "leaf smut", "brown spot", and "bacterial leaf blight", were collected from nearby rice fields during the harvesting season located on the outskirts of Dehradun. The photos are captured using a Canon EOS 1300D with 18–55 mm lens. A total of 160 images of each category are captured. The number of images for healthy leaves was also the same.

3.2 Pre-processing of Diseased Leaf Images

The captured images are pre-processed for better and accurate results. The images are refined and resized for uniformity and to reduce the computational burden. All the images are resized to a resolution of 200 × 200 pixels for precise results. Contrast enhancements and exposure refinements were made to images where the contrast and exposure needed amendments. Figure 2 contains the sample dataset of leaves after pre-processing.

3.3 Feature Extraction

Feature extraction mainly focuses on the identification of characteristics and features that are inherent in a given image such as shape [7, 9]. Several feature extractors are used, namely Inception v3, Visual Geometric Group 16, and Visual Geometric Group

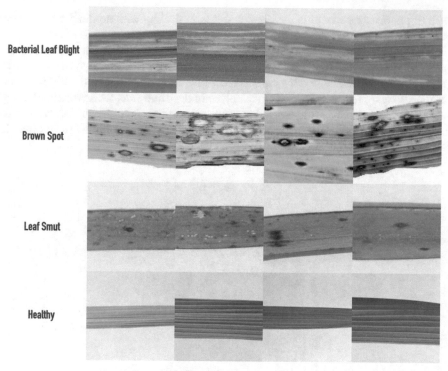

Bacterial Leaf Blight

Brown Spot

Leaf Smut

Healthy

Fig. 2 Sample dataset of leaves after pre-processing

19. Our work showed that the use of Inception v3 yields the best results for our sample images. Inception v3 is an image recognition model that is widely used and is based on a convolutional neural network. It is based on the paper by Szegedy et al. "Rethinking the inception architecture for computer vision" [8].

3.4 Detection

Our approach in this paper involves a two-stage classification process. The first stage includes detecting the presence of a disease on the plant. The second stage outputs if the plant is infested with any of the three diseases, namely: leaf smut, bacteria leaf blight, and brown spot. Out of the total number of images of diseased leaves, 80% are used as training model while the remaining 20% are used as testing models. Various state-of-the-art classification models are used to get the best results. We used support vector machine (SVM), random forest, KNN, naïve Bayes, and neural network for the classification of detected diseases. Among the mentioned classification models, support vector machine gave the best results with an average accuracy of 90% in classifying the three diseases, namely: bacteria leaf blight, leaf smut, and brown spot. Using SVM algorithms differentiation and classification of diseases in plants can be achieved [7]. The implementation of a neural network gave an accuracy of 88.3%. SVM is a machine learning algorithm that is based on statistical theory put forward by Vapnik in the 1990s

[10]. Support vector machine has a great ability in dealing with nonlinearity and small samples [2].

Using SVM a decision surface (H) is found, which is determined through some points of training set, known as support vectors taken between two-point classes. The training data ($''$, $''$) is divided by the surface without mistake, that is, the division of all points belonging to the same class takes place in the same side, while maximal margin is the minimum distance that exists between this surface and either of the two classes (Fig. 3). Using the solution of a problem of quadratic programming helps in obtaining this surface [11] (Table 1).

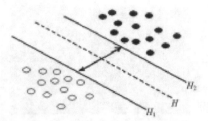

Fig. 3 Optimal plane of SVM

Table 1 Results for rice leaf classification

Model	AUC	CA	F1	Precision	Recall
kNN	0.978	0.833	0.833	0.833	0.833
SVM	0.987	0.908	0.909	0.918	0.908
Random forest	0.928	0.783	0.784	0.789	0.783
Neural network	0.982	0.883	0.884	0.885	0.883
Naïve Bayes	0.969	0.850	0.851	0.858	0.850

For linearly separable points, we can estimate the function of the surface as (1):

$$f(m) = sgm\left(\sum_{i=1}^{n} \alpha_i^* n_i (m_i \cdot m) + b^*\right) (m_i, n_i) \in R^N \times \{-1, 1\} \tag{1}$$

where Lagrange multiplier is α_i^* and bias is b^*.

For nonlinearly separable classes, we give the function of the surface as (2):

$$f(m) = sgn\left(\sum_{i=1}^{n} \alpha_i^* n_i k(m, n) + b^*\right) \tag{2}$$

where $k(m, n)$ is a kernel function.

The kernels that were commonly used are:
Linear function:

$$k(m, n) = m \cdot n \tag{3}$$

Polynomial function:

$$k(m, n) = (1 + mvn)^q, q = 1, 2, \ldots, N \tag{4}$$

Radial basis function:

$$k(m, n) = \exp(-||m - n||)^2 \tag{5}$$

4   Results and Discussion

Implementation of different classification models gave different results and accuracy. In the analysis, it is observed that the convolutional neural network and kNN gave an average accuracy of 88.3 and 83.3%, respectively. The support vector machine (SVM) gave an accuracy of 90.8% in the identification and classification of the diseases which is the highest accuracy. For the respective diseases, after using SVM 97.5% of the images are classified correctly for bacterial leaf blight, 90.8% of the images are classified correctly for brown spot disease, and 90% of the images are classified correctly for leaf smut disease. Figures 4, 5, and 6 represent the receiver operating characteristic (ROC) curve of bacterial leaf blight, the ROC curve of brown spot, and the ROC curve of leaf smut, respectively.

5   Conclusion

A system for detecting and classifying the rice leaf diseases has been proposed in this paper. Comparison of some very effective and state-of-the-art classification models has shown that SVM yields the best accuracy for detection and classification of the diseases like bacterial leaf blight, leaf smut, and brown spot. With some optimization, this system can also be used for the detection of other rice leaf diseases in the future. With the accuracy of 90%, thus it can be implemented and used for the detection of diseases in rice fields that can prove fruitful for the farmers.

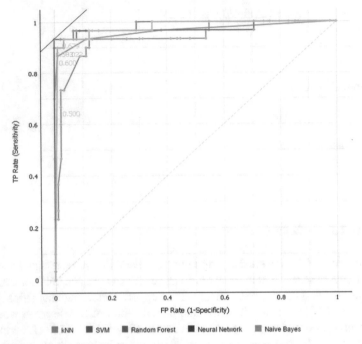

Fig. 4 ROC curve for bacterial leaf blight

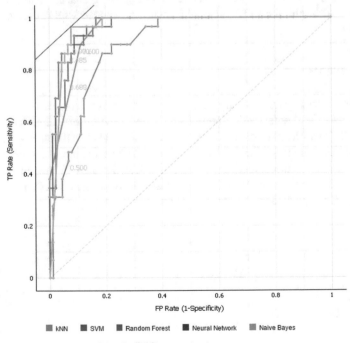

Fig. 5 ROC curve for brown spot

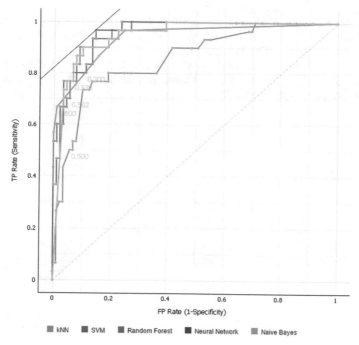

Fig. 6 ROC curve for leaf smut

References

1. Sladojevic, S., Arsenovic, M., Anderla, A., Culibrk, D., Stefanovic, D.: Deep neural networks based recognition of plant diseases by leaf image classification. In: Hindawi Publishing Corporation Computational Intelligence and Neuroscience, vol. 2016, p. 11. Article ID 3289801
2. Yao, Q., Guan, Z., Zhou, Y., Tang, J., Hu, Y., Yang, B.: Application of support vector machine for detecting rice diseases using shape and color texture features. In: 2009 International Conference on Engineering Computation
3. Phadikar, S., Sil, J., Das, A.K. Rice diseases classification using feature selection and rule generation techniques. Comput. Electron. Agri. **90**(2013), 76–85
4. Phadikar, S., Sil, J.: Rice disease identification using pattern recognition techniques. In: Proceedings of 11th International Conference on Computer and Information Technology (ICCIT 2008)
5. Sanyal, P., Patel, S.C.: Pattern recognition method to detect two diseases in rice plants. Imaging Sci. J. **56** IMAG mp171# RPS (2008)
6. Shah, J.P., Prajapati, H.B., Dabhi, V.K.: A Survey on Detection and Classification of Rice Plant Diseases. IEEE (2016)
7. Mainkar, P.M., Ghorpade, S., Adawadkar, M.: Plant leaf disease detection and classification using image processing techniques. Int. J. Innov. Emerg. Res. Eng. **2**(4) (2015)
8. Szegedy, C., Vanhoucke, V., Ioffe, S., Shlens, J., Wojna, Z.: Proceedings of the IEEE Computer Society Conference on Computer Vision and Pattern Recognition, vol. 2016, pp. 2818–2826. IEEE Computer Society (2016)

9. Kurniawati, N.N., Abdullah, S.N.H.S., Abdullah, S., Abdullah, S.: Texture analysis for diagnosing paddy disease. In: 2009 International Conference on Electrical Engineering and Informatics 5–7 August 2009, Selangor, Malaysia
10. Vapnik, V.: The Nature of Statistical Learning Theory. Springer, New York (1995)
11. Yue, S.H., Li, P., Hao, P.Y.: SVM classification: its contents and challenges. Appl. Math. J. Chinese Univ. Ser. B **18**(3), 332–334 (2003)

Transportation Spend Portfolio Optimization Using Multi-level, Multi-criteria Decision Analysis

Arti Deshpande[1]([✉]), Mamta R. Kumar[2], and Jonathan Chakour[3]

[1] Thadomal Shahani Engineering College, Mumbai, India
arti.deshpande@thadomal.org
[2] Cobotics Business Services LLC, Duluth, GA, USA
mamta.kumar@coboticsllc.com
[3] Logistics Ops Expert, Cumberland, RI, USA
jchakour@gmail.com

Abstract. An efficient supply chain is the backbone of any successful business. Such efficient supply chain relies on the performance of its suppliers and partners. Alignment of priorities is critical for strategic partnership. An enterprise can create greater value for their products and services by ensuring that their service providers and partners share common priorities and are willing to adjust their operations to meet the common goals. Careful partner evaluation based on quantifiable performance indicators is necessary. Conflicting criteria are typical in such partner evaluations. The AI-based solution explained in this paper leverages machine's ability to perform complex multi-level, multi-criteria-based ranking decisions. It allows decision makers to configure business criteria which are important to meet their goals. Behind the scene multi-criteria decision-making (MCDM) algorithms perform partner and quote ranking based on the latest uploaded request for quote (RFQ) responses and user-defined configuration setting of important business criteria, their relative importance and monotonicity. The proposed system also simulates your spend portfolio by changing the business distribution across various partners and service levels to achieve most optimal results prior to temporary or final award. The solution proposed optimizes your spend portfolio, strengthen your partner relationship and create greater value for your products and services.

Keywords: Machine learning · Ranking · Sourcing and costing · Logistics and transportation cost optimization

1 Introduction

Lack of systematic analysis to explicitly evaluate the multiple conflicting criteria impacts the decision making, especially when stakes are high. Conflicting criteria are typical in evaluating options like cost or price, which is usually one of the main criteria, and some measure of quality is typically another criterion that is easily in conflict with the cost. Both of these conflicting criteria influence the current and future state of the businesses.

© Springer Nature Singapore Pte Ltd. 2020
B. Iyer et al. (eds.), *Applied Computer Vision and Image Processing*,
Advances in Intelligent Systems and Computing 1155,
https://doi.org/10.1007/978-981-15-4029-5_11

Owing to lack of any unbiased and systematic analysis, we usually weigh multiple criteria implicitly and find out the consequences of such decisions that are made based on only intuition afterwards. When stakes are high, it is important to properly structure the problem and explicitly evaluate multiple criteria without cognitive biases.

Costing and sourcing organization of major enterprises, especially in transportation industry, are immensely pressurized due to e-commerce-driven changing business priorities and customer expectations. Traditional rule-based, cost-driven negotiation strategy no longer helps to meet the customer's expectations, which demand both quality and agility at lowest cost. It is imperative for businesses to adapt to dynamic multiple criteria-based decision-making processes. Managing and optimizing logistics spend and customer portfolio across complex and voluminous worldwide transportation network and multiple carriers is both resource and time intensive.

The solution for this is to exploit machine's ability to perform complex multi-level, multi-criteria-based ranking decisions for a complete evaluation at both strategic and operational perspective prior to sourcing and awarding. The technique proposed ranks the carrier partners and their quoted rates in-depth, including both transportation and accessorial cost elements, which help to eliminate unintentional cognitive bias and false-consensus effect from awarding and business decision making. The proposed model learns from historic bid and award decision to predict future cost trends by mode, partner, lane and service level across all price models. It leverages data insights to proactively predict and manage transportation spend portfolio.

In blog [1] on spend management strategies for optimized freight logistics, the author explained business approach freight spend management to achieve the results. According to the Federal Highway Administration, average overall logistics costs break down into three major expenses: transportation (63%), inventory-carrying costs (33%) and logistics administration (4%). So, the transportation plays a major role in costing and need optimization.

In paper [2], various challenges for transportation sourcing are given like management risk, uncertainty and so on. The author has elaborated four major additional factors to be considered while awarding business as capacity constraints, quantity discounts, historic service levels and carbon footprint.

2   Multi-criteria Decision-Making (MCDM) Approach

Applying single MCDM methods to real-world decisions, the progression of technology over the past couple of decades has allowed for more complex decision analysis methods to be developed [3]. There are many methods available like multi-attribute utility theory, analytic hierarchy process, fuzzy set theory, case-based reasoning, simple multi-attribute rating technique, simple additive weighting and so on. MCDM is divided into multi-objective decision-making (MODM) and multi-attribute decision-making (MADM) [4].

The technique used for ranking is MADM and the problem is associated with multiple attributes having different measuring units. Attributes are also referred to as "goals" or "decision criteria". If the number of attributes is large or more, then those can be arranged in a hierarchical manner or in terms of different level. Therefore, the concept of multi-level ranking is used while ranking. There are some major attributes which are associated with several sub-attributes and each sub-attribute may be associated with several sub-sub-attribute and so on [4].

2.1 Simple Additive Weighting (SAW)

SAW algorithm is used for ranking partners. All values are to be normalized before applying SAW [5] method. Min–max normalization [6, 7] is used to normalize all values in a range of [0, 1]. Benefit and cost criteria functions used are given in Eqs. 1 and 2, respectively.

$$f_{nk} = \frac{f_k - k_{\min}}{k_{\max} - k_{\min}} \tag{1}$$

$$f_{nk} = \frac{k_{\max} - f_k}{k_{\max} - k_{\min}} \tag{2}$$

Weight for each criterion is considered based on domain knowledge of business [8] in such a way that sum of weights of all criteria is equal to 1 as per the domain knowledge and business preference, and weight can be assigned to each criterion. Calculate rank for each partner using SAW [5] by applying formula given in Eq. 3.

$$A_n : \frac{\sum_{k=1}^{K} w_k \cdot f_{nk}}{\sum_{k=1}^{K} w_k} \tag{3}$$

where A_n: final rank of partner, w_k: weight given to criteria, and f_{nk}: normalized value of attribute. Final ranks for all partners are calculated by using the multi-level ranking.

2.2 Multi-level Ranking

MCDM criteria used for decision making are shown in Fig. 1. It illustrates the overall request for quote (RFQ) and business award process where we embed the MCDM ranking for the optimized business award. Response data against the RFQ is ingested and based on configurable business priorities and their respective monotonicity partners are ranked. Based on the ranking, business awards are simulated and carried out. This iterative process continues until the final RFQ round.

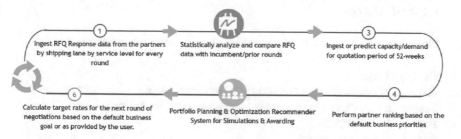

Fig. 1 MCDM approach used for ranking

Cost is considered as one of the criteria while calculating the overall partner ranking. This involves various implicit costs as shown in Fig. 2.

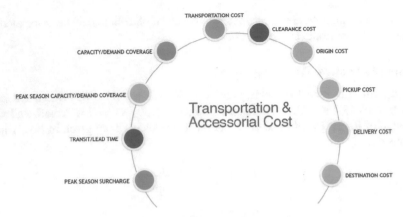

Fig. 2 Cost criteria

2.3 Proposed Algorithm

So the overall cost is ranked based on the values, monotonicity and weightage given to each cost element and sub-element. Multi-level ranking is performed. First level evaluates various service providers (partners) based on their quoted transportation, assessorial cost elements and sub-elements by lane id or shipping routes and service levels. Subsequently, ranked partners are further evaluated across various lane id or shipping routes that fall under a given lane grouping. One lane group can have multiple lane id or shipping routes and it is critical to evaluate the partners and their quotations at both level for more optimized results.

Major parameters considered for ranking are:

1. Transit time
2. Transportation cost
3. Assessorial cost
4. Available peak capacity/kg
5. Available non-peak capacity/kg
6. Airport to airport (ATA) or door to door (DTD) or airport to door (ATD) or door to airport (DTA) cost.

Further, transportation and assessorial cost has several sub-components like

1. Pick up charge
2. Average of origin charges amount by weight categories
3. Transportation charge by weight categories (airport to airport)
4. FSC surcharge amount
5. Security surcharge amount
6. Peak season surcharge amount
7. Destination charges amount
8. Clearance charges amount
9. Average of delivery charges by weight categories.

Two-additional calculated variables were used as described below:

1. Ratio Airport to Destination = Transportation Charge by Weight Categories (Airport to Airport)/Destination Charges Amount
2. Ratio Airport = Transportation Charge by Weight Categories (Airport to Airport)/(Pick Up Charge + Origin Charges Amount).

Pseudocode for proposed algorithm is given below:

Input: Transactional dataset D, n = Number of RFQ response parameters, m = Number of participating partners, Weightage for each criterion considered with domain knowledge is $W_{1 \ldots n}, p_{1 \ldots n}$ are input parameters, and $P_{1 \ldots m}$ are participating partners.

1. For each lane L and service level SL group within D

   a. Find minimum Min(p) and maximum Max(p) value of every input parameter $p(1)\ldots p(n)$ across all partners $P(1)\ldots P(m)$
   b. Calculate normalization function for each parameter $p(1)\ldots p(n)$ based on the business-defined monotonicity.

      i. For maximization, subtract the minimum value Min(p) from each parameter value $p(1)\ldots p(n)$ and divide it by the difference of the maximum value Max(p) and minimum value Min(p)
      ii. For minimization, subtract each parameter value $p(1)\ldots p(n)$ from the maximum value Max(p) and divide it by the difference of the maximum value Max(p) and minimum value Min(p)

2. End For
3. For each parameter p in D

   a. Add all sub-cost elements together to calculate overall normalized value for the cost element.
   b. Multiply it with the respective weightage $W_{1\ldots n}$ to calculate final score.

4. End For.

Output: Partner rank calculations based on the total score using the algorithm on multiple quotes provided by lane and service level

3 Transportation and Assessorial Cost-Based Ranking

Initially, using ranking algorithm SAW, best possible transportation cost is calculated considering both transportation and assessorial sub-components and then the cost-based ranking is used for hierarchical ranking of service providers. SAW, WSM (weighted sum model) [9, 10], weighted linear combination method [11] and SM (scoring method) calculate the overall score of each partner by calculating the weighted sum average of all the attribute values. SAW calculates an evaluation score for every alternative by multiplying the relative importance weights directly with the normalized value of the criteria for each alternative assigned by the user. The obtained product value is then summed up. The alternative service with the highest score is selected as the best service provider.

SAW-based cost ranking results will be the input for hierarchical ranking along with additional parameters or criteria. Initially, rank or score value for cost is calculated as many costs are involved in existing data, and then that score value is considered for final (hierarchal) ranking of service providers. As the measuring unit of attributes or criteria considered for ranking is different, Min_Max normalization is used to convert each value between 0 and 1 based on whether it's a minimization function or maximization function.

4   Experimental Results and Analysis

For experiment purpose transportation quotation response data received from various partners are considered.

Transit time and various cost and sub-cost elements are set as minimization functions (negative monotonicity) as less value is preferred. Available peak capacity/kg and available non-peak capacity/kg are set as maximization functions (positive monotonicity) as more available capacity can help reduce cost by avoiding higher spot market rates and adds flexibility to operations. Snapshot for data considered is given in Table 1. The actual table has totally 37 rows and 24 columns.

After applying Min_Max normalization on the data, total score obtained across multiple criteria is as shown in Table 2. Highest scored partner and respective quote for each lane and service-level combination is then recommended for potential business award.

Business priorities against each criterion can be controlled by weight assignment. Weights can be assigned manually to each criterion by the user to give the importance of each as per the business requirements. By default, equal weights are considered. For example, if the business priority is more minimizing ATA (airport-to-airport) cost [12] but at the same time wishes to consider and optimize all the rest of assessorial cost, transit time and capacity as well, the weights across each parameter can be adjusted. Table 3 presents the weights used for this experiment.

5   Time Series Forecasting for Demand Prediction (Capacity Prediction)

Once optimized cost is available based on partner ranking by lane and service level, it requires total demand or 52-week forward looking capacity in order to calculate the total future 52-week transportation spent. This will allow to award business contract for next year to best partner with most optimal quote to fulfill all the requirements to meet business priorities and goals. In order to predict 52-week forward looking capacity, time series forecasting algorithms [13, 14] are used.

Assume that t represents current time (end of week t), h forecast horizon and $F(t, h)$ quantity/weight forecast generated at the end of week t for week $t + h$. For example, $F(t,$

Table 1 Snapshot of sample input data (total clusters 8,604 for RFQ round 1, total 6-RFQ rounds were evaluated)

RFQ summary lane	BKK-BLR	BKK-BLR	BKK-BLR	BKK-BLR	BKK-BLR	BKK-BLR
Service level	STD	DEF	EXP	STD	DEF	EXP
Transit time	4	5	3	7	8	6
Service provider	Partner 1	Partner 1	Partner 1	Partner 6	Partner 6	Partner 6
Available capacity	14,000	21,000	10,500	0	0	0
Needed capacity	25,000	25,000	25,000	25,000	25,000	25,000
Capacity coverage	11,000	4000	14,500	25,000	25,000	25,000
PU minimum cost	90	90	90	79.76	79.76	79.76
PU Max 45 kg	0.08	0.08	0.08	0.13	0.13	0.13
PU Max 100 kg	0.08	0.08	0.08	0.13	0.13	0.13
PU Max 300 kg	0.08	0.08	0.08	0.13	0.13	0.13
PU Max 500 kg	0.08	0.08	0.08	0.13	0.13	0.13
–	–	–	–	–	–	–
37 parameters						

Table 2 Snapshot of partner rank based on the total score using the algorithm on multiple quotes provided by lane and service level

RFQ summary lane	Service level	Service provider	Total
BKK-BLR	STD	Partner 7	0.79675
BKK-BLR	STD	Partner 6	0.79475
BKK-BLR	DEF	Partner 7	0.78975
BKK-BLR	STD	Partner 4	0.75075
BKK-BLR	STD	Partner 6	0.74525
BKK-BLR	STD	Partner 1	0.72425
BKK-BLR	STD	Partner 2	0.70525
BKK-BLR	DEF	Partner 4	0.691
BKK-BLR	DEF	Partner 8	0.6845
BKK-BLR	STD	Partner 6	0.68275
BKK-BLR	STD	Partner 8	0.67625
BKK-BLR	DEF	Partner 2	0.6635

Table 3 Assigned weights to various input RFQ response parameters

RFQ response parameter	Weightage
Capacity gap	0.025
TT	0.025
PUMin	0.025
PU45	0.025
PU100	0.025
PU300	0.025
PU500	0.025
PU1000	0.025
PU1000Plus	0.025
ORMin	0.025
OR	0.025
ATAMin	0.05
ATA45	0.05
ATA100	0.05
ATA300	0.05

$h = 1$) represents actual plan/ship value for week $t + 1$. As shown in Table 4, plan/ship value for week $t + 1$, 52 forecasts are generated (forecasts are being sequentially updated) [15]:

Table 4 Illustration—Plan/ship week $t + 1$, 52 forecasts

Notations:

t = current time
h = forecast horizon
F(t,h) = Forecast generated at the end of week t for week t+h (see time line below)

						Plan/Ship week									
Week	t-51	t-50	t-49	…	t-1	t	t+1	t+2	t+3	t+4	t+5	t+6	t+7	t+8	…
t-52	F(t-52,h=1)	F(t-52,h=2)	F(t-52,h=3)	…	…	F(t-52,h=52)									
t-51		F(t-51,h=1)	F(t-51,h=2)	…	…	…	F(t-51,h=52)								
t-50			F(t-50,h=1)	…	…	…	…	F(t-50,h=52)							
…															
t-2					F(t-2,h=2)	F(t-2,h=3)	…	F(t-2,h=52)							
t-1					F(t-1,h=1)	F(t-1,h=2)	F(t-1,h=3)	…	F(t-1,h=52)						
t						F(t,h=1)	F(t,h=2)	F(t,h=3)	…	F(t,h=52)					
t+1							F(t+1,h=1)	F(t+1,h=2)	F(t+1,h=3)	…	F(t+1,h=52)				
t+2								F(t+2,h=1)	F(t+2,h=2)	F(t+2,h=3)	…	F(t+2,h=52)			
t+3									F(t+3,h=1)	F(t+3,h=2)	F(t+3,h=3)	…	F(t+3,h=52)		
…															

$F(t - 51, h = 52)$, $F(t - 50, h = 51)$, $F(t - 49, h = 50)$, … $F(t - 3, h = 4)$, $F(t - 2, h = 3)$, $F(t - 1, h = 2)$, $F(t, h = 1)$,

where the last value $F(t, h = 1)$ is generated at the end of week t (most recent week); for the next week's ship/plan value, $t + h = t + 1$. The goal is to determine max h (denoted by h^*) such that any transformation of forecasts up to h^* would give accurate enough plan/ship week's estimate. Various time series models were evaluated and their predictive accuracy is compared using mean absolute percentage error (MAPE) [16].

Historic 3-year shipment data are used to train the time series algorithm in order to generate forecast for the next 52 weeks from t ($h = 52$, $t = t + 1…t + 52$). The output is depicted in Figs. 3 and 4.

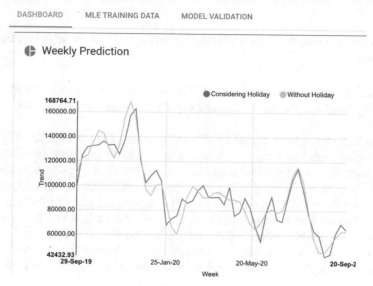

Fig. 3 Weekly prediction

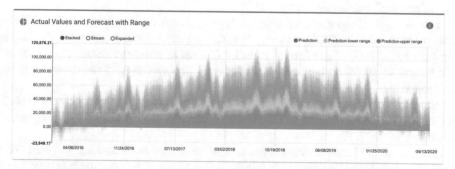

Fig. 4 Forward looking 52-week capacity prediction using time series forecasting algorithm for each lane

Figure 3 illustrates the predicted weekly tonnage generated by the time series models that are trained using historic tonnage for any given shipping lane. There are two models, first model considers impact of regional public holidays and significant business events on the predicted tonnage (output shown as "considering holidays") and the second model do not consider any external impact (output shown as "without holidays").

Figure 4 Illustrates information from Fig. 3, that is the 52-week forward looking prediction combined with the historic tonnage data which were used for training the models. This provides a complete time series and overall trend for the shipping lane. In addition, the models provide a range of predictions: floor (pessimistic or lower range) and ceiling (optimistic or upper range). All the three: lower, average and upper ranges of weekly predicted tonnage (a.k.a. capacity) are depicted in the above graph.

6   Business Award Distribution, Spend Prediction and Portfolio Planning

After ranking partners and predicting the 52-week forward looking capacity requirements, the spend for the next 52 weeks can be calculated using the formulae given below:

Formula_1 (Incumbent Spend Prediction) = Sum of (Quoted Rate) * ((52-week Capacity (i.e. needed capacity) * (% award from Rate sheet)/100) for all incumbent partners currently awarded with business for the selected lane(s) and price model.

Formula_2 (User Defined Ranking Based Spend Predictions) = Sum of (Quoted Rate of all selected Partners) * ((52-week capacity i.e. needed capacity) * (%award provided by user)/100) for all selected partners with %award value assigned by the users.

Formula_3 (Machine Learning Recommended Ranking Based Spend Predictions) = Sum of (Quoted Rate of MLE recommended top Partners) * ((52-week capacity i.e. needed capacity) * (%award calculated)/100) for MLE recommended high ranked partner(s).

Using the above calculations, business awarder can easily simulate and optimize their spend portfolio. They can change the weightage across various RFQ response parameters to re-rank the partners and best quotes to find the relative impact on their portfolio and loss of potential gain against each.

For example, if top-ranked partner from partner ranking algorithm can fulfill 100% of the needed capacity, 100% of the business can be awarded to that partner. If top partner can only fulfill 80% of the needed capacity (shown as coverage delta), then % award will be split to 80–20 and remaining 20% should be awarded to the next partner in the rank manually.

Snapshot of a sample award distribution based on ranks of partners is given in Table 5. For HKG-GDL lane, total 52 weeks predicted capacity is 2,373,633.84 kg and overall spend is 8,182,322.27.

Prior to MCDM ranking algorithms, simple cost-based stack ranking was used. That method awards 100% business to Partner_3 without considering that partner's available capacity, transit time or other RFQ response parameters that are mentioned in Sect. 4, Table 3 reflecting current business priorities.

MCDM algorithms perform real-time, dynamic partner-quote ranking based on the latest uploaded RFQ responses and user-defined configuration setting of important business criteria, their relative importance and monotonicity.

User can change the relative importance of predefined business criteria and simulate their impact on overall spend portfolio and find the relative impact on their portfolio and loss of potential gain against each.

Table 5 Example of award distribution based on ranks of partners to optimize overall spend

Service provider	ATA	Available capacity	Coverage delta	Spend	Distribution (%)
Partner_4-DEF	3.45	256,487.20	1,917,146.64	8,182,322.27	100
Partner_3-DEF	2.25	0.00	2,373,633.84	5,351,908.45	0
Partner_8-DEF	3.42	608,649.60	1,764,984.24	8,115,611.11	0
Partner_6-DEF	4.02	3,956.22	2,369,677.62	9,542,016.16	0

7 Conclusion and Future Work

Use of MCDM and time series forecasting algorithm allows enterprises to predict and optimize their logistics spend portfolio and provide long-range demand visibility to their partner network to plan proactively. While this solution makes use of internal business parameters to improve operational efficiency and cost savings, there is still a major gap to understand and leverage external market indicators while planning the transportation spend portfolio. Inability to quantify and time the combined impact of various ongoing economic and market influencers on current and future state of the businesses can be costly and result in complete deviation from the plan. Due to above limitations, most businesses are unable to continuously monitor and hedge their portfolio against the economic and market trends and thus are mostly dependent on their external logistic carrier partners. Future ongoing work by Cobotics Business Services LLC is dedicate toward an AI-enabled solution which combines enterprise data with external market influencers to personalize and quantify their current and future impacts on enterprise portfolio and partner strategy.

References

1. USCCG Homepage (2016). http://www.usccg.com/blog/spend-management-freight-logistics/
2. Steve Carr: Transportation Sourcing Optimization: a whitepaper by Elemica -Copyright © 2016 Elemica
3. Velasquez, M., Hester, P.T.: An analysis of multi-criteria decision-making methods. Int. J. Oper. Res. **10**(2), 56–66 (2013)
4. Triantaphyllou, E., Shu, B., Sanchez, S.N., Ray, T.: Multi-criteria decision making: an operations research approach. Encycl. Electr. Electron. Eng. **15**(1998), 175–186
5. Afshari, A., Mojahed, M., Yusuff, R.M.: Simple additive weighting approach to personnel selection problem. Int. J. Innov. Manage. Technol. **1**(5), 511 (2010)
6. Rao, R.V.: Decision Making in the Manufacturing Environment: Using Graph Theory and Fuzzy Multiple Attribute Decision Making Methods. Springer Science & Business Media (2007)
7. Patro, S., Sahu, K.K.: Normalization: A Preprocessing Stage. In: arXivpreprint arXiv:1503.06462 (2015)
8. Deshpande, A., Mahajan, A., Thomas, A.: Pattern discovery for multiple data sources based on item rank. In: Int. J. Data Min. Knowl. Manage. Process (IJDKP) **7**(1) (2017)

9. Miljković, B., Žižović, M.R., Petojević, A., Damljanović, N.: New weighted sum model. Filomat **31**(10), 2991–2998 (2017)
10. Moestopo, Universitas Prof Dr, S. T. I. K. O. M. TunasBangsaPematangsiantar.: Comparison of weighted sum model and multi attribute decision making weighted product methods in selecting the best elementary school in Indonesia. Int. J. Softw. Eng. Appl. **11**(4), 69–90 (2017)
11. Drobne, S., Lisec, A.: Multi-attribute decision analysis in GIS: weighted linear combination and ordered weighted averaging. Informatica **33**(4) (2009)
12. https://www.freightos.com/freight-resources/air-freight-rates-cost-prices/
13. Brockwell, Peter J., Davis, Richard A.: Introduction to Time Series and Forecasting, 2nd edn. Springer, USA (2002)
14. Montgomery, D.C., Jennings, C.L., Kuahci, M.: Introduction to Time Series Analysis and Forecasting. Wiley Publication (2008)
15. Shumsky, R.A.: Optimal updating of forecasts for the timing of future events. Manage. Sci. **44**(3), 321–335 (1998)
16. Statistics How to Homapage. https://www.statisticshowto.datasciencecentral.com/mean-absolute-percentage-error-mape/

The Design and Development of Marathi Speech Corpus from Native and Non-native Speakers

Shital S. Joshi[1](✉) and Vaishali D. Bhagile[2]

[1] Dr. Babasaheb Ambedkar, Marathwada University, Aurangabad 431002, Maharashtra, India
shitaljoshi1000@gmail.com

[2] Deogiri Institute of Technology & Management Studies, Aurangabad 431005, Maharashtra, India

Abstract. This paper describes development process of a creation of speech database from native and non-native speakers of Marathi language. Reason for choosing this topic is people may feel hesitation in speaking language other than their mother tongue. So, for working on native and non-native Marathi speech recognition, it is the basic and essential step to develop speech database. It has been used to capture various variations of the visiting non-native speakers having the non-Marathi mother tongue. A total of 50 native and 50 non-native speakers have been tested to collect the sample data. In this analysis each speaker is supposed to repeat the text corpus approximately three times, which acquires 112 isolated words and has maximum capacity of 33,000 words.

Keywords: Native Marathi speech recognition · Non-native Marathi speech recognition · Marathi speech recognition · Text corpus

1 Introduction

1.1 Speech

Speech is a biological method to express and share our ideas with other human beings [1]. It is natural and easy way of communication. Humans can easily interact with the help of speech. Therefore, speech recognition is becoming the area of interest of researchers. They are trying to develop such systems that can be operated with speech inputs. Our today's smart systems like ATMs, phones assume some sort of literacy from user. But with the help of speech-based interactive systems, illiterate users can also easily interact with systems as they can speak. Also speech-based interactive systems can be helpful to physically handicap people those who can speak. The language technology can bridge the gap between technically illiterate people to join the mainstream and be part of digital India [2].

Database not yet developed for native and non-native Marathi speech recognition. There is a scope to develop database to capture variations of native and non-native Marathi speaker. Speech does not only carry acoustic and verbal information and message but it also carries information about emotional state, intension and so on. Some people may hesitate to speak new language. So, for working on native non-native Marathi speech recognition, it is the basic and essential step to develop speech database.

© Springer Nature Singapore Pte Ltd. 2020
B. Iyer et al. (eds.), *Applied Computer Vision and Image Processing*,
Advances in Intelligent Systems and Computing 1155,
https://doi.org/10.1007/978-981-15-4029-5_12

1.2   Marathi Language

Marathi is a language of Indo-Aryan language family. It is mother tongue of Maharashtra. It is spoken in complete Maharashtra state which covers a vast geographic area consisting of 35 different districts [3]. Marathi is also spoken by non-native speakers of Maharashtra, who migrate from different states. As being regional language they are trying to adopt and speak Marathi. But there are some variations in their pronunciation.

Importance of Marathi

1. Marathi is mother tongue of Maharashtra. So they feel some sort of emotional attachment with Marathi.
2. Native speakers of Maharashtra prefer Marathi language in their day-to-day life. It is their thinking that language represents our culture.
3. If any concept or system is developed in mother tongue, then it will be easily understood. System developed in Marathi can be easily handled by native speakers of Maharashtra.
4. Marathi is one of the 23 recognized languages by the Constitution of India written in Devnagari script [4].

2   Literature Review

Lot of work is done with respect to speech recognition. But very little amount of work has been done for Indian languages. Regarding Marathi language, research is going on to reach critical level so that it can be used in day-to-day life. It is also observed that various databases in Marathi language are available for agriculture purpose, travel purpose, numerals, stuttering Marathi speech database and continuous sentences speech database [2]. There is a scope to work on native and non-native Marathi speech. Variations in pronunciations by non-native speakers, effect of accents and dialects can be observed and compared with native Marathi speaker's speech data.

3   Development of Text Corpus

For developing a speech database, preliminary step is to develop text corpus. We designed text corpus by selecting Marathi alphabets and numerals from Script Grammar for Marathi Language prepared by Technology Development for Indian Languages (TDIL) Programme of Department of Information Technology, Government of India (DIT, GoI) in association with Center for Development and Advanced Computing (CDAC) [5]. Isolated words are selected from basic Marathi books of primary school. Text corpus was recorded from 50 native and 50 non-native Marathi speakers. There were 12 vowels in Marathi but in year 2018, two more vowels are added in Marathi vowel list. They are— ॲ(ae)and ऑ(ao). By considering these, there are 14 vowels in Marathi. Therefore, we have considered these in our text corpus.

There are 46 consonants in Marathi, but in the decade of 1990–2000, one more consonant was considered in Marathi language and that was ऱ्र(tra). This consonant ऱ्र(tra)is actually omitted from Marathi Grammar Script prepared by Technology Development for Indian Languages (TDIL) Programme of Department of Information Technology, Government of India (DIT, GoI) but it is still spoken by the native Marathi speakers. It is pronounced as त्र(tra). We have included it in our text corpus.

Our text corpus consists of 10 Marathi numerals, 51 Marathi alphabets and 51 Marathi isolated words, which are described in Table 1.

Table 1 Text corpus

Type of Data	No. of words
1) Marathi numerals (शून्य to नऊ)	10
2) Marathi alphabets (अ to ज्ञ)	51
3)Marathi isolated words(अननस to ज्ञानदेव)	51
Total size of text corpus	**112**

The English translation and International Phonetic Alphabets (IPA) are as follows (Tables 2 and 3).

Table 2 English transliteration and IPAs of Marathi numerals

Marathi Numerals	English Transliteration	IPAs
शून्य	śunya	/ʃuːnjə /
एक	eka	/ekə/
दोन	dona	/donə/
तीन	tīna	/tiːnə/
चार	cāra	/ tɕɑrə/
पाच	paca	/pətɕə/
सहा	sahā	/səɦə/
सात	sāta	/satə/
आठ	āṭha	/aʈʰə/
नऊ	naū	/nəuː/

4 Data Collection

Designed text corpus was recorded by 50 native and 50 non-native Marathi speakers. It is tried to capture more variations in the non-native speech data. We have tried to visit people having mother tongue except Marathi, like Guajarati, Haryanvi, Marwadi, Sindhi, Tamil, Malayalam, Hindi, Tulu and Arabic. Speech samples were recorded using

Table 3 English transliteration and IPAs of Marathi alphabets and corresponding isolated words

Marathi Alphabets	English Transliteration	IPAs	Marathi Isolated Word	english Transliteration	IPAs
अ	A	/ə/	अननस	ananas	/ənənəsə/
आ	Ā	/a/	आई	āī	/ai:/
इ	I	/i/	इमारत	imārat	/imərət/
ई	Ī	/i:/	ईडलिंबू	īḍalimbu	/i:ḍəlimbu:/
उ	U	/u/	उपवन	upavan	/upəvən/
ऊ	Ū	/u:/	ऊस	ūsa	/u:sə/
ए	E	/e/	एडका	eḍakā	/eḍəkə/
ऐ	ai	/əi/	ऐरण	airaṇa	/əirəṇə/
ओ	O	/o/	ओठ	oṭha	/oṭʰə/
औ	au	/əu/	औषध	auṣadha	/əuṣədʱə/
अं	aṃ	/əm/	अंगठा	angaṭhā	/əngəṭʰa/
आः	aḥ	/əɦə/	आःहा	āḥhā	/əɦɦa/
ॲ	ae	/æ/	ॲना	aena	/ænə/
ऑ	ao	/ɔ/	ऑस्कर	aoskara	/ɔskərə/
ऋ	ṛ	/ru/	ऋषी	ṛuśī	/ruṣi:/
क	ka	/kə/	कप	kappa	/kəpə/
ख	kha	/kʰə/	खटारा	khaṭārā	/kʰəṭara/
ग	ga	/gə/	गणपती	ganapaṭī	/gənəpəti:/
घ	gha	/gʱə/	घर	ghara	/gʱərə/
ङ	ṅa	/ŋə/	वाङ्मय	vaṅamaya	/vəŋəməjə/
च	ca	/tɕə/ or /tsə/	चमचा	camacā	/tsəmətsa/
छ	cha	/tɕʰə/	छत्री	chañī	/tɕʰəji:/
ज	ja	/dzə/ or /dʑə/	जहाज	jahāja	/dzəɦadzə/
झ	jha	/dʑʱə/ or /dzʱə/	झबले	jhabale	/dzʱəbəle/
ञ	ña	/jə̃/	त्राण	ñāṇa	/janə/
ट	ṭa	/ʈə/	टरबूज	ṭarabūja	/ʈərəbu:dzə/
ठ	ṭha	/ʈʰə/	ठसा	ṭhasā	/ʈʰəsa/
ड	ḍa	/ɖə/	डबा	ḍabā	/ɖəba/
ढ	ḍha	/ɖʱə/	ढग	ḍhaga	/ɖʱəgə/
ण	ṇa	/ɳə/	बाण	bāṇa	/baɳə/
त	Ta	/tə/	तलवार	talavāra	/tələvarə/
थ	tha	/tʰə/	थडगे	thaḍage	/tʰəɖəge/
द	da	/də/	दौत	dauta	/dəutə/
ध	dha	/dʱə/	धनुष्य	dhanuṣya	/dʱənuṣjə/
न	na	/nə/	नळ	naḷa	/nəɭə/
प	pa	/pə/	पतंग	patanga	/pətəngə/
फ	pha	/pʰə/ or /fə/	फणस	phaṇasa	/pʰəṇəsə/
ब	ba	/bə/	बदक	badaka	/bədəkə/

(continued)

Table 3 (*continued*)

Marathi Alphabets	English Transliteration	IPAs	Marathi Isolated Word	english Transliteration	IPAs
भ	bha	/bʰə/	भटजी	bhaṭajī	/bʰətədzi:/
म	ma	/mə/	मका	makā	/məka/
य	ya	/jə/	यज्ञ	yajña	/jəɡɳə/
र	ra	/rə/	रथ	rath	/rətʰə/
ल	la	/lə/	लसूण	lasūṇ	/ləsu:ɳə/
व	va	/ʋə/	वजन	vajan	/ʋədzənə/
श	śa	/ʃə/	शहामृग	śahāmṛūga	/ʃəɦamru:gə/
ष	ṣa	/ʂə/	षट्कोन	ṣaṭkon	/ʂəʈəkonə/
स	sa	/sə/	ससा	sasā	/səsa/
ह	ha	/ɦə/	हरीण	harīṇ	/ɦəri:ɳə/
ळ	ḷa	/ɭə/	कमळ	kamaḷ	/kəmələ/
क्ष	kṣa	/kɛə/	क्षत्रिय	kṣañiya	/kɛəjɔ̃ijə/
ज्ञ	jña	/ɡɳə/	ज्ञानदेव	jñanadev	/ɡɳanədeʋə/

PRATT from 100 speakers and stored in .wav format. Standards for speech data capturing and annotation are strictly followed while collecting data from different native and non-native speakers. These standards are prescribed by Linguistic Data Consortium for Indian Languages (LDC-IL) [6].

4.1 Speaker Selection

As speech characteristics changes according to the age of a person, age group variation plays important role [7]. We have recorded speech samples from different age groups, that is, 11–20, 21–30, 31–40, 41–50, 51–60 for native and non-native speaker. Data is recorded from 50 male speakers and 50 female speakers. We have selected 5 males and 5 females from each age group for native as well as non-native speech data collection. Hence the data is age-balanced and gender-balanced.

4.2 Recording Environment

Speech samples are recorded in a closed room without noise to achieve high-quality speech samples. It was not possible to bring each speaker to the lab for recording. So, we have visited each speaker at their home and recorded their speech in closed room without noise of fan, AC and so on.

4.3 Recording Device

Redgear cosmo 7.1 headphones with mic were used for recording. It was put at a distance of 2.5–3.5 cm from mouth of speaker. General quality headphones are used to balance better data quality and also cost of application.

4.4 Recording Software

PRAAT software is used for recording speech and was stored in .wav file format. One of the important features of PRAAT is its graphical user interface (GUI). PRAAT is developed by Paul Boersma and David Weenink of University of Amsterdam [8]. It also provides the functions like spectral analysis, pitch analysis, intensity and formant analysis.

Text corpus size of 112 isolated words was recorded from 100 speakers. Mono-channel speech sounds are recorded at sampling frequency of 16,000 Hz and bit rate was 16 bit. Each speaker is asked to repeat text corpus three times. So, total database contains 33,600 isolated words. It is described in Table 4.

Table 4 Total vocabulary size

Type of speaker	Number of Speaker	Size of text corpus	Utterances	Total size
Native	50	112	3	16,800
Non-native	50			16,800
Total	100			33,600

5 Annotation and Labeling of Speech Corpus

The term annotation of speech data covers any indicative, descriptive or analytic notations applied to recorded raw speech data. These added notations include information of various kinds. Such recorded speech corpora must be tagged or annotated adequately for both linguistic as well as acoustic criteria [9]. Database is being labeled and annotated. For annotation and labeling we are using PRAAT. PRAAT is a software developed to study phonetics using computer.

Information about boundaries and associated labels is stored in text file in a specific format in PRAAT.

6 Conclusion

This paper has been designed and developed in order to capture the speech corpus from native and non-native Marathi speakers. In future we will furnish this database for other researchers too. It consists of ample amount of variations in pronunciations by the non-native speakers. This work is an initiative taken to recognize native and non-native Marathi speech recognition. Database has various applications, like military, banking, education, speech therapy, railway ticket booking system, building smart devices and home appliances, machine learning and so on.

7 Future Work

A robust speech recognition system is to be developed that can recognize Marathi speech from both native and non-native speakers. Also, variations in the prosody features, pronunciation and acoustic variations are to be studied to develop real-life applications. This work can be extended to develop real-life applications used in banks, education and speech therapy.

References

1. Waghmare, S., Deshmukh, R.R., Kayte, S.N.: Analysis of fundamental frequency, jitter and shimmer in stuttered and non-stuttered speech of Marathi language. In: Jitter and Shimmer in Stuttered and Non-Stuttered Speech of Marathi Language (2019). Available at SSRN
2. Joshi, S.S., Bhagile, V.D., Deshmukh, R.R.: A Review on Marathi Speech Recognition (2019). Available at SSRN 3419224
3. Shrishrimal, P.P., Deshmukh, R.R., Waghmare, V.B.: Marathi isolated words speech database for agriculture purpose. Int. J. Eng. Innov. Res. 3(3), 248
4. Shrishirmal, P.P., Deshmukh, R.R., Waghmare, V.B., Borade, S., Janse, P.V., Janvale, G.B.: Development of Marathi Language Speech Database from Marathwada Region
5. Script for Marathi Grammar prepared by Technology Development for Indian Languages (TDIL) programme of Department of Indian Languages (DIL), Government of India (GoI), in association with Center for Development Advanced Computing (CDAC)
6. Standards for Speech Data Capturing and Annotation: Linguistic Data Consortium for Indian Languages (LDC-IL) (2008)
7. Gaikwad, S., Gawali, B., Mehrotra, S.: Creation of Marathi speech corpus for automatic speech recognition. In: 2013 International Conference Oriental COCOSDA held jointly with 2013 Conference on Asian Spoken Language Research and Evaluation (O-COCOSDA/CASLRE), pp. 1–5. IEEE (2013)
8. https://en.wikipedia.org/wiki/Praat
9. http://tdil.meity.gov.in/pdf/Vishwabharat/16/5.pdf

Speckle Noise Reduction of In Vivo Ultrasound Image Using Multi-frame Super-Resolution Through Sparse Representation

Suchismita Maiti[1(✉)], Sumantra Maiti[2], and Debashis Nandi[3]

[1] Narula Institute of Technology, Agarpara, Kolkata, West Bengal, India
suchismita2006@gmail.com
[2] Magma HDI, Ecospace, Rajarhat, Kolkata, West Bengal, India
sumantramaiti@gmail.com
[3] National Institute of Technology, M. G. Avenue, Durgapur, West Bengal, India
debashisn2@gmail.com

Abstract. An approach for speckle noise reduction in in vivo medical ultrasound image has been proposed. The proposed technique makes use of multi-frame-based super-resolution (SR) image reconstruction which further is performed through sparse representation. This approach makes use of dictionary learning using medical ultrasound images as training data. Furthermore, for the preparation of multiple *looks* or frames, LoG (Laplacian of Gaussian) filter has been used to make noisy frames which helps in preserving the typical texture of speckle noise generated in real medical ultrasound images. The sets of data include real ultrasound images displayed from in vivo raw RF ultrasound echo.

Keywords: Speckle noise reduction of in vivo ultrasound image · Super-resolution reconstruction · Sparse representation · Medical ultrasound image quality enhancement

1 Introduction

Ultrasound imaging is one of the most popular diagnostic imaging modalities owing to its cost efficiency, non-invasiveness, portability, and comparatively good imaging speed. The application areas include fetal health monitoring, early detection of cancer, etc. In spite of these advantages, visual quality and resolution of ultrasound image is severely affected for backscattering of ultrasound echo [1]. Hence the quality and resolution enhancement of medical ultrasound image is still regarded as an active research area.

Resolution enhancement is one of the major issues to be taken care of by ultrasound imaging system designers, since this issue is inherently co-related to imaging system features. Intrinsic resolution is limited by different parameters

© Springer Nature Singapore Pte Ltd. 2020
B. Iyer et al. (eds.), *Applied Computer Vision and Image Processing*,
Advances in Intelligent Systems and Computing 1155,
https://doi.org/10.1007/978-981-15-4029-5_13

like attenuation, scattering from biological tissues, wave frequency, number of transducer elements in the probe or sensor density, etc. Extrinsic resolution is dependent on the point spread function (PSF) and live tissue movements in case of in vivo imaging. Hence enhancement of intrinsic resolution has been focused in research. Corresponding hardware-centric methods of solution include increasing the transmitted wave frequency, the number of transducer array elements, and the number of scan lines which are costly, non-portable, and prone to high attenuation at higher depth of body. Past research works show that SR reconstruction can be used to overcome inherent resolution issues in image processing [2–5]. Many of these application areas involve medical image processing [6–8].

Single-frame and multi-frame-based SR are two categories of SR image reconstruction depending on the number of LR frames are used [9]. In multi-frame SR reconstruction, multiple images of a particular scene or *looks* obtained with sub-pixel misalignment provide several complementary information which is further used for higher resolution (HR) image reconstruction. The LR *looks* of a stationary scene may be obtained by positioning multiple cameras at different spatial locations. If the shifting or rotation parameter values are known within sub-pixel accuracy, it is possible to generate a high-resolution image by combining LR frames or images [9–12]. However there are some application areas where multiple LR frames may not be available. In those cases HR image is recovered using limited available information from a single LR image which is defined as single-frame SR reconstruction [13–16].

Single-image SR reconstruction is an inverse and ill-posed problem having many consistent solutions. On the other hand conventional and potential interpolation-based techniques increase the blur or aliasing effect massively. Moreover the capability of prediction about lost information is not feasible in interpolation-based techniques. In comparison to that, intuitively increasing the number of LR samples, obtained by capturing multiple shifting or rotation (at sub-pixel level) of the image frame, provides more information, hence that capability of providing predictive information. In the present paper, we propose an ultrasound image resolution enhancement technique through speckle noise reduction using multi-frame SR reconstruction using sparse representation.

In Sect. 2 overall SR reconstruction technique using sparse representation has been discussed briefly. Section 3 contains the details about the proposed approach. Section 4 contains experimental results and other relevant discussion. Section 5 contains the conclusive comment.

2   SR Reconstruction Using Sparse Representation

In the single-image SR problem an HR image, I_h, is to be recovered using a single LR image. Two identified constraints for the aforementioned problem are reconstruction constraint and sparsity prior [15]. Reconstruction constraint mainly focuses on the consistency of the recovered image with input image while keeping an eye on the observation model. The observed LR image I_l is basically a blurred and down-sampled version of the corresponding HR image I_h using the following formula.

$$I_h = F_{dw}F_{bl}I_l \qquad (1)$$

Here, F_{bl} and F_{dw} denote the blurring filter and down-sampling operator, respectively. As there can be more than one HR images for a single input LR image satisfying the aforementioned reconstruction constraint, regularization process has been done here via sparsity prior on small patches p_H of I_h. The assumption is that the HR patches can be sparsely represented in an appropriately chosen overcomplete dictionary and the sparse representations of the HR patches can be recovered from the LR observation. The image patches, I_{ph}, from I_h are hereby represented as a sparse linear combination in a dictionary, D_{HR}, trained from the HR image patches, sampled from training images. Mathematically it is stated as follows.

$$p_H \approx D_{rh}\,\phi \text{ for some } \phi \in \mathbb{R}^K \text{ with } \|\phi\|_0 \ll K \qquad (2)$$

Here, ϕ denotes the sparse representation which is to be recovered by the LR image patches p_L with the help of an LR dictionary D_{LR}. D_{LR} is obtained by joint training with the high-resolution dictionary D_{HR} [17].

Dictionary learning is another important step in sparse representation-based image processing where the representation of a signal is performed by combining a set of linear combination of atoms from a pair of overcomplete dictionaries—$D_{HR} \in \mathbb{R}^{M \times K}$ and $D_{LR} \in \mathbb{R}^{N \times K}$ consisting of K atoms (where $K > N > M$). Using a set of training examples $X = \{x_1, x_2, \ldots, x_t\}$ and following the learning formula the dictionary D is learned [18]. The optimization steps of dictionary learning is discussed later in Sect. 3.

The concept of single-frame SR reconstruction has been extended to implement multi-frame SR reconstruction via sparse representation by Wang et al. [17]. Several LR patches are generated by degrading HR patch to recover the sparse coefficients for the SR reconstruction using proper multi-frame LR observation model.

3   Proposed Approach

In the present work, a novel multi-frame SR reconstruction framework using sparse representation for denoising-based resolution enhancement of medical in vivo ultrasound images has been proposed. The standard multi-frame observation model has been chosen to relate the original HR image to the observed LR images in the current approach [2]. Conventional interpolation-based technique of SR reconstruction schema has been further amalgamated hereby with sparse representation-based SR reconstruction technique.

In the proposed technique while generating LR frames with rotation and translation parameters, LoG filter have been used to enhance the noise texture in stead of simply adding up random multiplicative noise. A closer look shows that the texture of simply simulated speckle noise and that of typical speckle noise in medical ultrasound images are quite different. Hence the process have been modified to increase the noisy pixel values along with translation and rotation parameters. Non-local mean-based approach helps to exploit the information from all of the degraded LR frames. The rest of the proposed algorithm is similar to that of the SR reconstruction algorithm proposed by Yang et al. [15].

Initially registration parameters of LR frames are computed using the frequency domain approach. Afterwards these pixel values are projected upon an HR grid. In this phase registration is performed using previously computed parameters. Afterwards the weighted features from previously generated low-pass filtered LR patches are calculated. Finally, the SR reconstruction has been executed via sparse representation to obtain final reconstructed image. The steps of the proposed technique are stated below. *Registration parameters* denote the sub-pixel shifts and rotations of LR frames with respect to a reference frame. The projection of the pixels from LR frame to HR grid requires estimated values of registration parameters to ensure correctness. Among several available parameter estimation techniques, a frequency domain approach has been deployed in the proposed approach [19]. For the estimation of rotation parameter, the Fourier transforms in polar coordinates of two spatial domain images, $f_1(I)$ and $f_2(I)$, are to be derived and have been denoted by $F_1(k)$ and $F_2(k)$, respectively. The rotation angle θ between $|F_1(k)|$ and $|F_2(k)|$ is computed such that the Fourier transform of the reference image $|F_1(k)|$ and the rotated Fourier transform of the image to be registered $|F_2(R_\theta k)|$ would possess maximum correlation. The frequency content, h, as a function of angle α is calculated by integrating over the range of radial lines as stated below.

$$h(\alpha) = \int_{\alpha - \frac{\Delta\alpha}{2}}^{\alpha + \frac{\Delta\alpha}{2}} \tag{3}$$

Finally the estimation of shift parameter the parallel shifting of an image to the plane is expressed as follows:

$$F_2(k) = \iint_I f_2(I)e^{\ j2\pi k^T I}dI = \iint_I f_1(I + \delta I)e^{-j2\pi k^T I}dI = e^{-j2\pi k^T \delta I}f_1(k) \tag{4}$$

Shift estimation parameter, δI, is computed from $\angle\frac{(F_2(k))}{(F_1(k))}$. In the next step the projection of LR frame pixels on an HR grid takes place and a large noisy image having the information of all the LR frames is produced. Input noisy image has been moved through a low-pass filter and further grid interpolated. The image patches are extracted afterwards and prepared for further phases of processing.

In the next step, feature selection and extraction is performed through feature transformation operator, O_F, which helps to confirm that the compatibility of the coefficients are good enough to fit to the most pertinent part of the LR image. As human eyes are more sensitive to high-frequency components, the selected features must emphasize to predict about desired high-frequency components. Thus high-pass filter has been considered as the feature operator. First-order and second-order derivatives have been used for feature extraction in both directions [15]. Four one-dimensional filters have been used to extract the derivatives as stated below.

$$f_1 = [-1, 0, 1], \quad f_2 = f_1^T, \quad f_3 = [1, 0, -2, 0, 1], \quad f_4 = f_3^T \tag{5}$$

Here f_1^T and f_3^T denote the transpose of vectors f_1 and f_3, respectively. Thus per LR patch, four feature vectors are obtained by applying these feature vectors. These four vectors are further concatenated to obtain a single feature vector, O_F,- the final feature level representation of that LR patch.

The aforementioned transformation operator sometimes fails to perform satisfactorily while the LR frames remain noisy owing to some amount of high-frequency noise components remaining in it. So noise reduction from these feature vectors would obviously help to obtain better performance. In the proposed approach, this issue has been taken care of minutely to control the noise and conserve desirable high-frequency components. For this purpose relative weight of feature vector is computed from the patches of previously low-pass filtered and grid-interpolated image. This added filtering does the initial screening for noise which further helps in noise reduction. The formula to calculate the weight of nth patch is stated below.

$$W(i, j) = \frac{1}{\sqrt{2\pi\sigma^2}} \exp\left(\frac{-(X_c^n - X^n(i, j))^2}{2\sigma^2}\right) \tag{6}$$

where $X^n(i, j)$ denotes the pixel value of the noisy n^{th} patch at (i, j). X_c^n and σ^2 denote the central pixel value of filtered n^{th} patch and the nth patch variance of the noisy image, respectively. The weighted feature vector, O_{FG}, is computed by multiplying both the vectors O_F and G in an element-by-element manner.

After computing feature vector, sparse representation vector ϕ is determined by basic SR reconstruction algorithm using the following minimization statement.

$$\min_\phi \|\tilde{\mathbf{D}}\phi - \tilde{\mathbf{y}}\|_2^2 + \lambda\|\phi\|_1 \tag{7}$$

where $\tilde{\mathbf{D}} = \begin{bmatrix} \mathbf{0}_F & \mathbf{D}_l \\ \beta P & \mathbf{D}_h \end{bmatrix}$ and $\tilde{\mathbf{y}} = \begin{bmatrix} \mathbf{0}_{FG} & \mathbf{y} \\ \beta & \mathbf{W}. \end{bmatrix}$

Minimization performed along with regularization computes the sparse vector representation of the patch.

Finally, estimated HR image, (I_{est}), is computed from the input LR image, I_L, which is obtained by Eq. 1 and by calculating the projection of I_H on the solution space of the following equation. The proposed algorithm has been stated in algorithm 1 (Fig. 1).

$$(I_{est}) = \arg \ \min_I \|I_L - F_{dw}F_{bl}I_H\|_2^2 + \lambda_0\|I_H - I_{HR}\|_2^2 \tag{8}$$

In the sparse representation-based signal processing, the linear system $D\phi = I_H$ is regarded as the reconstruction model to construct the HR image I_H. In this model, D is termed as the dictionary of atoms which is basically a full rank matrix whose every column is a possible signal in $\mathbb{R}^n$. An image is basically a linear combination of the atoms (or the columns) of the dictionary which basically refers to a dictionary D having dimension $Q \times R$, $R >> Q$.

Here a pair of overcomplete dictionaries $D_{HR} \in \mathbb{R}^{Q \times K}$ and $D_{LR} \in \mathbb{R}^{R \times K}$ have been used. K_A number of atoms have been obtained in the pair of dictionaries, where $K > R > Q$, by a joint learning process using a set of medical ultrasound images, $I = \{I_1, I_2, \ldots, I_t\}$. The learning process of the pair of dictionaries has been previously proposed by Yang et al. and is stated below [18] (Fig. 2).

$$\mathbf{D} = \arg \min_{\mathbf{D}, \phi} \|X - \mathbf{D}\phi\|_2^2 + \lambda \|\phi\|_1 \quad \text{s. t.} \quad \|\mathbf{D}_i\|_2^2 \leq 1, i = 1, 2, \ldots k \quad (9)$$

In the above equation, l^2-norm works as fitting term whereas l^1-norm acts as the regularization term. The process of further optimization is as follows.

1. Initialize the dictionary, $\mathbf{D}$, using random Gaussian distributed matrix performing column normalization.
2. Update ϕ (keeping $\mathbf{D}$ fixed) using the equation stated below.

$$\phi = \arg \min_{\phi} \|I - \mathbf{D}\phi\|_2^2 + \lambda \|\phi\|_1 \quad (10)$$

3. Update $\mathbf{D}$ (keeping ϕ fixed) using the following equation.

$$\arg \min_{\mathbf{D}} \|I - \mathbf{D}\phi\|_2^2 \quad \text{s. t.} \quad \|\mathbf{D}_i\|_2^2 \leq 1, i = 1, 2, \ldots k \quad (11)$$

4. Repeat 2 and 3 till convergence.

These aforementioned optimization steps are required to reach the convergence.

Algorithm 1: Multi-frame-based SR reconstruction of ultrasound image using sparse representation

Input: P number of generated LR frames, D_{HR} and D_{LR} (two learned dictionaries)
Output: Reconstructed Image I_{est}
1. Estimate the values of Registration Parameters.
2. Generate grid-interpolated image with the help of the estimated registration parameters using LoG filter.
3. Create the copy of the image by using low-pass filtering and grid interpolation.
4. Extract features from the patches of using four filters.
5. Repeat the following steps for each patch.
   (a) Calculate the mean value m of patch p_L.
   (b) Estimate the weight vector by considering Eq. 3 from each patch and .
   (c) Perform element-by-element multiplication between the weight vector W and each feature vector.
   (d) Solve the optimization problem according to Eq. 7
   (e) Reconstruct high-resolution patches. Generate HR image from the HR patches by implementing $p_H = D_{HR} \phi + m$.
   (f) Generate final estimated image close to original HR image using Eq. 8.

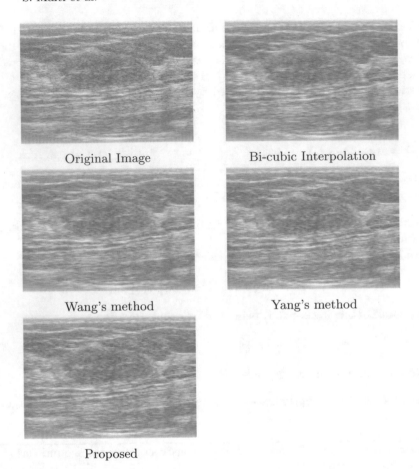

Fig. 1. Benign lesion in female breast (OASBUD data): (1) original; (2–5) denoised by Bicubic interpolation, Wang's method, Yang's method, and proposed method, respectively

In the present approach, the pair of dictionaries D_{HR} and D_{LR} are trained through joint learning technique [18]. Patch pairs formed from the taken sample training images are $\{P^H, P^L\}$ where HR patches, $P^H = \{h_1, h_2, \ldots, h_n\}$ and LR patches, $P^L = \{l_1, l_2, \ldots, l_n\}$. Here we aim to train and learn both the dictionaries for HR and LR patches in such a way so that the sparse representation of the HR image patch happens to be the same as the sparse representation of corresponding LR patch [17]. The optimization problems for sparse coding in case of the HR and LR patch spaces may be separately stated as follows.

$$D_{HR} = \arg\min_{\{D_{HR},\phi\}} \|P^H - D_{HR}\phi\|_2^2 + \lambda\|\phi\|_1 \tag{12}$$

and

$$D_{LR} = \arg\min_{\{D_{LR},\phi\}} \|P^l - D_{LR}\phi\|_2^2 + \lambda\|\phi\|_1 \tag{13}$$

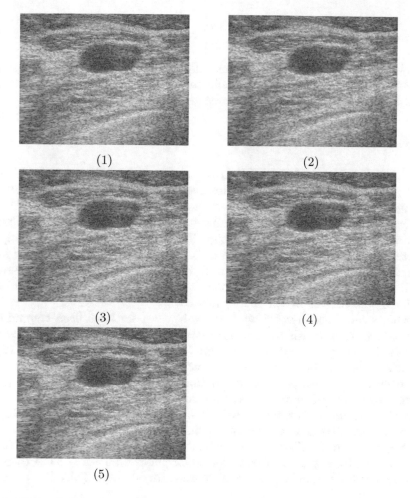

Fig. 2. Benign Fibroadenoma in female breast (Insana lab data): (1) original; (2–5) denoised by Bicubic interpolation, Wang's method, Yang's method, and proposed method, respectively

By combining these objectives the following equation is obtained.

$$\min_{\mathbf{D}_{HR}, \mathbf{D}_{LR}, \phi} \frac{1}{Q} \|\mathbf{X}_h^t - \mathbf{D}_{HR}\phi\|_2^2 + \frac{1}{R} \|(P^L)^t - \mathbf{D}_{LR}\phi\|_2^2 + \lambda\left(\frac{1}{Q} + \frac{1}{R}\right) \|\phi\|_1$$

$$(14)$$

Here Q and R are the dimensions of the HR and LR image patch vectors, respectively. λ is the Lagrange multiplier for regularization. Equation 15 may be rewritten in the following form.

$$\min_{\mathbf{D}_h, \mathbf{D}_l, \phi} \|\hat{P} - \hat{D}\phi\|_2^2 + \hat{\lambda}\|\phi\|_1$$

$$(15)$$

where $\hat{P} = \begin{bmatrix} \frac{1}{\sqrt{Q}}P_H \\ \frac{1}{\sqrt{R}}P_L \end{bmatrix}$, $\quad \hat{D} = \begin{bmatrix} \frac{1}{\sqrt{Q}}D_{HR} \\ \frac{1}{\sqrt{R}}D_{LR} \end{bmatrix}$ and $\hat{\lambda} = \lambda\left(\frac{1}{Q} + \frac{1}{R}\right)$.

Combined approach of dictionary learning helps in accelerating the speed of computation as well as acquiring a nonlinear correlation between two dictionaries by using input LR image patch features.

4   Experimental Result

In this work, two sets of raw RF ultrasound data have been used which are freely available from the website of Insana Laboratory of University of Illinois, Chicago [20] and Open Access Series of Breast Ultrasonic Data (OASBUD) [21]. The first set has been obtained from Insana Lab, University of Illinois, Chicago which consists of female breast benign fibroadenoma (precisely non-palpable tumors). The second dataset is a part of OASBUD and has been recorded in the Department of Ultrasound, Institute of Fundamental Technological Research Polish Academy of Sciences which consists of female breast lesion. Patient data files have been further by the biopsy reports.

Other than visual perception, two quality metrics have been computed to compare the results with three other methods of SR reconstruction—methods proposed by Wang et al., Yang et al., and Bicubic Interpolation (BI) [15,17]. As the techniques have been tested on real ultrasound images, two quality metrics have been used for performance evaluation—Effective number of looks (ENL) and speckle index (SI) have been calculated for input and output images for the comparison. Other performance evaluation metrics like PSNR, EKI, MSE, etc. require a ground truth image for calculation. For evaluation of the performance of noise reduction algorithms, there is no such ground truth available. ENL and SI do not require any ground truth for performance evaluation. Thus these two evaluation metrics have been selected to evaluate our work Table 1).

5   Conclusion

In the present paper a novel technique using multi-frame SR reconstruction through sparse representation has been proposed for speckle noise reduction of in vivo medical ultrasound images. LoG filter has been used to generate LR frames for preserving the typical medical ultrasound speckle texture. The training images used for dictionary learning are in vivo medical ultrasound images. Experimental data consists of real ultrasound images generated from in vivo raw RF echo. Other than the visual comparison, output image has been compared to four other methods using ENL and SI metrics. The comparison shows that the proposed method produces better result than the other four methods.

Table 1. Comparison of quality metric values

Image: Female breast fibroadenoma; source: Insana Lab			
Method	ENL_{in} ENL_{out}	SI_{in}	SI_{out}
Wang	0.0101 0.0126	0.2624	0.2481
Yang	0.0101 0.0133	0.2624	0.2452
Bicubic	0.0101 0.0142	0.2624	0.2414
Proposed	0.0101 0.0176	0.2624	0.2286

Image: Female breast lesion; source: OASBUD			
Method	ENL_{in} ENL_{out}	SI_{in}	SI_{out}
Wang	0.061 0.0837	0.1586	0.1463
Yang	0.061 0.0789	0.1586	0.1424
Bicubic	0.061 0.0868	0.1586	0.1449
Proposed	0.061 0.1026	0.1586	0.1390

References

1. Michailovich, O.V., Tannenbaum, A.: Despeckling of medical ultrasound images. IEEE Trans. Ultrason., Ferroelectr., Freq. Control **53**(1), 64–78 (2006)
2. Nandi, D., Karmakar, J., Kumar, A., Mandal, M.K.: Sparse representation based multi-frame image super-resolution reconstruction using adaptive weighted features. IET Image Process. **13**(4), 663–672 (2019)
3. Trinh, D.-H., Luong, M., Dibos, F., Rocchisani, J.-M., Pham, C.-D., Nguyen, T.Q.: Novel example-based method for super-resolution and denoising of medical images. IEEE Trans. Image Process. **23**(4), 1882–1895 (2014)
4. Fang, L., Li, S., McNabb, R.P., Nie, Q., Kuo, A.N., Toth, C.A., Izatt, J.A., Farsiu, S.: Fast acquisition and reconstruction of optical coherence tomography images via sparse representation. IEEE Trans. Med. Imaging **32**(11), 2034–2049 (2013)
5. Malczewski, K., Stasinski, R.: Toeplitz-based iterative image fusion scheme for MRI. In: 15th IEEE International Conference on Image Processing 2008, pp. 341–344. IEEE (2008)
6. Guan, J., Kang, S., Sun, Y.: Medical image fusion algorithm based on multi-resolution analysis coupling approximate spare representation. Futur. Gener. Comput. Syst., Elsevier **98**, 201–207 (2019)
7. Liu, Y., Chen, X., Ward, R.K., Wang, Z.J.: Medical image fusion via convolutional sparsity based morphological component analysis. IEEE Signal Process. Lett. **26**(3), 485–489 (2019)
8. Tan, L., Yu, X.: Medical image fusion based on fast finite shearlet transform and sparse representation. Comput. Math. Methods Med. **2019**, 485–489 (2019)
9. Park, S.C., Park, M.K., Kang, M.G.: Super-resolution image reconstruction: a technical overview. IEEE Signal Process. Mag. **20**(3), 21–36 (2003)
10. Tekalp, A.M., Ozkan, M.K., Sezan, M.I.: High-resolution image reconstruction from lower-resolution image sequences and space-varying image restoration. In: IEEE International Conference on Acoustics, Speech, and Signal Processing 1992, pp. 169–172. IEEE (1992)

11. Rajan, D., Chaudhuri, S.: Generation of super-resolution images from blurred observations using Markov random fields. In: IEEE International Conference on Acoustics, Speech, and Signal Processing 2001, pp. 1837–1840. IEEE (2001)
12. Chaudhuri, S.: Super-resolution imaging. Springer Science & Business Media (2001)
13. Glasner, D., Bagon, S., Irani, M.: Super-resolution from a single image. In: IEEE International Conference on Computer Vision 2009, pp. 349–356. IEEE (2009)
14. Kim, K.I., Kwon, Y.: Single-image super-resolution using sparse regression and natural image prior. IEEE Trans. Pattern Anal. Mach. Intell. **32**(6), 1127–1133 (2010)
15. Yang, J., Wright, J., Huang, T.S., Ma, Y.: Image super-resolution via sparse representation. IEEE Trans. Image Process. **19**(11), 2861–2873 (2010)
16. Freeman, W.T., Jones, T.R., Pasztor, E.C.: Example-based super-resolution. IEEE Comput. Graph. Appl. **22**(2), 56–65 (2002)
17. Wang, P., Hu, X., Xuan, B., Mu, J., Peng, S.: Super resolution reconstruction via multiple frames joint learning. In: 2011 International Conference on Multimedia and Signal Processing 2011, pp. 357–361. IEEE (2011)
18. Yang, J., Wright, J., Huang, T., Ma, Y.: Image super-resolution as sparse representation of raw image patches. In: IEEE conference on computer vision and pattern recognition 2008, pp. 1–8. IEEE (2008)
19. Vandewalle, P., Süsstrunk, S., Vetterli, M.: A frequency domain approach to registration of aliased images with application to super-resolution. EURASIP J. Adv. Signal Process. **2006**(1), 233 (2006)
20. Insana, Lab.: University of Illinois. Patient and gelatin phantom RF echo data from siemens antares ultrasound system - download data. http://ultrasonics.bioengineering.illinois.edu/index.asp (2007)
21. Piotrzkowska-Wrblewska, H., Dobruch-Sobczak, K., Byra, M., Nowicki, A.: Open access database of raw ultrasonic signals acquired from malignant and benign breast lesions. Med. Phys. https://doi.org/10.1002/mp.12538

On the Novel Image Algorithms Filling Material Index over the Finite-Difference Time-Domain Grid for Analysis of Microstrip Patch Antenna Using MATLAB

Girish G. Bhide[✉], Anil B. Nandgaonkar, and Sanjay L. Nalbalwar

Dr. Babasaheb Ambedkar Technological University, Lonere, Raigad 402103, India
ggb_rtn@yahoo.co.in

Abstract. This paper presents a simple but effective technique to fill out the material index grid for the FDTD analysis of microstrip patch antenna. To solve any electromagnetic problem, the physical space is discretized in three dimensions. The grid points are used for the evaluation of electric and magnetic fields. Further, these grid points are also assigned the various properties of the material, such as conductivity, permittivity, and permeability. The performance of FDTD analysis mainly depends upon the accurate filling of the material indexes over the grid. It becomes a tedious job when the patch layout is consisting of circles or triangles. Two compact algorithms are proposed here for the creation of shapes close to circle and triangle. The algorithm for plotting circular shape is based on overlapping rectangles. The algorithm for plotting triangular shape uses the area comparison technique. The algorithms are quite useful for the creation of presence as well as an absence of material. The patch layouts produced by these algorithms are matching with the physical shapes in the design.

Keywords: Image algorithm · Material index grid · FDTD · Microstrip patch antenna

1 Introduction

The finite-difference time-domain analysis is one of the widely used computational electromagnetics methods. It finds applications in electromagnetic wave scattering models for computation of near steady-state fields, far fields, radar cross-section [1]. There are numerous applications of FDTD, but that is not the focus of this work hence not mentioned here. FDTD is widely used in the analysis of microstrip patch antennas [2, 3]. FDTD is a method wherein the direct solution of Maxwell's time-dependent curl equations is obtained without employing potentials. Maxwell's curl equations in the magnetic loss-free region are as given below.

$$\nabla \times \bar{E} = -\frac{\partial \bar{B}}{\partial t} \tag{1}$$

© Springer Nature Singapore Pte Ltd. 2020
B. Iyer et al. (eds.), *Applied Computer Vision and Image Processing*,
Advances in Intelligent Systems and Computing 1155,
https://doi.org/10.1007/978-981-15-4029-5_14

$$\nabla \times \bar{H} = \mathrm{J} + \frac{\partial \bar{D}}{\partial t} \tag{2}$$

The constitutive relations necessary to support Maxwell's equations are given below.

$$\bar{D} = \epsilon \, \bar{E} \tag{3}$$

$$\bar{B} = \mu \bar{H} \tag{4}$$

The vector equations (1) and (2) can be decomposed each into three scalar equations to get six equations in the cartesian coordinate system (x, y, z) as given hereunder.

$$\frac{\partial E_x}{\partial t} = \frac{1}{\varepsilon} \left(\frac{\partial H_z}{\partial y} - \frac{\partial H_y}{\partial z} \right) \tag{5}$$

$$\frac{\partial E_y}{\partial t} = \frac{1}{\varepsilon} \left(\frac{\partial H_x}{\partial z} - \frac{\partial H_z}{\partial x} \right) \tag{6}$$

$$\frac{\partial E_z}{\partial t} = \frac{1}{\varepsilon} \left(\frac{\partial H_y}{\partial x} - \frac{\partial H_x}{\partial y} \right) \tag{7}$$

$$\frac{\partial H_x}{\partial t} = \frac{1}{\mu} \left(\frac{\partial E_z}{\partial y} - \frac{\partial E_y}{\partial z} \right) \tag{8}$$

$$\frac{\partial H_y}{\partial t} = \frac{1}{\mu} \left(\frac{\partial E_x}{\partial z} - \frac{\partial E_z}{\partial x} \right) \tag{9}$$

$$\frac{\partial H_z}{\partial t} = \frac{1}{\mu} \left(\frac{\partial E_y}{\partial x} - \frac{\partial E_x}{\partial y} \right) \tag{10}$$

The FDTD algorithm dates back to 1966 when Yee originated a set of finite-difference equations for the time-dependent Maxwell's curl equations system [4]. These equations are shown in discrete form, both in space and time, using a central difference formula. The electric and magnetic field components are sampled at discrete positions both in time and space. This FDTD technique divides the three-dimensional problem geometry into cells to form a grid. Figure 1 shows a Yee cell consisting of various field components.

The FDTD algorithm samples and calculates the fields at discrete time instants 0, $\Delta t, 2\Delta t, \ldots n\Delta t$. The material parameters, such as permittivity ε, permeability , electric conductivity σ, and magnetic conductivity σ_m, are distributed over the FDTD grid. These are associated with field components. They are, therefore, indexed the same as their respective field components. Solutions are obtained by adequately terminating the computational domain with absorbing boundary conditions [5].

The FDTD method can solve a variety of problems. The important steps of this method are as given hereunder.

- Discretization of problem space,
- Filling material constants/indices at all the nodes,
- Application of the boundary condition (ABC) such as Mur's ABC, Berenger's PML, uniaxial PML, and CPML,
- Calculating FDTD coefficients,

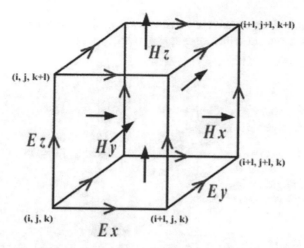

Fig. 1 Field components on a Yee cell indexed as (i, j, k)

- Running the FDTD time marching loop for evaluation of electric and magnetic field components,
- Postprocessing, etc.

One of the important parts of this method is to fill in the material indices over the discretized structure of the problem space. In this work, a microstrip patch antenna structure is created using the proposed algorithms of creating various shapes like circle, triangle, rectangle, etc.

1.1 Motivation for the Work

There is a lot of literature available wherein microstrip patch antenna is applied for imaging applications, such as human head imaging, breast cancer imaging, high-resolution RF imaging radar applications, medical applications, and human existence detections applications [6–14]. However, the reverse, i.e., application of image for antenna analysis, is not available, especially for FDTD. The MATLAB antenna toolbox allows us to accept an image of the patch for antenna design, but it employs the Method of Moments from computational electromagnetics [15]. It was felt that image in binary format can be used to prepare patch layout or even other layers of a microstrip patch antenna; hence this work is undertaken.

1.2 Problem Statement

The patch layouts are laid over the dielectric substrate having different shapes made up of a combination of rectangular, circular, semicircular, triangular, etc. These shapes represent conducting material (copper). Further, when slots are created, we need to remove copper material in that portion of the patch. Thus, we can say that these shapes may have a presence or absence of the copper material. We need to draw the shapes for both presence and absence and are considered as positive or negative shapes. To help understand this statement following Fig. 2 is useful.

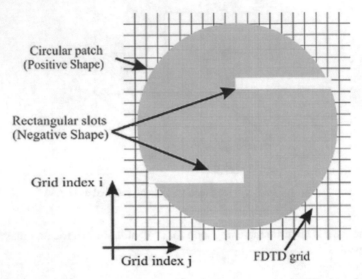

Fig. 2 Sample Patch layout over the grid to be implemented in FDTD Algorithm

Let the index values of various materials involved in the structure of a microstrip patch antenna are as given in Table 1.

Table 1 Index values of materials in FDTD implementation of microstrip patch antenna structure

Sr.	Type of material/structure	Index value assigned
1	Conducting	2
2	Vacuum	1
3	Dielectric	3
4	Matched load	4

Now, these index values can be assigned when we have the correct patch layout confined to the FDTD grid. In this work, we propose image creation algorithms for the creation of a patch layout above the substrate layer. Once the substrate size along x- and y-direction and the relative locations of various shapes like circle, triangle, rectangle, slots, etc., are fixed, we are ready to invoke the image creation algorithms.

2   Image Creation Algorithm

The circle plotting algorithm and triangle plotting algorithms are both based upon the rectangle plotting algorithm. As we know that no curve is displayed as a smooth curve on the computer screen; instead, it is in the form of staircase approximation; the same is the approach used in these algorithms. When the number of pixels is more, we do not feel as if there is a staircase approximation. Before we deal with the circle plotting algorithm and triangle plotting algorithm, it would be better if we go through the rectangle plotting code.

The rectangle plotting is not a great deal. It simply needs a lower limit along the x-direction (x_{min}), the upper limit for x-direction (x_{max}) as outer "for loop" indices. Similarly, for y-direction, the limits are (y_{min}) and (y_{max}) as indices of inner "for loop." A simple MATLAB code section for the creation of a rectangular image is given below in Table 2.

Table 2 MATLAB code section for plotting rectangle

```
imszx = input ('enter x size of image');
imszy = input ('enter y size of image');
  for i = 1: imszx
    for j = 1: imszy
      b (i, j) = 200;
    end
  end
xmin = input ('top left corner x coordinate');
ymin = input ('top left corner y coordinate');
xmax = input ('bottom right corner x coordinate');
ymax = input ('bottom right corner y coordinate');
  for i = xmin: xmax
    for j = ymin: ymax
      b (i, j) = 0;
    end
  end
```

In the above code index, i is measured from the top of the image, while j is measured from the left edge of the image. The same conventions are followed in all the subsequent discussion. To invert the shape from positive to negative the values assigned to $b(i, j)$, i.e., 200 and 0 are swapped.

2.1   Circle Plotting Algorithm

There are various algorithms available to draw a circle; however, none of them is suitable for filling the material indices over the FDTD grid. Therefore, there was a need for a novel algorithm for the creation of shapes in the form of images so that filling material indices becomes quite easy.

The circle is plotted using overlapping rectangles developed in every run of the control loop in such a way that the shape resembles the circle. Varying numbers of overlapping rectangles can control accuracy. The following Table 3 shows the algorithm.

Table 3 Circle plotting algorithm

Sr.	Action
1.	Read the size of the image along x and y diretion.
2.	Fill all the indices in this range with zeros. b(i ,j) = 0; for all (i, j) within image.
3.	Read the radius of the circle to be drawn (odd integer.
4.	Read the center coordinates say 'Cx' and 'Cy'.
5.	Read the number of steps of accuracy, say 'st'.
6.	Calculate the span of a rectangle in x and y directions by using following formulae Spanx(k) = 1 + round (Sin(k*90/st) * radius); Spany(k) = 1 + round (Cos(k*90/st) * radius);
7.	Fill in the rectangle by using above spans on either side of 'Cx' and 'Cy'. b(i, j) = 0; for all (i, j) in the span.
8.	Check whether all steps are over. If No, then go to step 6, else proceed.
9.	Display the image.

The output of the algorithm: A MATLAB code is developed based on this algorithm. The sample outputs are shown for an image size of 500 × 500 pixels. The radius for plotting the circle is 151. The center coordinates of the circle (Cx, Cy) are (250, 250). The circles have been plotted for the different number of steps of accuracy, as shown in Fig. 3. One can easily notice that the staircasing is more when an accuracy setting is of 15 steps, and it decreases as the number of steps is increased to 30.

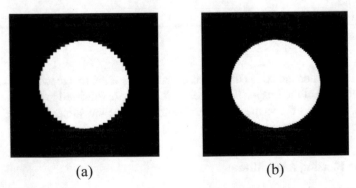

(a)   (b)

Fig. 3 Sample outputs—negative shape with accuracy **a** 15 steps **b** 30 steps

The higher accuracy settings certainly improve the quality of the image, but at the same time, it increases the number of FDTD cells while implementing the FDTD algorithm. A moderate accuracy is good enough to implement FDTD code.

2.2 Triangle Plotting Algorithm

The algorithm uses the mathematical principle, which says that "*A point lies inside the triangle if the area of that triangle is equal to the sum of three triangles formed by selecting two vertices at a time and the third vertex is the point under test.*" We need to calculate the areas of the triangle to be plotted and the three triangles for deciding each point within the region of interest. The following Table 4 shows the algorithm.

Table 4 Triangle plotting algorithm

Sr.	Action
1.	Read the size of the image along x and y direction.
2.	Read the vertices of the triangle. (x1, y1), (x2, y2), and (x3, y3)
3.	Calculate the area using the following formula. Area = abs(0.5*(x1*(y2-y3)+x2*(y3-y1)+ x3*(y1-y2)))
4.	Determine maximum of x1, x2, and x3 Fill indices from 1 to maxx and 1 to maxy with value 100. b(i, j) = 100; ∀ (i, j) ≤ (maxx, maxy).
5.	For every (x, y) in the above range, calculate the areas of the three triangles by using two of the three vertices and (x, y) as the third vertex.
6.	A1 = abs(0.5*(x1*(y2-j)+x2*(j-y1)+i*(y1-y2))); … (x3 and y3 replaced by i, j in above). A2 = abs(0.5*(i*(y2-y3)+x2*(y3-j)+x3*(j-y2))); … (x1 and y1 replaced by i, j in above). A3 = abs(0.5*(x1*(j-y3)+i*(y3-y1)+x3*(y1-j))); … (x2 and y2 replaced by i, j in above). sumA123 = round (A1+A2+A3);
7.	If Area >= sumA123 then b(i, j) = 0; (point is inside)
8.	Check whether all steps are over. If No, then go to 6. If Yes, then proceed.
9.	Display the image.

The output of the algorithm: A MATLAB code is developed based on this algorithm. The sample outputs are obtained to represent the presence of copper (positive) are shown in Fig. 4.

Similarly, three triangles are plotted to represent the absence of copper (negative) and shown in Fig. 5.

The higher accuracy settings certainly improve the quality of the image, A moderate accuracy is good enough to implement the FDTD code using the rectangular grid.

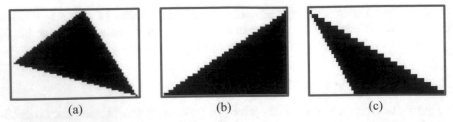

(a)   (b)   (c)

Fig. 4 Sample outputs—positive type triangle **a** acute angle **b** right angle **c** obtuse angle

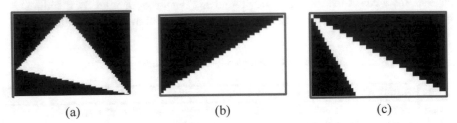

(a)   (b)   (c)

Fig. 5 Sample outputs-negative type triangle **a** acute angle **b** right angle **c** obtuse angle

3 Performance Analysis

The performance analysis of all the above three algorithms is done by comparing the area of the developed shape with the area of the desired shape as per the geometric formula. The following Table 5 displays the performance analysis.

There is lot of literature available wherein material indexes are filled directly in the FDTD code using the coordinates for solving the electromagnetic problems [2, 3, 16]. A comparison between existing method for filling material index and proposed algorithms is shown in the Table 6 given below.

4 Implementation

The MATLAB code consisting of the above algorithms, i.e., for rectangle, circle, and triangle, is developed and used for preparing the top layer of the patch antenna. Figure 6 shows a sample layout for a circular shape antenna with L shape slots where the dimensions are in mm [17]. It consists of a circle (positive), rectangles (positive as well as negative). The circle has a radius of 30 mm having center coinciding with substrate center. The circle represents copper and hence is in the form of a positive shape. The feed line is made up of copper hence it is a positive shape having length of 10 mm. The L shape slots are also implemented using rectangles (negative). The slots have vertical length of 30 mm while horizontal lengths of 15 mm.

Table 5 Performance analysis of the three-image algorithms

Sr.	Type of shape	Number of accuracy steps	Area of output shape (sq. mm) for a target area of 100 sq. mm	% Difference of areas	Qualitative remark
1	Rectangle	NA	100	0.0	Exact, obvious
2	Circle				
		15	99.10	−0.90	Slightly undersize
		30	100.95	0.95	Slightly oversize
		60	101.83	1.83	Oversize, smooth
3	Triangle				
	– Acute angle	NA	102.05	2.05	Oversize
	– Right angle	NA	103.55	3.55	Oversize
	– Obtuse angle	NA	102.73	2.73	Oversize

5 Conclusion

The proposed algorithms are straightforward to understand and adopt as compared to the existing method. Any complicated microstrip patch antenna layout, comprising basic shapes, like a circle, triangle, and rectangle, can be implemented using these algorithms. Union and subtraction of shapes are easily achieved, provided the coordinates of the center and radius of a circle, vertices of a triangle, etc., are appropriately chosen. Thus, these algorithms can contribute and may serve as a suitable candidate in the implementation of the FDTD method.

Table 6 Comparison with existing method

Sr.	Parameter	Existing method	Proposed algorithms
1	Approach	Contiguous range of coordinates representing grid nodes	Using geometric properties of desired shape
2	Comfort	Cumbersome	Easy
3	Accuracy	As per the grid setting	As per the grid setting

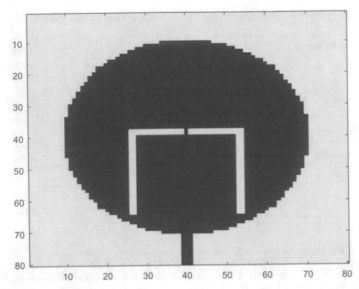

Fig. 6 Sample patch layout of circular microstrip patch antenna with L shape slots

References

1. Umashankar, K.R., Taflove, A.: A novel method to analyze electromagnetic scattering of complex objects. IEEE Trans. Electromagn. Compat. **24**(4), 397–405 (1982)
2. Gao, S., Li, L.W., Leong, M.S., Yeo, T.S.: A broad-band dual-polarized microstrip patch antenna with aperture coupling. IEEE Trans. Antennas Propag. **51**(4), 898–900 (2003)
3. Kar, M., Wahid, P.F.: The FDTD analysis of a microstrip patch antenna with dual feed lines. In: Proceedings IEEE Southeastcon '98 'Engineering for a New Era', pp. 84–86, Orlando, FL, USA (1998)
4. Yee, K.: Numerical solution of initial boundary value problems involving Maxwell's equations in isotropic media. IEEE Trans. Antennas Propag. **14**(3), 302–307 (1966)
5. Katz, D.S., Thiele, E.T., Taflove, A.: Validation and extension to three dimensions of the Berenger PML absorbing boundary condition for FD-TD meshes. IEEE Microw. Guided Wave Lett. **4**(8), 268–270 (1994)
6. Rokunuzzaman, M., Samsuzzaman, M., Islam, M.T.: Unidirectional wideband 3-D antenna for human head-imaging application. IEEE Antennas Wirel. Propag. Lett. **16**, 169–172 (2017)
7. Hidayat, M.V., Apriono, C.: Design of 0.312 THz microstrip linear array antenna for breast cancer imaging application. In: 2018 International Conference on Signals and Systems (ICSigSys), pp. 224–228, Bali, Indonesia (2018)
8. Aggarwal, N., Gangwar, V.S.: M-shaped compact and broadband patch antenna for high resolution RF imaging radar applications. In: 2014 IEEE International Microwave and RF Conference (IMaRC), pp. 356–359, Bangalore (2014)
9. Iyer, B., Pathak, N.P., Ghosh, D.: Dual-input dual-output RF sensor for indoor human occupancy and position monitoring. IEEE Sens. J. **15**(7), 3959–3966 (2015)
10. Iyer, B., Garg, M., Pathak, N.P., Ghosh, D.: Contactless detection and analysis of human vital signs using concurrent dual-band RF system. Elsevier Procedia Eng. **64**, 185–194 (2013)
11. Iyer, B., Pathak, N.P., Ghosh, D.: Concurrent dualband patch antenna array for non-invasive human vital sign detection application. In: 2014 IEEE Asia-Pacific Conference on Applied Electromagnetics (APACE), pp. 150–153 (2014)

12. Iyer, B.: Characterisation of concurrent multiband RF transceiver for WLAN applications. Adv. Intell. Syst. Res. **137**, 834–846 (2017)
13. Iyer, B., Pathak, N.P., Ghosh, D.: RF sensor for smart home application. Int. J. Syst. Assur. Eng. Manag. **9**, 52–57 (2018). https://doi.org/10.1007/s13198-016-0468-5
14. Iyer, B., Abegaonkar, M.P., Koul, S.K.: Reconfigurable inset-fed patch antenna design using DGS for human vital sign detection application. Comput. Commun. Signal Process. **810**, 73–80 (2019)
15. Antenna Toolbox—MATLAB-Mathworks. https://in.mathworks.com/company/newsletters/articles/using-a-photo-for-full-wave-antenna-analysis.html
16. EI Hajibi, Y., EI Hamichi, A.: Simulation and numerical modeling of a rectangular patch antenna using finite difference time domain (FDTD) method. J. Comput. Sci. Inf. Technol. **2**(2), 01–08 (2014)
17. Lu, J.-H.: Broadband dual-frequency operation of circular patch antennas and arrays with a pair of L-shaped slots. IEEE Trans. Antennas Propag. **51**(5), 1018–1023 (2003)

A Simulation Model to Predict Coal-Fired Power Plant Production Rate Using Artificial Neural Network Tool

Ravinder Kumar[1], Keval Nikam[1,2(⊠)], and Ravindra Jilte[1]

[1] School of Mechanical Engineering, Lovely Professional University, Phagwara 144411, Punjab, India
ravchauhan8@gmail.com, nikamkeval26@gmail.com
[2] Faculty of Mechanical Engineering, Dr. D. Y. Patil Institute of Engineering, Management and Research, Savitribai Phule Pune University, Akurdi, Pune 411044, India

Abstract. At present, the performance evaluation of a coal-fired power plant is highly required to enhance its efficiency. The present paper deals with the power prediction of a 600 million watts (MW) typical subcritical coal-fired power plant situated in North India using real operational data at different load conditions. The entire complex thermal system comprises various systems like boiler, turbine, condenser, re-heater, deaerator, boiler feed pump, etc. A simulation model is prepared to predict the performance of the power plant. The Artificial Neuron Network tool is used to validate the simulation model for known input and output data. The selected performance criterion and network error is found satisfactory after the analysis of computational results. The coefficient of determination was calculated as 0.99787 which gives an idea of the close relation of one output variable with several different input variables. This paper will definitely help out to those interdisciplinary engineers who are dealing with combining technology with the conventional process.

Keywords: Subcritical coal-fired power plant · Artificial neuron network · Boiler · Turbine

1 Introduction

A coal-based power plant is a complex and highly sophisticated system. Heavy finance is involved annually in the operation and maintenance of such complex power production plants to achieve the desired availability level. Performance prediction of such a thermal system is of great concern because of increasing cost, competition, and public demand in one way, while the risk of failure on the other. In India, there is a heavy load on coal-fired power plants where coal is generally used as a common fuel for electricity generation. This generated electricity plays an important role to raise the modern economy for industry, agriculture, transport, and household for any nation. But, the growing energy demand has increased energy consumption in the entire world. As energy demand is continuously increasing, the improvement in operating characteristics of electrical power

© Springer Nature Singapore Pte Ltd. 2020
B. Iyer et al. (eds.), *Applied Computer Vision and Image Processing*,
Advances in Intelligent Systems and Computing 1155,
https://doi.org/10.1007/978-981-15-4029-5_15

generation plants has become a mandate in this era. According to the data published by the Central Electricity Authority (CEA) of India, reported that per capita consumption has increased from 15 kWh in 1950 to about 1,122 kWh (kilowatts hours) in the year 2016–17. Out of 597,464 census villages, 592,135 villages (99.25 %) have been electrified as on 31.03.2017. Efforts are been taken by the government of India to fulfill the needs of the citizen by increasing generation about 5.1 billion units in 1950 to 1,242 BU (including imports) in the year 2016–17. Coal-fired power plants are ruling over the power sector from past decades and will continue to remain on top position in India due to the easy availability of coal. The installed capacity of India as on 31.3.2017 was 326,833 MW comprising 59 % of coal-fired power plants as compared to other power sources as shown in Fig. 1.

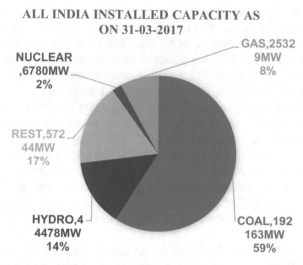

Fig. 1 All India installed capacity as on 31-03-2017 [1]

    The subcritical power plants are contributing at the major level in electricity generation in India. Installed unit of a subcritical power plant in India as per data published by CEA were 74 units working under Private companies, 33 units are under State government and 23 units are operating under the Central government. The exact thermodynamic analysis of the subcritical power plant is done using suitable assumptions. Without suitable assumptions, thermodynamic analysis of the subcritical power plant will lead to more number of nonlinear equations and the solution of such equations will lead to an increase in computational time. To overcome this issue, an Artificial Neural Network (ANN) is used to analyze the systems for operational input and output patterns. Using the previously stored operational data, the neural network can be modeled to simulate the power plant operation. ANN models are more dominant than physical models as these can be trained with live operational data. By comparing previous ANN models with the latest model, factors affecting the output can be studied and suitable maintenance majors would be taken in advance. The response time of the ANN model is quick [2]. Some researchers have presented a brief review of applications of ANN in different energy systems [3]. Some of them have proposed energy and exergy analysis using ANN for heat evaluation [4]. Exergy efficiency can also be predicted by using ANN techniques [5]. Multiple regression techniques are also used to predict the energy performance of power generating plants [6]. The use of soft computing techniques like Genetic Algorithm, Particle Swarm Optimization, Fuzzy Logic, and ANN for simulation and prediction of thermodynamic systems is presented in past research work [7, 8]. Specifically, the performance of power generation turbines can also be modeled by using ANN [9]. Researchers nowadays are focusing on carbon capture technology to understand the emissions produced from existing power plant [10]. Emission contents like NO_x (Nitrogen oxide) and CO_2 (Carbon dioxide) from coal-fired power plants can also be predicted by using the ANN technique along with the Genetic Algorithm [11, 12]. It is particularly useful in system modeling while implementing complex mapping, fault detection, and prediction of output [13, 14]. Only a few researchers developed a simulation model for predicting equipment maintenance and their priorities using probabilistic approaches [15]. Several authors have reported the capability of ANN to replicate an established correspondence between points of an input domain and points of an output domain to interpret the behavior of phenomena involved in energy conversion plants [16, 17]. However, ANN models can be developed with definite objective and training with data from existing plants with lesser effort but with great utility. Prediction of output with nearly zero error helps to monitor the performance of power plants [18]. Plant operators can decide to change the water and steam flow rate, temperature and pressure values in accordance with usage of ANN model [19]. Online tracking of the generator output will increase operator awareness, and also plant efficiency and stability in parallel [20]. This kind of prediction method helps in the prediction of amount of energy used and required to obtain the desired output. Recently researchers are implementing ANN to the ultra-supercritical power plant for developing innovative modeling for controlling parameters [21].

In this paper, ANN was trained with the operational data from the subcritical power plant of 600 MW. The objective of the model was to predict mass flow rate, specific enthalpy, pressure and temperature of steam exiting various inlet and outlet of the components such as boiler, high-pressure turbine, intermediate pressure turbine, low-pressure turbine, condenser, re-heater, deaerator, boiler feed pump, etc. In order to have control of the plant with the operator, components should able to communicate based on the history with it. This will increase the plant utilization capacity and finally increase the operating hours. The motivation to select this integration of steam power plant components with the latest technology such as ANN is to improve the overall performance of the subcritical power plant.

2 Experimental Facility

2.1 Typical Plant Description

The unit of 600 MW subcritical power plant was selected for the modeling. The basic description of a subcritical coal-fired thermal power plant is shown in Fig. 2. It uses coal energy to convert it into mechanical energy through the expansion of steam in steam turbines. Coal received from collieries in the rail wagons is routed by belt conveyors. After crushing the coal in coal mills, it is then pushed to the boiler furnace (1), which is comprised of water tube walls all around through which water circulates. The chemically treated water through the boiler walls is converted into high-temperature steam. This steam is further heated in the super-heaters (3). The thermal energy of this steam is utilized to produce mechanical work in high pressure, intermediate pressure, and low pressure turbines. The steam blend collected from the turbine extractions is returned to feedwater heaters. The output of the turbine rotor is coupled to the generator (17) to produce electric energy. The steam after doing useful work in the turbine is condensed to water in condenser (8) for recycling in the boiler. For our study, we had not considered the air preheated circuit and coal feeding system.

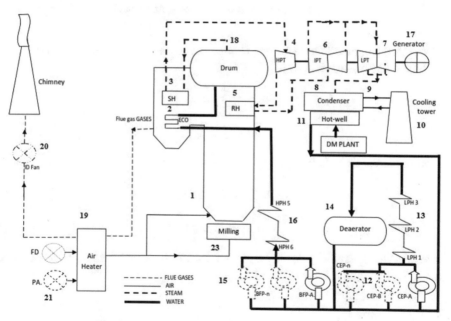

Fig. 2 Typical subcritical coal-fired power plant [22]

2.2   Design of Neural Network Model for 600 MW Plant

The neural network technique is developed from the behavior of the human brain. ANN gets trained by using the previous operational data and develop the relationship between the performance affecting input and output variables. The feed-forward transform function is used to predict the output for known operational inputs parameters. It consists of an input layer, two hidden layers, and an output layer as shown in Fig. 3.

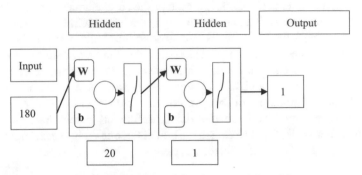

Fig. 3 Typical neuron for the proposed model

Weighted summed is transformed to predict the output in the neural network tool [9]. Hidden Layer Calculations [4]

$$\text{net}_i = \sum x_i w_{ij} \tag{1}$$

$$y_i = f(\text{net}_i) \tag{2}$$

Output Layer Calculations

$$\text{net}_x = \sum y_i w_{jk} \tag{3}$$

$$O_k = f(\text{net}_k) \tag{4}$$

where x_i is the input data, y_i is the output obtained from layer 1, and O_k is the output of layer 2. W is weight, b is bias.

2.3   Measurements

Operational data of 600 MW subcritical power plants were taken for each component. These readings were then imported to the ANN tool present in MATLAB software for constructing a predictive model. The following are the parameters (see Table 1) which were considered while taking a reading.

Table 1 Parameters considered and inputs for predicting output

Measured Parameters	Particulars	Reading	Particulars	Reading	Particulars	Reading	Particulars	Reading	Particulars	Reading	Particulars	Reading	Particulars	Reading
h(Kcal/Kg)	Before steam Generator	269.5	Blend 2 after IPT	728.9	Before Condenser	568.5	After LPH 1	50.5	Blend 1	677	After Boiler Feed Pump	174.5	Blend of HPH 2 send back to Dearator	218.2
C(Temp)		259		297.4		0.9141		50.3		183.9		170.1		210
T/H(Mass F)		1979.4		190.8		1241.3		1541.03		78.53		1979		223.9
AT		207.26		8.17		0.105		11.07		2.729		209.6		0
h(Kcal/Kg)	After Steam Generator/ Before Super heater	419.8	After IPT	728.9	After Condenser	46.3	Blend 3	613.6	Before LPH4	127.5	Before HPH(After 1 BFP)	174.5	Before SG	269.5
C(Temp)		359.3		297.7		46.4		0.969		127		170.1		259
T/H(Mass F)		0		1458.4		1547		64.478		2646		1979		1979
AT		188.89		8.32		0.105		0.44		9.39		209.6		207.3
h(Kcal/Kg)	After Superheater/ Before HPT	810.7	Blend 1	677	Before Condenser Extra.Pump	46.3	Before LPH2	613.8	After LPH4	728.9	Blend 1 after IPT	781.9	After SG	419.8
C(Temp)		587		184.3		46.3		0.9701		297.4		409.9		359.3
T/H(Mass F)		1979.4		78.528		1547		64.418		190.8		103.1		0
AT		170		2.878		0.105		0.416		8.17		18.16		188.9
h(Kcal/Kg)	Blend 1 after HPT	729.9	Blend 2	638.5	After Condenser Extra.Pump	46.9	After LPH2	73.5	Blend 2 after IPT	728.9	After HPH 1	210.5		
C(Temp)		342.8		99.3		46.6		74.3		296.6		204.7		
T/H(Mass F)		223.88		63.678		1547		1547.03		105.4		1979		
AT		49.24		1.119		20.89		10.83		7.48		208.4		
h(Kcal/Kg)	Blend 2 after HPT /Before Reheater	729.9	Blend 3	729.9	Before Gland Steam Condenser	46.9	Blend 2	638.5	Before Dearator	638.5	Blend of HPH 1 send back to Dearator	177.5		
C(Temp)		342.8		342.8		46.6		99.498		99.498		175.4		
T/H(Mass F)		1749.6		64.478		1547		63.678		63.678		327		
AT		49.24		49.24		20.89		1.119		1.119		0		
h(Kcal/Kg)	After Reheater/B efore IPT	842.2	Blend 4	64.1	After Gland Steam Condenser	47.3	Before LPH3	638.5	After Dearator	638.5	Blend 1 after HPT	729.9		
C(Temp)		537		64		47.1		98.7		98.7		342.6		
T/H(Mass F)		1749.6		11.831		1541		63.678		63.678		223.9		
AT		48.35		0.242		11.71		1.059		1.059		48.9		
h(Kcal/Kg)	Blend 1 after IPT	781.7	Blend 5	568.5	Before LPH 1	47.3	After LPH3	98.1	Before Boiler Feed Pump	168.1	After HPH 2	269.5		
C(Temp)		410		0.9141		47.1		97.9		166.6		259		
T/H(Mass F)		103.11		1241.3		1541		1547.03		1979		1979		
AT		18.26		0.105		11.71		10.33		7.33		201.9		

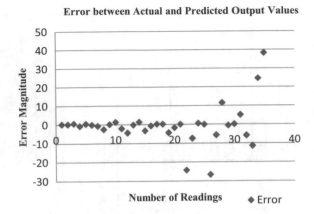

Fig. 4 Graphical representation of error for different readings during training

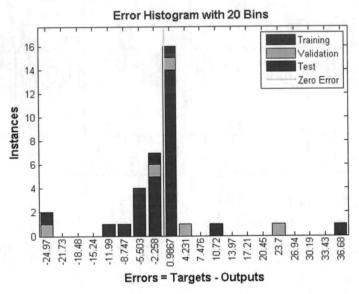

Fig. 5 Error considering 20 neurons between target value and output

3   Results and Discussions

Graphical representation of error between actual and predicted output values for different readings during training is shown in Fig. 4. During the training of neurons, minimum error magnitude was evaluated as 0.11182 for 35 readings as predicted from the ANN model.

As per the discussion of reading in Table 1, the actual output of the plant was observed to be 600 MW. The same said output was predicted from ANN Model with 20 neurons in the hidden layer. It is concluded that considering 20 neurons predicted values come to be true value where the error is equal to zero as shown in Fig. 5. Different neuron combination strategy was used to predict the output as shown in Table 2.

Table 2 Computational table at different combination of neurons

Itr.	Neurons	Training	Validation	Test	ALL R	Predicted values (MW)
1	5	0.9998	0.9959	0.9958	0.9958	661.19
2	10	0.9899	0.9899	0.9541	0.9850	660.51
3	15	0.9983	0.999	0.9911	0.9943	654.00
4	20	1	0.9990	0.9933	0.9978	660
5	25	0.9959	0.9961	0.9733	0.9952	660.76
6	30	0.9983	0.9994	0.9951	0.9943	665.54
7	35	0.9962	0.9992	0.9957	0.9964	656.07
8	40	0.9992	0.9478	0.9933	0.9958	656.48
9	45	0.9984	0.9963	0.9912	0.9972	663.58
10	50	0.8030	0.7904	0.6863	0.7797	780
11	55	0.9985	0.9801	0.9902	0.9959	658.52

In addition to the above work, energy and exergy analysis of power plant is performed and exergetic efficiency is determined as shown in result Table 3. It is clear from Table 3 that ANN is positively applied to existing plants to increase the performance. The exergetic efficiency of components namely High-Pressure Turbine (HPTr), Intermediate Pressure Turbine (IPTr), Low-Pressure Turbine (LPTr), Condensate Extraction Pump (CEP), Boiler Feed Pump (BFP), High-Pressure Heater 1 (HPHeater1), High-Pressure Heater 2 (HPHeater2), Low-Pressure Heater 1 (LPHeater1), Low-Pressure Heater 2 (LPHeater2), Low-Pressure Heater 3 (LPHeater3), Low-Pressure Heater 4 (LPHeater4).

Energetic η and Exegetic ψ efficiency from the sample reading was evaluated as follows:

$$\eta = 41.2\% \quad \text{and} \quad \psi = 39.23\%$$

The small difference in efficiencies is due to chemical exergy of coal being greater than its specific energy measured by its high heating value. From the exegetic efficiency, it is clearly seen that waste heat emissions from the condenser although greater in quantity are low in quality (i.e., have little exergy as compared with other) because of temperature near to surrounding temperature. So improvement in condenser will slightly increase the overall exergy efficiency. From Fig. 6 with minimum deviation from actual reading a straight line is fitted with $R = 0.99787$.

Table 3 Comparison of calculated exergetic efficiency with reference paper

Sr.No.	Components	Calculated exergetic efficiency	Ref. [23]	Ref. [24]	Ref. [25]
1	HPTr	95.04	73.5	72.66	92.11
2	IPTr	94.8	–	–	–
3	LPTr	68.77	–	–	–
4	CEP	67.18	–	–	54.91
5	BFP	88.83	82.5	81.51	–
6	HPHeater1	95.9	97.4	97.65	91.58
7	HPHeater2	92.9	95.3	96.95	86.12
8	LPHeater1	89.99	89.5	89.06	85.41
9	LPHeater2	86.94	67.3	82.79	82.65
10	LPHeater3	86.91	–	–	82.65
11	LPHeater4	85.93	–	–	82.65

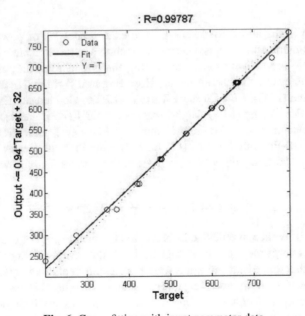

Fig. 6 Curve fitting with input parameter data

4 Conclusion

ANN model was developed with the help of operational data of the existing power plant with different inputs and corresponding outputs in the form of power generated. The main objective of this model was to predict power output for known input parameters. On comparing the error in prediction, it has been found that the ANN model with 20 neurons yields and minimum value of error in actual value recorded. Satisfactory coefficient of determination "R" was found to be 0.99787 which gives an idea of the close relation of one output variable with several different input variables. With this model linked with the control system of the power plant, operators can do changes accordingly in input parameters to achieve the desired power output. As a result of this efficiency and stability of a plant is continuously observed and necessary action could be taken to avoid the losses at different sections of coal-fired power plant.

References

1. Verma, V.S.: Adoption and introduction of supercritical technology in the power sector and consequential effects in operation, efficiency and carbon dioxide emission in the present context. In: Goel, M., Sudhakar, M. (eds.) Carbon Utilization, Green Energy and Technology, pp. 35–43. Springer, Singapore (2017)
2. Ianzhong Cui, X., Shin, K.G.: Contributed paper application of neural networks to temperature control in thermal power plants. Eng. Appl. Artif. Intell. **5**(6), 527–538 (1992)
3. Azadeh, A., Ghaderi, S.F., Anvari, M., Saberi, M.: Performance assessment of electric power generations using an adaptive neural network algorithm. Energy Policy **35**, 3155–3166 (2007)
4. Anead, H.S.: Evaluation and improvement performance of a boiler in a thermal power plant using artificial neural network. Eng. Technol. J. **36**(6), 656–663 (2018)
5. Acır, A.: Application of artificial neural network to exergy performance analysis of coal fired thermal power plant. Int. J. Exergy **12**(3), 362–379 (2013)
6. Kumar, R., Jilte, R., Mayank, B., Coal, Á.: Steady-State Modelling and Validation of a Thermal Power Plant. Springer, Singapore (2019)
7. Naserabad, S.N., Mehrpanahi, A., Ahmadi, G.: Multi-objective optimization of HRSG configurations on the steam power plant repowering specifications. **159** (2018)
8. Qi, J., Zhou, K., Huang, J., Si, X.: Numerical simulation of the heat transfer of superheater tubes in power plants considering oxide scale. Int. J. Heat Mass Transf. **122**, 929–938 (2018)
9. Salim, H., Faisal, K., Jawad, R.: Enhancement of performance for steam turbine in thermal power plants using artificial neural network and electric circuit design. Appl. Comput. Intell. Soft Comput. **2018** (2018)
10. Kumar, R., Jilte, R., Nikam, K.: Status of carbon capture and storage in India's coal fired power plants: a critical review. Environ. Technol. Innov. **13**, 94–103 (2019)
11. Tunckaya, Y., Koklukaya, E.: Comparative prediction analysis of 600 MW coal-fired power plant production rate using statistical and neural-based models. J. Energy Inst. **88**(1), 11–18 (2015)
12. Shi, Y., Zhong, W., Chen, X., Yu, A.B., Li, J.: Combustion optimization of ultra supercritical boiler based on artificial intelligence. Energy **170**, 804–817 (2019)
13. Chandrasekharan, S., Panda, R.C.: Statistical modeling of an integrated boiler for coal fired thermal power plant. Heliyon 3(October 2016), e00322 (2017)
14. Nurnie, N., Nistah, M., Lim, K.H., Gopal, L., Basim, F., Alnaimi, I.: Coal-fired boiler fault prediction using artificial neural networks. Int. J. Electr. Comput. Eng. **8**(4), 2486–2493 (2018)
15. Kumar, R., Tewari, P.C.: Markov approach to evaluate the availability simulation model for power generation system in a thermal power plant. Int. J. Ind. Eng. Comput. **3**(3), 743–750 (2013)
16. De, S., Kaiadi, M., Fast, M., Assadi, M.: Development of an artificial neural network model for the steam process of a coal biomass cofired combined heat and power (CHP) plant in Sweden. Energy **32**, 2099–2109 (2007)

17. Basu, S.: Modelling of steam turbine generators from heat balance diagram and determination of frequency response. **2**(1), 1–15 (2018)
18. Kumar, R., Jilte, R., Ahmadi, M.H., Kaushal, R.: A simulation model for thermal performance prediction of a coal-fired power plant, pp. 1–13 (2019)
19. Gurusingam, P., Ismail, F.B., Gunnasegaran, P.: Intelligent monitoring system of unburned carbon of fly ash for coal fired power plant boiler. In: MATEC Web of Conferences, 2017, vol. 02003, pp. 0–5 (2017)
20. Mikulandri, R., Cvetinovi, D., Spiridon, G.: Improvement of existing coal fired thermal power plants performance by control systems modifications zen Lon. Energy **57**, 55–65 (2013)
21. Hou, G., Yang, Y., Jiang, Z., Li, Q., Zhang, J.: A new approach of modeling an ultra-super-critical power plant for performance improvement. Energies **9**(310), 1–15 (2016)
22. Kumar, R.: Performance evaluation of a coal-fired power performance evaluation of a coal-fired power plant. Int. J. Perform. Eng. **9**(4), 455–461 (2013)
23. Aljundi, I.H.: Energy and exergy analysis of a steam power plant in Jordan. Appl. Therm. Eng. **29**(2–3), 324–328 (2009)
24. Hasti, S., Aroonwilas, A., Veawab, A.: Exergy analysis of ultra super-critical power plant. Energy Procedia **37**, 2544–2551 (2013)
25. Topal, H., et al.: Exergy analysis of a circulating fluidized bed power plant co-firing with olive pits: a case study of power plant in Turkey. Energy (2017)

Motion Object Tracking for Thermal Imaging Using Particle Filter

Jayant R. Mahajan[1]([✉]), Neetu Agarwal[2], and Chandansingh Rawat[3]

[1] Department of ETE, Pacific University, Udaipur, India
mahjayant@gmail.com
[2] Department of Computer Applications, Pacific University, Udaipur, India
neetu.agarwal1508@gmail.com
[3] Department of ETE, Vivekanand Institute of Technology, Chembur, Mumbai, India
csrawat3@gmail.com

Abstract. Measurement monitoring has been applied in many applications such as scrutiny, the advancement in the system to assist the driver (ADAS), contumacious biometrics, environment generated by computer, etc. In the computer learning apparatus, video scrutiny has a decisive topic for doing research. Current research in this field includes the creation of a robust and reliable tracking system.

Keywords: Motion tracking · Particle filter · Thermal imaging

1 Introduction

Human following is a crucial space that ought which may applied in many use of happening like scrutiny, combatant, amusement etc. The analysis of traffic, unrestricted biometry, ADAS having additionally and necessary applications in occurrent situations, wherever sovereign driving rushes the step within the self-propelled industries. Ongoing endeavour during this field is to create the motive force help system conformable and correct in some situations. In computer vision world the object pursuit is the greatest challenge. The execution is liable to numerous parametric quantity like barricade, background disorder, amendment in lighting and fluctuation. The improvement scale of personal computer, a straight forward creation of correct constancy and inexpensive device internal representation and increase the capture of machine-driven, video investigation for the aim of video perception analysis of objects. There are a couple of crucial phases, location of the cutting edge in motion, apprehend (uninterrupted determination of the position through visual communication) of the prospect of the image, where internal representation following continuous type and illation of the complete of the foreground position within the scene to understand the behavior of the article. The primary objective of motion following is to investigate the optical illusion in internal representation sequences. Secondary to the ocular image, recently thermic image has been victimised in applications like motion apprehend and face identification. This has attracted the eye of computer vision, researchers to the present newfound representational process modality. There are some tries to incorporate thermic pictures at the side

© Springer Nature Singapore Pte Ltd. 2020
B. Iyer et al. (eds.), *Applied Computer Vision and Image Processing*,
Advances in Intelligent Systems and Computing 1155,
https://doi.org/10.1007/978-981-15-4029-5_16

of the camera visible through information unification methods. However, if just one of the logical relation is employed as tested results are acquired with it, the system would get pleasure from value savings as well as quicker process speed.

In the concern for the outside closed-circuit television wherever the background temperature is drastically completely different from the mostly citizenry. Thermal image will play a vital role in characteristic the person in motion and following them. This truth has motivated us in the main to use solely thermal imaging to trace the movement of beings. Secondly human, thermal imaging cannot understand the darkness or lightweight elucidation, which are typically a obstruct in most motion apprehend schemes supported ocular wavelength. Therefore, thermic imaging becomes more acceptable for motion apprehend in outside encircle throughout the daylight similarly in the dark, wherever shadows, uneven light weighting and low light (nighttime) are dominating parts, creating the lot of complicated following. Grayscale image that ends up in less processing compared to the visible color camera output of the thermal imaging camera solely.

2   State of Art

The vital task for determinate the movement apprehend is that the emplacement of the destination within the structure of the internal representation concatenation or in the box. Counting on the angle of the deciphering devices, there are most likely 2 kinds of movement or the prevalence of apprehend, which are still observed the camera or the innovative monitoring movement. Depending on the fluctuation of the article, it divided into 2 categories on the markers (utilized in two dimensions and box-shaped animations) [1], while not markers. Ascertained by the strategies accustomed begin the chase movement, 2 kinds of formula detection (MT) movement are used, the acceptance dependent detection and therefore the prevalence of the detection. Depending on the sort of prevalence to be caterpillar-tracked in observation manlike occurrences [2], the observation of prevalence use can be organized into 2 categories.

The chase of various objects (MNO) [3, 4] was essential in several applications, police work within the same method. Clarification of the varied motion chase technologies are created within the literature, nominal because the Kalman Filter (KF) [5], Extended KF (EKF), Particle Filter (PF), FREE KF-Fragrance (USKF), Hidden Markov Model (HMM), transformation of Gabor. A sure prevalence detection (MT) system consists of an internal representation theme. An improvement internal representation formula furnish the position of the articles as well as a tool for demonstrating the trail of the object and the signal-forming principle to spot the outcome. The fundamental recursive formulation consists of initial part of target perception, which might achieved by subtracting the focusing internal representation between successive frames. If the {required|the mandatory} morphological operation is used; might require segmentation.

The prevalence detection (MT) that uses the approximation of occurrence implies AN silver-tongued correspondence of objects around its position. There are the subsequent approaches to the localization of the units (computationally complex), supported regions and mesh (triangle, polygonal shape supported the content) more practical based on pixels based on the correspondence of the blocks [1, 6, 7]. The most effective technique of the variation unit for occurrence vector estimation consider a base verified by a similarity measures between adjacent element. This search window is sometimes fifteen × 15 element, thus nice computing power is needed. To measure back this advanced process, there are several improvement methods at intervals the literature. Search logarithms in 2D [8], TSS [9], NTSS [10], FSS [11], 2 section have backed up several endemic winners [12], therefore explore for married management [13]. The Kalman filter [5, 14] and conjointly the particle filter [15] are the 2 frequent practical application within the field of motion search.

3   Particle Filter (PF) [16–22]

PF have considered one of the way for employ a Bayesian recursive filter exploiting the resolving power. The particle filter (PF) is beneficial as compared to the Kalman filter, especially once the quantity is process options a multi modal arrangement and which is not mathematical. In the particle filter (PF), the significant roaring point is to represent the winning concentration acting requisite work from a bunch of haphazard samples with the related to weights then compute the supported estimation for these sampling and weights. The particle filter (PF) epistemology is employed to trait the difficulty of additive filtering. Depending on the aim of reading math and casuistry conception, particle filters consist to the class of friendly divergent/genetic algorithms and synergistic strategies of middle particles.

The representation of those particle methods counts on the sphere of study. Within the functional calculation of the method, the middle genetic methods are usually utilized as heuristics and artificial search algorithms. In machines and molecular chemistry to trait the solution of segregation of the Feynman-Mac path, or within the calculation of Boltzmann-Gibbs measurements, the most own values and also the primary states of the manipulator of Schrodinger. The use of Particle filters within the life science and bio scientific discipline of the biological process of a population of individuals or sequence, the PF is depicted in several contexts.

The important purpose of a particle filter (PF) to approximation of the adjacent density of the variable observation state variables at interval of filter. The particulate informs has been intended for an Andrei Markov Hidden Model (HMM), at intervals that the scheme belongs to hidden and distinguishable covariant. A big covariant (measuring process) is related to invisible covariant (state operation) finished associate in nursing known sensible sympathetic. Likewise, the impulsive system describing the development of state covariant is to boot probabilistically known.

The flow sheet of the fundamental knowledge particle filter is shown within the following Fig. 1.

PF is intended to use the theorem computer. Apart from the delimited set of weighted particles or samples, the theorem computer provides a periodic approximation of succeeding distribution. Once victimization the non-linear and non-Gaussian system, several researchers have used the theorem computer to unravel the estimation downside. Primarily for PF, the Monte Carlo modelling is that the fundamental estimation, within which an particle established with attached artifact is approximated by succeeding concentration. Forecasting and change having two primary steps once a theorem computer uses it to style the particle filter. Victimization the system model, the sample estimate

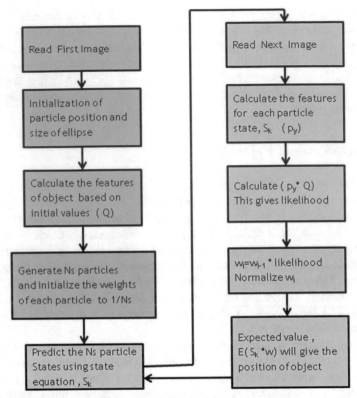

Fig. 1 Accumulation flow particle filtering technique for target/object tracking [21]

transfer are going to be performed. Betting on the observation model, the update are going to be performed to live the artifact of every sample.

The performance of the particle filter technique will be delineated as given below.

1. Read the 1st image within the successiveness and create a restricted accumulation of samples or weighted particles.
2. The particles which are weighted note particle is performed within the opening, the position and size of the conic section particle is initialized.
3. Next, betting on the initial worth, the calculation of the characteristics of the article.
4. Based on the newfound measurements and therefore the preceding state, the contemporaries of the latest samples are going to dispensed for an adequate arrangement denoting from the sample N. Each particle should be initialized respective to weight of $1/Ns$.
5. By using the transformation model of the casuistry system, particles will be calculable by k. The Ns prediction interval of the particles created victimization the state equation and therefore the propagation of every particle relative to the model or transition system.

$$S_{k+1} = f_k(S_k, \omega_k) = p(S_k|S_{k-1}) \tag{1}$$

6. The subsequent image will be reading.
7. By reading the subsequent image, calculate characteristics of each particle state in relevance model or biological process scheme.

$$S_{k+1} = f_k(S_k, \omega_k) = p(S_k|S_{k-1}) \tag{2}$$

8. For getting continuous samples, new sampling weights should be established.
9.

$$Q_k^i = Q_k^i \frac{p(y_k|x_k^i)p(x_k^i|x_{k-1}^i)}{q(x_k|x_{1\cdot k-1}, y_{1\cdot k})} \tag{3}$$

This will be done exploitation the likelihood measuring of every sample that depends on the appliance framework.
10. After reaching the probability of samples, establish the samples for wi, wherever looking on the new weights of the champion are generated for consecutive steps that are known as re sampling. Exploitation the sampler re-phase, the quantity of the high weight sample will increase, also because the low weight sample can decrease the probabilities. Exploitation normalized weights, that is,
11. Next, the expected particle values are approximate

$$Ep(x_k|y_{1\cdot k}) \cong \sum_k^i \bar{Q}_k^i x_k^i \tag{4}$$

4 Results and Discussion

MATLAB is employed to develop the algorithmic program, and therefore the use of the OTC BVS site is employed for the info. The OSU color and thermal info was used, that may be a third info. Six sequences of pictures enclosed during this info, every sequence has concerning 1500 frames (the precise range can vary from the sequence). The primary three sequences of an area with respective to pedestrians. The remainder of the sequences were interpreted from the non-identical position. To judge PF, the initial sequence is employed as colour (to convert to grayscale) and thermic.

The parameters of PF are listed below

Initial distribution/redistribution of particle weights—uniform distribution

range of particles, N two hundred and four hundred

distribution Threshold no/10

Particle form—parallelogram

Uncontrolled pursuit threshold: ten frames.

RGB and HSV performance for PF is depicted in Table 1. Here we tend to thought of 2 differing kinds of motion development to investigate the results of the illustration of the RGB and HSV entity on the particle filter and, additionally, there's a regular non-luminous condition because of the existence of the structure wall on one facet and therefore the open garden on the opposite region. This provides an awfully reliable example for testing detection performance (Fig. 2).

Table 1 Measurements of performance of particle filter [21]

Video name/Person ID	Initialization, orientation = pi			Particles	RGB or HSV	Controlled tracking length	(MSE, C-MSE)	Sampling redistributions
	(Start frame No.	(Width, height)/2	X and Y locations					
2 V/P1	1	2, 13.5	239, 116	400	RGB	356	110/8.37	166
					HSV	232	19/1.04	58
2 V/P1	1	3.5, 11	238, 120	400	RGB	122	19.21/1.36	147
					HSV	38	92.01/11.62	52
2 V/P1	1	2.5, 16.5	245, 114	400	RGB	122	19.21/1.36	179
					HSV	38	92.01/11.62	225

5 Conclusion

In fashionable digital progression and in an exceedingly video device, they need semi-conductor diode to recently discovered applications like management, advanced driver support, non-cooperative biometry, computer game, etc. data regarding the color which will be pictured within the bar graph. The feature vector of every particle is predicated on the distribution of pixels within the RGB and HSV color planes. RGB illustration in particle filtering has been discovered to behave higher than HSV representation. The superior performance of the RGB set up remains severally in keeping with the amount of particles utilized in the particle filter. The longer term work is to investigate multiple styles of motion methods in videos and additionally to get the foremost economical illustration of particles.

References

1. Van Beek, P.J.L., Murat Tekalp, A., Puri, A.: 2-D mesh geometry and motion compression for efficient object-based video representation. In: Proceedings of International Conference on Image Processing, vol. 3, pp. 440–443. IEEE (1997)
2. Haritaoglu, I., Harwood, D., Davis, L.S.: W sup 4: real-time surveillance of people and their activities. IEEE Trans. Pattern Anal. Mach. Intell. 22(8), 809–830 (2000)
3. Cai, Q., Aggarwal, J.K.: Tracking human motion in structured environments using a distributed-camera system. IEEE Trans. Pattern Anal. Mach. Intell. 21(11) 1241–1247 (1999)
4. Feng, P., et al.: A robust student's-t distribution PHD filter with OCSVM updating for multiple human tracking. In: 2015 23rd European Signal Processing Conference (EUSIPCO), pp. 2396–2400. IEEE (2015)
5. Bishop, G., Welch, G.: An introduction to the Kalman filter. Proc. SIGGRAPH Course 8, 27599–3175 (2001)
6. Altunbasak, Y., Murat Tekalp, A., Bozdagi, G.: Two-dimensional object-based coding using a content-based mesh and affine motion parameterization. Int. Conf. Image Process. Proc. 2, 394–397 (1995)
7. Badawy, W., Bayoumi, M.: A mesh based motion tracking architecture. In: IEEE International Symposium on Circuits and Systems (ISCAS 2001), vol. 4, pp. 262–265 (2001)

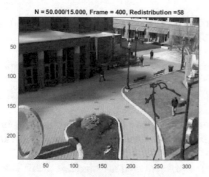

(a) X-Y Position:- 239,116 and Width and Height :- 2,13.5

(b) X-Y Position:- 238,120 and Width and Height :-3.5,11

(c) X-Y Position:- 245,114 and Width and Height :- 2.5,16.5

Fig. 2 Tracking results of particle filter for two subjects from two different videos with $N = 400$. Left: HSV; Right: RGB [21]

8. Jain, J., Jain, A.: Displacement measurement and its application in interframe image coding. IEEE Trans. Commun. **29**(12), 1799–1808 (1981)

9. http://www.ece.cmu.edu/~ee899/project/deepak_mid.html

10. Li, R., Zeng, B., Liou, M.L.: A new three-step search algorithm for block motion estimation. IEEE Trans. Circuits Syst. Video Technol. **4**(4), 438–442 (1994)

11. Po, L.-M., Ma, W.-C.: A novel four-step search algorithm for fast block motion estimation. IEEE Trans. Circuits Syst. Video Technol. **6**(3), 313–317 (1996)

12. Hsieh, H.-H., Lai, Y.-K.: 2A novel fast motion estimation algorithm using fixed subsampling pattern and multiple local winners search. In: IEEE International Symposium on Circuits and Systems, ISCAS 2001, vol. 2, pp. 241–244 (2001)

13. Srinivasan, R., Rao, K.: Predictive coding based on efficient motion estimation. IEEE Trans. Commun. **33**(8), 888–896 (1985)

14. Stauffer, C., Grimson, W.E.L.: Learning patterns of activity using real-time tracking. IEEE Trans. Pattern Anal. Mach. Intell. 747–757 (2000)

15. Arulampalam, S., Maskell, S., Gordon, N.J., Clapp, T.: A tutorial on particle filters for on-line non-linear/non-Gaussian bayesian tracking. IEEE Trans. Signal Process. **50**(2), 174–188 (2002)

16. Sugandi, B., Kim, H., Tan, J.K., Ishikawa, S.: Object tracking based on color information employing particle filter algorithm. AIP Conf. Proc. (Artif Life Robotics) **1159**(1), 39–42 (2009)

17. Zhang, X., Li, C., Hu, W., Tong, X., Maybank, S., Zhang, Y.: Human Pose estimation and tracking via parsing a tree structure based human model. IEEE Trans. Syst. Man Cybern. Syst. **44**(5), 580–592 (2014)

18. Feng, P., Wang, W., Naqvi, S.M., Dlay, S., Chambers, J.A.: Social force model aided robust particle PHD filter for multiple human tracking. In: IEEE International Conference on Acoustics, Speech and Signal Processing (ICASSP), Shanghai, pp. 4398–4402 (2016)

19. Yang, G.: WiLocus: CSI based human tracking system in indoor environment. In: Eighth International Conference on Measuring Technology and Mechatronics Automation (ICMTMA), Macau, pp. 915–918 (2016)

20. Jayaraman, S., Esakkirajan, S., Veerakumar, T.: Book on Digital-Image Processing, p. 747. Tata McGraw Hill Education (2009). ISBN-10:0070144796, ISBN-13: 978-0070144798

21. Mahajan, J.R.: Tracking Object/Human for Surveillance Using Computational Intelligence. Ph.D. thesis submitted, Department of Electronics and Telecommunication Engineering, Pacific University, Udaipur (2019)

22. Mahajan, J.R., Rawat, C.S.: Color based motion particle filter for object tracking. In: Processing of 2nd International Conference on Electronics, Communication and Aerospace Technology (ICECA 2018), at RVS technical campus, Coimbatore, Tamil Nadu, IEEE conference record # 42487. IEEE Xplore (2018). ISBN: 978-1-5386-0965-1

A Report on Behavior-Based Implicit Continuous Biometric Authentication for Smart Phone

Vinod P. R[1](✉), Anitha A.[1], and Usha Gopalakrishnan[2]

[1] Department of Computer Science and Engineering, Noorul Islam Centre for Higher Education, Kumaracoil, Tamilnadu, India
vinod_p_r@yahoo.co.uk, anidathi@yahoo.co.in
[2] Department of Computer Science and Engineering,
Musaliar College Of Engineering And Technology, Pathanamthitta, Kerala, India
writeto_usha@yahoo.com

Abstract. Continuous authentication mechanism is gaining popularity in the recent days as the conventional mechanism of authentication is difficult for the user to memorize, such as the passcode and pattern, and for the device and data, it is vulnerable to different types of attacks. The continuous authentication mechanism has the advantage that, at any point of time, without the user's permission the device can check the genuineness of the user credentials, thereby ensuring that the registered owner is using the device and the stored data. This technique uses the behavioral biometric data which is difficult for an impersonator to mimic and ensure the security of the data as well as to the device. This paper proposes a method based on continuous authentication by exploring the possibilities of the signals generated from micro-electro-mechanical systems (MEMS) that are inbuilt in the smartphone. Signals collected which are generated by the movement of the hand, orientation of the device, and the holding behavior of the user are analyzed to get a unique parameter, which will be used for the authentication purposes.

Keywords: Continuous authentication · HMOG · Biometric signals · Gait characteristics · F1 score

1 Introduction

Recently, smartphones have been playing a ubiquitous role in the life of human beings as they have become an inevitable part of life. The last decade has seen the rapid development in the field of smartphone technology and other handheld devices which makes humans to heavily depend on these devices. These devices can store huge amount of person-centric data, especially bank details, family photos, and other personal information; users other than the actual owner must

B. Iyer et al. (eds.), *Applied Computer Vision and Image Processing*,
Advances in Intelligent Systems and Computing 1155,
https://doi.org/10.1007/978-981-15-4029-5_17

be blocked from using the phone which makes authentication a necessary and inevitable prerequisite [1]. The simple and popular user authentication is using a password, PIN, or pattern, which demands the user to memorize these information and reproduce at the time of login. But such authentication mechanism has got some drawbacks: (1) the authentication data can be lost or forgotten, (2) it can be illegally shared or exposed to the attacker and can be used for unauthorized access, and (3) these types of mechanisms are applied only once, generally in the beginning of the activity, once passed the legal owner as well as the intruder can continue the use of smartphone without any further authentication check. Repeated entry of the authentication key during an activity session is annoying to the user and may distract the ongoing activity . In practical environments, the user who has initially logged into the system may not be the same person who is currently working on the smartphone [2]. Other authentication mechanisms are PIN and pattern input [3]. PIN and patterns are very small and can be leaked or guessed very easily and is more prone to attacks, which further attribute to the risk of information disclosure. This also suffers the drawback of one-time authentication as in the case of a password. Thus to authenticate a genuine smartphone user, some continuous authentication mechanisms have to be evolved to ensure security of data. Previous research in this direction [4,5] claimed 70% of users are not interested in protecting their smartphones using either PIN or passcodes.

For authentication purpose in smartphone, watchword is the most frequently used technique. Shoulder surfing, profile spoofing, wiretapping, dictionary attacks, etc., are some of the vulnerabilities of this technique.The system can be made more secure by using long and fanciful passwords, however, these long passwords will eat up the precious resource in the handheld device. If the user selects short or easy to recollect passwords, that will further enhance the system vulnerability [6,7]. These passwords can be easily guessed or applied to brute-force technique to break. The choice of graphical patterns have their own disadvantage of forgetting the pattern. Yet another authentication mechanism is using biometrics such as fingerprints or iris scan [8,9]. Here, the Identification process takes much time and also the cost is high compared to other approaches. Eventhough several graphical positive identification schemes are used as substitutes, nowadays, most of them have many disadvantages [10].The graphical positive identification schemes done against shoulder surfing have major disadvantages like usability problems or take much longer time for users to login [11,12].

User behavior-dependent authentication mechanism is an old technique evolved in 1990s and is a behavior-based biometric authentication used in personal computing scenarios and was derived from user behavior such as mouse interaction and keyboard handling and is difficult to be mimicked [11]. This authentication mechanism derives the general profile of an authorized user from the usage statistics. If the authentication mechanism detects a wide variation in the profile that was created earlier and the current user activities, then interprets it as an unauthorized access [13]. The advantage of behavior-based authentication is that, it can authorize a user not only at the time of login but also during the entire system usage. Another advantage of such a system is the absence of unnecessary dedicated electronic devices. Research in the direction of behavior-based authentication mechanism for handheld devices have gained momentum due to the advancements in MEMS technology. Eventhough a new field of research, the authentication based on behavior-based mechanism have progressed well [14,15] as in keystroke-based, accelerometer and other sensor-based and touch screen-based biometrics. The gait characteristics of a human can be collected using high-accuracy sensors present in the smartphone. Since login process may not differ from one user to another user, most of the research process is bounded to user's first login process. The activities performed after login process may be different for every user, making it harder to identify them. Hence the area of behavior-based continuous user authentication after login is a tricky situation, which is the motivation behind this work [16].

Modern smartphones are equipped with advanced MEMS, which can be used for biometric authentication and in a continuous manner which are referred to as sensor-based authentication [17]. The different gait characteristics of a human can be collected and analyzed for authentication purpose in this type of mechanisms using the sensors in the smartphone [18–22]. Eventhough a lot of mechanisms have been proposed and are in use, they all have their own drawbacks and limitations. Most of them need frequent user interaction, which annoys or disturbs the user. Above all, the computation complexity and energy management needs to be improved a lot. Thus this paper suggests an automatic way of authenticating the user in a continuous mode after login without frequent user interaction. The paper is organized as follows. Section 2 reviews the major works that are carried out and Sect. 3 details the data collection and preprocessing techniques, Sect. 4 presents the different feature extraction methods, Sect. 5 explains the various classifiers used, Sect. 6 is the comparative study of the recent works, Sect. 7 details the proposed work, and Sect. 8 is the conclusion followed by references.

2   Related Works

The continuous authentication and monitoring of smartphone users are done by means of utilizing the motion sensors such as accelerometer, gyroscope, and other sensors present in the smartphone. Data features like time domain, frequency domain, and wavelet domain were extracted from the signals generated from the motion sensors, and the characterization of the fine grains of user movements were accomplished using empirical feature analysis [3].

The behavioral features like movement of the hand, orientation and grasp (HMOG) derived from the sensor signals generated from the smartphone users can also be used as an authentication mechanism in an unceasing manner [23]. This approach was efficient in capturing the subtle micro-movement as well as the resulting orientation dynamics on the basis of the grasping behaviors, holding and tapping method of a smartphone. The data for authentication of the smartphones were gathered under two main conditions: (1) sitting and (2) walking.

The extracted features were evaluated on the basis of three different features like HMOG, keystroke, and tap. Further, the evaluation of the features were accomplished by authentication, biometric key generation (BKG), and energy consumption on smartphones.

AnswerAuth is another authentication mechanism for smartphone users on the basis of the behavioral biometric of the user in sliding and lifting movements of the device [24]. This approach was based on the frequent behavioral features that were utilized by the authorized user while sliding the screen lock button in order to unlock the phone and the way the user brings the phone toward the ear. These extracted features from the biometric behavior were derived from the records by utilizing the built-in smartphone sensors like accelerometer, gyrometer, gravity, magnetometer, and touch screen sensors. Further, the user-friendliness of the proposed scheme was verified by usability study based on Software Usability Scale (SUS).

Finger movements and pressure exerted on the screen can also be used as authentication parameters. Safeguard is a re-authentication system for accurate verification of smartphone users on the basis of on-screen finger movements and pressure parameters [25]. There is a transparency for the users about the back-end computation and processing on the key features. The unique features from each user were extracted from angle-based metrics and pressure-based metrics which will uniquely identify individual users.

Secure shaking of the smartphone in a controlled way can be used for secure access of the device. ShakeIn is such a mechanism with the aspire of unlocking the smartphone in a secured manner by means of shaking the phone [26]. The effective motion sensors embedded in the smartphone capture the distinct and dependable biometrical features of users' shaking behavior of the phone. Further, to enhance the authentication of the smartphones, ShakeIn had endowed the users with an utmost flexibility in operation, thereby permitting the users to customize the way of shaking the phone. Cao and Chang [27] formulated a novel new framework for obtrusive and continuous mobile user verification with the intention of diminishing the required frequency that it utilized by the user in order to feed the security token. Further, the customized Hidden Markov Models as well as sequential likelihood ratio test were employed to construct a cheap, readily accessible, anonymized, and multimodal smartphone data.

Energy efficiency needs to be considered while developing any authentication mechanism. An energy-efficient implicit authentication (IA) system on the basis of adaptive sampling in order to choose the dynamic sets of activities for user behavior in an automatic manner is proposed in [28]. The authentication features were obtained from various activities like the location of the user, usage of application, and motion of the user. The partially labeled Dirichlet allocation (PLDA) was employed for more accurate extraction of the features. For battery-powered mobile devices, the soft biometric-based authentication model was superior to the hard biometric-based authentication as well as password-based authentication in terms of energy efficiency and lack of explicit user action. Gasti et al. [1] have proposed an energy-preserving outsourced technique for continuous authentication by offloading the calculation overhead on an untrusted cloud. This type of continuous authentication is desirous in case of low latency situations.

The existing password/pattern authentication method can be enhanced by formulating an implicit authentication approach by employing an additional security layer [29]. In this approach, three times the security checks were made and that too in two steps: in the first step, the matching of the mobile angle that the user holds the mobile was carried out and in the second step, the time taken to draw the pattern as well as pattern check was performed. The result of the work shows 95% as maximum accuracy and 60% as the lower range. Privacy preserving protocol is an alternate way for the secure authentication of the smartphone [30]. The resources available in handheld devices are limited. Hence the verifiers namely, scaled Euclidean and scaled Manhattan suffer the drawback of limited memory in smartphones.

Touch dynamics is another area that can be exploited for user authentication in smartphones. The dynamic features of the touch dynamics are extracted for active authentication by applying different classification techniques achieving EER between 1.72 and 9.01% [2]. The work observed the increase in authentication accuracy with the increase in touch operation length. Yang et al. [9] protect the smartphone with the users' four different touch-based behavioral biometrics and the behavioral model was trained using the one-class Support Vector Machine and isolation Forest (iForest) algorithm.

Camera is another mechanism that can be exploited for authentication in smartphones. An innovative user authentication by distinctively identifying the camera with its high-frequency components of photo response and nonuniformity of the optical sensor from the captured images is presented in [16]. It is a server-based verification mechanism using adaptive random projections and an innovative fuzzy extractor using polar codes. The probability of two cameras having the same photo response nonuniformity in their imaging sensor is negligible.

Authentication based on the bodily activity pattern of individual user can be performed on signals from accelerometer, gyroscope, and magnetometer sensors located in the smartphone and SVM, decision tree, and KNN are the machine learning classifiers that are used for the individual recognition and authentication [13]. Alghamdi et al. [7] conducted an experiment to collect data for the gestures like tapping, scrolling, dragging, and zooming and then machine learning classifiers were applied to achieve authentication and reached a good equal error rate.

Biometric authentication in smartphones can be carried out by putting the signature in the air by carrying the phone in the hand [8]. The signals generated by the accelerometer are recorded and sent to the server for authentication. The server cross-correlates the received data with the templates that are generated in the learning phase and a satisfying threshold is used to authenticate the user (Table 1).

Table 1. Device, sensors, action context, and number of participants

Reference #	Participants	Action context	Sensors	Device
[18]	51	Gait	Accelerometer	Google G1
[53]	36	Walking, jogging, ascending, and descending stairs	Accelerometer	Android
[54]	10	Answering or placing a phone call	Accelerometer and gyroscope	Android
[19]	36	Gait	Accelerometer	Motorola
[55]	30	Gait	Accelerometer	Google Nexus
[56]	8	Gait	Accelerometer, gyroscope, magnetometer	Google Nexus
[57]	35	Gait	Accelerometer	Sony Xperia
[58]	100	Arm gesture	Accelerometer and gyroscope	Samsung S4

3   Data Collection and Preprocessing

Shen et al. [31] used 10 subjects with five different actions in five different positions and four different contexts to evaluate the influence of phone placement in human activity. They used Savitzky–Golay filter for accelerometer data filtering and Kalman filter for filtering gyroscope data. The filtered data is then segmented using windowing with a window segment size of one second and with an overlapping level of 50%.

Muazz et al. [32] use accelerometer to capture the data using 35 participants keeping the phone in the trouser's front pocket and walking for 2–3 min, and then it is mean normalized. The constant sampling rate is obtained by linear interpolation and to remove the random noise, Savitzky–Golay smoothing filter is used.

Lee et al. [33] collected data from wearable and handheld devices and the time and frequency domain features are extracted. The feature vector used for authentication are the context feature vector and authentication feature vector. Kernel ridge regression algorithm has been used for classification process.

Lee et al. [34] uses a multi-sensor authentication procedure and the data are captured from magnetometer, orientation sensor, and accelerometer and the dimensionality of the data set is reduced by averaging with a suitable window size (Table 2).

Table 2. Device Positioning and Preprocessing Methods

Reference #	Device position	Preprocessing
[35]	Right hand side of the hip	Interpolation, zero normalization, piecewise linear approximation
[36]	In a small purse fastened on the right side of the hip	Noise removal, interpolation, normalization
[37]	Right back pocket	Signal vector magnitude calculation, linear time interpolation, zero normalization
[38]	Smartwatch	Time interpolation, low-pass filtering
[39]	Phone is facing outwards and oriented vertically and placed in the right pocket	Piecewise cubic spline interpolation, resampling and normalization
[40]	Right front pocket	Cubic spline interpolation, low-pass FIR
[41]	Front left pocket	Principal eigenvector calculation

4   Feature Extraction Methods

The preprocessed data has to be exploited to derive the features that are significant for the system performance and is the basis of feature extraction. Many categories of feature extraction techniques exists in the literature like time domain, frequency domain, wavelet domain, heuristic search, etc. [25]

Time domain technique is a simple statistical metric to collect information from raw sensor data. The different measures in this category are mean, median, variance, standard deviation, RMS, correlation, and cross-correlation. The other functions in this category are signal vector magnitude and zero crossings. These measures are used as input to a classifier algorithm or to some thresholding algorithm.

Frequency domain techniques are used to find the repetition that correlates the periodic nature of an activity. Fourier transform of a time-based signal gives the dc component and dominant frequency component present in it. Another recent approach is converting the sensor signals to discrete form by converting into a window of approximate size, calculating the average and mapping to a symbol. Then this coded form is analyzed to find the known pattern or to classify the user activity (Table 3).

Table 3. Feature extraction methods

Reference #	Time domain	Frequency domain	Code based
[3]	✓	✓	✓
[24]	✓		
[25]	✓		
[26]	✓	✓	
[2]	✓		
[15]	✓		

5   Classifiers

Classification is the final stage in the gait recognition system. A lot of classifiers are used in the literature. A few of them are Decision Tree (DT), Naive Bayesian (NB), k-Nearest Neighbor (KNN), Logistic Regression (LR), Support Vector Machines (SVM), Neural Networks, and Random Forest.

Decision tree is a simple and powerful tool that uses a tree-like structure to model decisions and its consequences. Each node in the tree contains nodes that are originated from a root that contains the name of the field which is also called object of analysis. The input field values are used to calculate the likely value of the expected result. Neural network is a parallel computing system with large number of nano-processors with a huge number of interconnections between them. As in the case of real life, each interconnection has different weights attached to it which shows the importance of the information processed by the previous neuron. The product of input and the weight is fed to a mathematical function which determines the neuron activation.

k-Nearest Neighbor (KNN) algorithm is based on a similarity measure by storing all available data and calculating the position of the new one based on this similarity test, i.e., the classification of new instance is done by assigning a label which is closest to the k-training samples. The value of k can be chosen randomly and the best choice of it depends on the data and the best value of k can be achieved by a heuristic technique called cross-validation.

Support Vector Machine (SVM) derives a pattern or model from the input data and fits it to a hyperplane to differentiate classes. A radial-based function is used for training. The model will be validated using a fivefold cross-validation in each training iteration. Bayesian network is used to show how long a window stays opened without changing the status to be closed. Bayes classifier takes all properties of an object to come to the conclusion in an independent way. Thus it considered to be easy and fast and it uses the principle of maximum likelihood method in applications.

In Random Forest classifier, the strong classifier is derived from weak classifiers. A random subset of features was selected and searches for the best feature before splitting a node rather than searching for the most important feature. This leads to a large diversity that typically results in a more robust model (Table 4).

Table 4. Classifiers used for feature recognition

Reference #	Decision tree	Bayesian	SVM	Random Forest	KNN	HMM	KRR
[42]		✓					
[31]			✓		✓		
[43]			✓			✓	
[44]						✓	
[33]							✓
[45]					✓		
[46]			✓				
[34]			✓	✓	✓		
[47]			✓				
[48]							✓
[49]	✓		✓		✓		
[50]	✓						

6 Performance Measures

The effectiveness of a biometric authentication can be evaluated using a lot of metrics available in information security. False Acceptance Rate (FAR) is the probability that the system erroneously identifies an attacker as an authorized user. False Rejection Rate (FRR) indicates the probability of the system to reject a genuine user from accessing the resources. Equal Error Rate (EER) indicates the rate at which FRR and FAR are same. The accuracy of the system is higher for lower value of ERR. True Positive Rate (TPR) is a proportion of all identification attempts for the correct identification. False Positive Rate (FPR) is the probability of incorrectly rejecting the invalid hypothesis for a particular check. True Negative Rate (TNR) is the ratio of all recognition attempts to the subjects identified incorrectly. Failure to Capture Rate (FCR) is the likelihood that the system fails to sight a biometric input once given properly. False Match Rate (FMR) is the probability of the system to falsely match input to a dissimilar value. False non-match Rate (FNMR) is the probability of a failure to verify any similarity between given value and required template.

7 Problem Definition

The literature has come out with several techniques for behavior-based authentication of smartphones as per Table 5. However, they require more improvements

Table 5. Performance evaluation parameters

Reference #	Measures used
[2]	FAR, FRR
[3]	EER, FAR, FRR
[6]	Accuracy, precision, recall, F-measure, error rate
[7]	FAR, FRR, EER
[23]	EER
[24]	TAR
[25]	FAR, FRR
[28]	Accuracy, precision
[37]	ERR
[42]	TPR, FPR, F1 score, accuracy

because of the lack of several features in behavioral authentication. In Support Vector Machine (SVM) and k-Nearest Neighbors (K-NN), the accuracy in identification is high and it utilizes the one-class classifier as the indicator to differentiate the authorized and the unauthorized users, hence the discrimination accuracy is high. Apart from this, it suffers the drawbacks like high EER and it is infeasible to action-aware context and the placement-aware placement process that takes place in authentication. Further, with HMOG [23], EER is low while walking as well as sitting and in addition to that the sampling rate is also low. Apart from this, it is not suitable under certain constraints like a) walking or jogging at unusual speeds; (b) using the smartphone in different climatic environments; and (c) applications which do not need any typing input. Then, AnswerAuth [24] has the advantages like high TAR, and it is robust to the possible mimicry attacks. Yet, it suffers the disadvantages like lack of consideration on seamless detection of the users' current activity and the training of AnswerAuth is much complex. Then, SVM [25] has the advantages like low FAR and low FRR. Further, the accuracy of authentication is much lower and this identification process consumes more time. In ShakeIn [26], the average error rate is low and the proposed model can work under different modes of transport. In contrast to these advantages, it suffers the drawbacks like high FPE and has no consideration on the physiological and behavioral characteristics. Then, in customized Hidden Markov Models (HMM) and sequential likelihood ratio test (SLR) [27], the detection rate is high and there is a high trade-off between the security and usability. But, this method has low effectiveness and efficiency and it too consumes high time in identification and makes use of much battery resources. Moreover, in PLDA [28], there is more accuracy and precision in feature extraction and light weight authentication. This method is unable to provide a solution to the deviation problem of user behavior. Then, in Markov-based decision procedure, the accuracy in identification as well as authentication is high and the identification process consumes much time. Thus, the necessity of designing an optimal authentication model becomes essential (Table 6).

8 Methodology

In the current scenario, the commonly available smartphone authentication mechanisms like fingerprint images, PINs, and graphical passcodes are limited in terms of security. In case of the fingerprint, iris-based biometric scans, etc., there is high level of probability in spoofing. Apart from this, the fundamental limitation behind the other authentication technique like PINs, passwords, and pattern draws are that they are susceptible to be guessed and the other channel attacks like the smudge, reflection, and video capture attacks have high probability of occurrence. This contributes them inefficient, when the smartphone's access is

Table 6. Features and challenges of behavior-based smartphone authentication

Reference #	Adopted methodology	Features	Challenges
[3]	KNN and SVM	High authentication and re-authentication accuracy. High discrimination accuracy	High equal error rates (EER) Feasible in placement-aware placement and the action-aware context
[23]	HMOG	Low EER while walking and sitting. Low sensor sampling rate	Not applicable under stringent constraints. Lack of cross-device interoperability.
[24]	AnswerAuth	High acceptance rate (tar). Highly robust against the possible mimicry attacks.	Not considered speed and seamless detection of the users current activity. The training of AnswerAuth is difficult.
[25]	SVM	Low false rejection rate (FRR) and low false acceptance rate (FAR).	Low accuracy Tedious process
[26]	ShakeIn	Low average equal error rate. Shows high reliability and works well under various transportation modes	Under shoulder-surfing attacks, possesses high false positive errors (FPE). No consideration on both physiological and behavioral characteristics
[27]	HMM and SLR	Detecting illegitimate users rate is high. High trade-off between usability and security	Low effectiveness and efficiency. High time consumption and high battery-power usage
[28]	PLDA	More accurate and precious feature extraction, high compatibility	Behavior deviation is left unsolved Time consumption is high
[29]	Markov-based decision procedure	High accuracy Low False positives (FP) and False negatives (FN) rate	More expensive Slow identification process

gained by an attacker after login. Active biometric authentication or continuous biometric authentication addresses these challenges by constantly and indistinguishably authenticating the user via signals generated through behavioral biometric, such as voice, phone location, touch screen communications, hand movements, and gait. So, to override the challenges faced by the traditional smartphone authentication techniques, the proposed smartphone authentication model is based on two main phases, viz., (i) Feature Extraction and (ii) Classification. The initial stage behind this research is feature collection and they are gathered from the Hand Orientation, Grasp, and Movement (HMOG) for continuous monitoring of the authenticated smartphone users. The main contributors of HMOG is the gyroscope, magnetometer, and accelerometer readings with the intention of unobtrusively capturing fine-drawn micro-movements of hand and orientation patterns produced when a user taps on the screen. The proposed feature extraction model uses the HMOG features, where the main intention will be on converting those HMOG features to another domain and the transformed optimal HMOG features are extracted, which need to be classified. The classification of the selected features is done by employing a suitable machine learning tool. Hence, the classified output shall recognize whether the respective user is authorized or not.

9   Conclusion

The proposed behavior-based smartphone authentication model has to be carried out in Android smartphones and the experimented outcome will be investigated. The performance of the proposed model will be analyzed by determining Type I and Type II measures. Here, Type I measures are positive measures like Accuracy, Sensitivity, Specificity, Precision, Negative Predictive Value (NPV), F1-score, and Mathews Correlation Coefficient (MCC), and Type II measures are negative measures like False Negative Rate (FNR), False Discovery Rate (FDR), and False Positive Rate (FPR).

References

1. Gasti, P., Edenka, J., Yang, Q., Zhou, G., Balagani, K.S.: Secure, fast, and energy-efficient outsourced authentication for smartphones. IEEE Trans. Inf. Forensics Secur. **11**(11), 2556–2571 (2016)
2. Shen, C., Zhang, Y., Guan, X., Maxion, R.A.: Performance analysis of touch-interaction behavior for active smartphone authentication. IEEE Trans. Inf. Forensics Secur. **11**(3), 498–513 (2016)
3. Shen, C., Chen, Y., Guan, X.: Performance evaluation of implicit smartphones authentication via sensor-behavior analysis. Inf. Sci. **430–431**, 538–553 (2018)
4. Jakobsson, M., Shi, E., Golle, P., Chow, R.: Implicit authentication for mobile devices. In: Proceedings of the 4th USENIX conference on Hot topics in security (2009)
5. Survey on password protection in mobiles: http://nakedsecurity.sophos.com/2011/08/09/ downloaded on January 2019

6. Ehatisham-ul-Haq, M., Azam, M.A., Naeem, U., Amin, Y., Loo, J.: Continuous authentication of smartphone users based on activity pattern recognition using passive mobile sensing. J. Netw. Comput. Appl. **109**, 24–35 (2018)
7. Alghamdi, S.J., Elrefaei, L.A.: Dynamic authentication of smartphone users based on touchscreen gestures. Arab. J. Sci. Eng. **43**(1), 789–810 (2018)
8. Laghari, A., Waheed-ur-Rehman, Memon, Z.A.: Biometric Authentication Technique using Smartphone Sensor, pp. 381–384. Applied Sciences and Technology (IBCAST), Islamabad (2016)
9. Nyang, D., Mohaisen, A., Kang, J.: Keylogging-resistant visual authentication protocols. IEEE Trans. Mob. Comput. **13**(11), 2566–2579 (2014)
10. Schaffer, K.B.: Expanding continuous authentication with mobile devices. Computer **48**(11), 92–95 (2015)
11. Lin, Y., et al.: SPATE: small-group PKI-less authenticated trust establishment. IEEE Trans. Mob. Comput. **9**(12), 1666–1681 (2010)
12. Alzubaidi, A., Kalita, J.: Authentication of smartphone users using behavioral biometrics. IEEE Commun. Surv. Tutor. **18**(3), 1998–2026 (2016)
13. Martinez-Diaz, M., Fierrez, J., Galbally, J.: Graphical password-based user authentication with free-form doodles. IEEE Trans. Hum. Mach. Syst. **46**(4), 607–614 (2016)
14. Galdi, C., Nappi, M., Dugelay, J., Yu, Y.: Exploring new authentication protocols for sensitive data protection on smartphones. IEEE Commun. Mag. **56**(1), 136–142 (2018)
15. Thavalengal, S., Corcoran, P.: User authentication on smartphones: focusing on IRIS biometrics. IEEE Consum. Electron. Mag. **5**(2), 87–93 (2016)
16. Valsesia, D., Coluccia, G., Bianchi, T., Magli, E.: User authentication via PRNU-based physical unclonable functions. IEEE Trans. Inf. Forensics Secur. **12**(8), 1941–1956 (2017)
17. Muaaz, M., Mayrhofer, R.: Smartphone-based gait recognition: from authentication to imitation. IEEE Trans. Mob. Comput. **16**(11), 3209–3221 (2017). https://doi.org/10.1109/TMC.2017.2686855
18. Derawi, M.O., Nickel, C., Bours, P., Busch, C.: Unobtrusive user-authentication on mobile phones using biometric gait recognition. In: Sixth International Conference on Intelligent Information Hiding and Multimedia Signal Processing, pp. 306–311 (2010). https://doi.org/10.1109/IIHMSP.2010.83
19. Nickel, C., Wirtl, T., Busch, C.: Authentication of smartphone users based on the way they walk using KNN algorithm. In: Eighth International Conference on Intelligent Information Hiding and Multimedia Signal Processing, pp. 16–20 (2012). https://doi.org/10.1109/IIH-MSP.2012.11
20. Patel, S.N., Pierce, J.S., Abowd, G.D.: A gesture-based authentication scheme for untrusted public terminals. In: Proceedings of the 17th Annual ACM Symposium on User Interface Software and Technology, UIST 04, pp. 157–160. ACM, New York, NY, USA (2004). https://doi.org/10.1145/1029632.1029658
21. Sun, B., Wang, Y., Banda, J.: Gait characteristic analysis and identification based on the iphones accelerometer and gyrometer. Sensors **14**(9), 17037–17054 (2014). https://doi.org/10.3390/s140917037
22. Tamviruzzaman, M., Ahamed, S.I., Hasan, C.S., Obrien, C.: EPET: when cellular phone learns to recognize its owner. In: Proceedings of the 2Nd ACM Workshop on Assurable and Usable Security Configuration, SafeConfig 09, pp. 13–18. ACM, New York, NY, USA (2009). https://doi.org/10.1145/1655062.1655066

23. Sedenka, J., Yang, Q., Peng, G., Zhou, G., Gasti, P., Balagan, K.S.: HMOG: new behavioral biometric features for continuous authentication of smartphone users. IEEE Trans. Inf. Forensics Secur. **11**(5), 877–892 (2016). May
24. Buriro, A., Crispo, B., Conti, M.: AnswerAuth: a bimodal behavioral biometric-based user authentication scheme for smartphones. J. Inf. Secur. Appl. **44**, 89–103 (2019)
25. Lu, L., Liu, Y.: Safeguard: user reauthentication on smartphones via behavioral biometrics. IEEE Trans. Comput. Soc. Syst. **2**(3), 53–64 (2015)
26. Zhu, H., Hu, J., Chang, S., Lu, L.: ShakeIn: secure user authentication of smartphones with single-handed shakes. IEEE Trans. Mob. Comput. **16**(10), 2901–2912 (2017)
27. Cao, H., Chang, K.: Nonintrusive smartphone user verification using anonymized multimodal data. IEEE Trans. Knowl. Data Eng.
28. Yang, Y., Sun, J., Guo, L.: PersonaIA: a lightweight implicit authentication system based on customized user behavior selection. IEEE Trans. Dependable Secure Comput. **16**(1), 113-126 (2019). https://doi.org/10.1109/TDSC.2016.2645208
29. Agrawal, A., Patidar, A.: Smart authentication for smart phones. Int. J. Comput. Sci. Inf. Technol. **5**(4), 4839–4843 (2014)
30. Ednka, J., Govindarajan, S., Gasti, P., Balagani, K.S.: Secure outsourced biometric authentication with performance evaluation on smartphones. IEEE Trans. Inf. Forensics Secur. **10**(2), 384–396 (2015). https://doi.org/10.1109/TIFS.2014.2375571
31. Chen, Y., Shen, C.: Performance analysis of smartphone-sensor behavior for human activity recognition. IEEE Access p. 1. https://doi.org/10.1109/ACCESS.2017.2676168
32. Muaaz, M., Mayrhofer, R.: Smartphone-based gait recognition: from authentication to imitation. IEEE Trans. Mob. Comput. **16**(11), 3209–3221 (2017). https://doi.org/10.1109/TMC.2017.2686855
33. Lee, W., Lee, R.B.: Sensor-based implicit authentication of smartphone users. In: 2017 47th Annual IEEE/IFIP International Conference on Dependable Systems and Networks (DSN), Denver, CO, pp. 309–320 (2017). https://doi.org/10.1109/DSN.2017.21
34. Chen, Y., Shen, C.: Performance Analysis of Smartphone-Sensor Behavior for Human Activity Recognition, 3536(c). https://doi.org/10.1109/ACCESS.2017.2676168
35. Muaaz, M., Mayrhofer, R.: An analysis of different approaches to gait recognition using cell phone based accelerometers. In: Proceedings of International Conference on Advances in Mobile Computing and Multimedia (MoMM '13), p. 293. ACM, New York, NY, USA (2013). https://doi.org/10.1145/2536853.2536895
36. Wasnik, P., Schafer, K., Ramachandra, R., Busch, C., Raja, K.: Fusing biometric scores using subjective logic for gait recognition on smartphone. In: International Conference of the Biometrics Special Interest Group (BIOSIG), pp. 1–5 . Darmstadt (2017). https://doi.org/10.23919/BIOSIG.2017.8053508
37. Ferrero, R., Gandino, F., Montrucchio, B., Rebaudengo, M., Velasco, A., Benkhelifa, I.: On gait recognition with smartphone accelerometer. In: 2015 4th Mediterranean Conference on Embedded Computing (MECO), Budva, pp. 368–373 (2015). https://doi.org/10.1109/MECO.2015.7181946
38. Al-Naffakh, N., Clarke, N., Haskell-Dowland, P., Li, F.: A Comprehensive Evaluation of Feature Selection for Gait Recognition Using Smartwatches. Int. J. Inf. Secur. Res. (IJISR). **6** (2017). https://doi.org/10.20533/ijisr.2042.4639.2016.0080

39. Juefei-Xu, F., Bhagavatula, C., Jaech, A., Prasad, U., Savvides, M.: Gait-ID on the move: pace independent human identification using cell phone accelerometer dynamics. In: 2012 IEEE Fifth International Conference on Biometrics: Theory, Applications and Systems (BTAS), Arlington, VA, pp. 8–15 (2012). https://doi.org/10.1109/BTAS.2012.6374552

40. Gadaleta, M., Rossi, M.: INet: Smartphone-based gait recognition with convolutional neural networks. Comput. Vision Pattern Recogn. Mach. Learn. https://doi.org/10.1016/j.patcog.2017.09.005

41. Crouse, M.B., Chen, K., Kung, H.T.: Gait recognition using encodings with flexible similarity measures. Int. J. Comput. Sci. Inf. Technol. **5**(4), 4839–4843 (2014)

42. Albayram, Y., Khan, M.M.H.: Evaluating smartphone-based dynamic security questions for fallback authentication: a field study. Hum. Cent. Comput. Inf. Sci. **6**, 16 (2016). https://doi.org/10.1186/s13673-016-0072-3

43. Shen, C., Li, Y., Chen, Y., Guan, X., Maxion, R.A.: Performance analysis of multi-motion sensor behavior for active smartphone authentication. IEEE Trans. Inf. Forensics Secur. **13**(1), 48–62 (2018). https://doi.org/10.1109/TIFS.2017.2737969

44. Gjoreski, H., Lustrek, M., Gams, M.: Accelerometer placement for posture recognition and fall detection. In: 2011 Seventh International Conference on Intelligent Environments, Nottingham, pp. 47–54 (2011). https://doi.org/10.1109/IE.2011.11

45. Brezmes, T., Gorricho, J., Cotrina, J.: Activity recognition from accelerometer data on a mobile phone. In: Proceedings of the International Working Conference Artificial Neural Networks (IWANN), pp. 796–799, Salamanca, Spain (2009)

46. Sun, L., Zhang, D., Li, B., Guo, B., Li, S.: Activity recognition on an accelerometer embedded mobile phone with varying positions and orientations. In: Proceedings of the 7th International Conference on Ubiquitous Intelligence and Computing (UIC 2010), pp. 548–562, Xian, China (2010)

47. Lee, W., Lee, R.B.: Multi-sensor authentication to improve smartphone security. In: 2015 International Conference on Information Systems Security and Privacy (ICISSP), Angers, pp. 1–11 (2015)

48. Wei-Han, L., Ruby, L.: Implicit smartphone user authentication with sensors and contextual machine learning, pp. 297–308 (2017). https://doi.org/10.1109/DSN.2017.24

49. Raziff, A., Rafiez, A., Sulaiman, M., Perumal, T.: Gait identification using smartphone handheld placement with linear interpolation factor, single magnitude and one-vs-one classifier mapping. Int. J. Intell. Eng. Syst. **10**, 70–80. https://doi.org/10.22266/ijies2017.0831.08

50. Derlatka, M., Ihnatouski, M.: Decision tree approach to rules extraction for human gait analysis. Int. Conf. Artif. Intell. Soft. Comput. 597–604 (2010)

51. Yang, Y., Guo, B., Wang, Z., Li, M.: BehaveSense: continuous authentication for security-sensitive mobile apps using behavioral biometrics. Ad Hoc Netw. **84**, 9–18 (2019)

52. Mirjalili, S., Lewis, A.: The whale optimization algorithm. Adv. Eng. Softw. **95**, 51–67 (2016)

53. Kwapisz, J.R., Weiss, G.M., Moore, S.A.: Cell phone-based biometric identification. In: 2010 Fourth IEEE International Conference on Biometrics: Theory, Applications and Systems (BTAS), pp. 1–7 (2010). https://doi.org/10.1109/BTAS.2010.5634532

54. Conti, M., Zachia-Zlatea, I., Crispo, B.: Mind how you answer me!: transparently authenticating the user of a smartphone when answering or placing a call. In: Proceedings of the 6th ACM Symposium on Information, Computer and Communications Security, in: ASIACCS 11, pp. 249–259. ACM, New York, NY, USA (2011). https://doi.org/10.1145/1966913.1966945

55. Primo, A., Phoha, V.V., Kumar, R., Serwadda, A.: Context-aware active authentication using smartphone accelerometer measurements. In: 2014 IEEE Conference on Computer Vision and Pattern Recognition Workshops, Columbus, OH, pp. 98–105 (2014). https://doi.org/10.1109/CVPRW.2014.20

56. Lee, W.H., Lee, R.B.: Multi-sensor authentication to improve smartphone security. In: International Conference on Information Systems Security and Privacy (ICISSP), pp. 1–11 (2015)

57. Muaaz, M., Mayrhofer, R.: Smartphone-based gait recognition: from authentication to imitation. IEEE Trans. Mob. Comput. **16**(11), 3209–3221 (2017). https://doi.org/10.1109/TMC.2017.2686855

58. Abate, A., Nappi, M., Ricciardi, S.: I-Am: implicitly authenticate me person authentication on mobile devices through ear shape and arm gesture. IEEE Trans. Syst. Man Cybern. Syst. 1–13 (2017). https://doi.org/10.1109/TSMC.2017.2698258

Adaptive Digital Image Watermarking Technique Through Wavelet Texture Features

Sarita P. Ambadekar[1]([⊠]), Jayshree Jain[2], and Jayashree Khanapuri[1]

[1] K. J. Somiaya Institute of Engineering & Information Technology, Mumbai, India
sarita.ambadekar@somaiya.edu
[2] Pacific Academy of Higher and Research University, Udaipur, India

Abstract. Remarkable advancement in social media websites and chats, sharing photos, audios, and videos has become very easy and dangerous. Digital image watermarking technique has the potential to address the issue of privacy, ownership, and authenticity of the media shared using such medium. Invisible digital image watermarking techniques are necessary which are imperceptible within host image and robust to common signal and image processing attacks. In this paper, we present a watermarking technique for digital images through adaptive texturization, statistical parameters, and Bhattacharya distance. The primary idea is to segment the host image into four different regions based on frequency distribution using discrete wavelet transform. Subsequently important statistical parameters mostly applied in image processing techniques such as mean, standard deviation, skew, kurtosis, and variance are calculated from the wavelet transform coefficients of each region of host and watermark image. These statistical parameters of segmented regions of host and watermark image are then applied to obtain Bhattacharya distance. Wavelet transform coefficients of watermark image are embedded into wavelet transform coefficients of one of the regions of carrier image with minimum Bhattacharya distance through embedding factor. Performance of the technique was tested on multiple host and watermark images under common image processing attacks, which yield better results.

Keywords: Digital image watermarking · Discrete wavelet transform (DWT) · Statistical parameters · Bhattacharya distance · Embedding factor

1 Introduction

The remarkable development in very high-speed local area network (LAN), wide area network (WAN), metropolitan area network (MAN), internet technology and social media websites and chats sharing of photos, audios and videos has become very easy and dangerous. Due to sharing of digital multimedia documents such as images, audios, and video sequences by uploading/downloading/circulating/forwarding intentionally or unintentionally on social media websites such as Facebook, Twitter, LinkedIn, and WhatsApp. Manipulating, copying, and moderating of data are very simple since one and all have got access to the data/information through several signal and image processing

© Springer Nature Singapore Pte Ltd. 2020
B. Iyer et al. (eds.), *Applied Computer Vision and Image Processing*,
Advances in Intelligent Systems and Computing 1155,
https://doi.org/10.1007/978-981-15-4029-5_18

algorithms. Furthermore an original copy of digital images, videos is not distinguishable from the inventive record. Hence copyright protection and owner authentication has been more and more challenging tasks for multimedia data [1–3]. Nowadays it has been extremely necessary to propose, innovate, design, and develop an image processing algorithm for copyright protection, protection against duplication, and authentication of digital multimedia contents. Digital watermark is a message/data/information which is exactly simpler to digital signature. This signature is then embedded into digital multimedia files that can be detected or extracted later to prove the authenticity of the owner or inventor [4–6]. The complete method of embedding/extracting digital watermark information in the form of signature (text/image) into digital multimedia content is termed as digital watermarking. Sharing/posting/uploading an image on website require resizing, contrast adjustments, cropping, and compression through JPEG or any other compression algorithms which may result in partial or complete damage to host image that requires sophisticated image processing and watermarking algorithms.

Digital image watermarking techniques have been the focus of many researchers since the last two/three decades. The major objective of the digital image watermarking techniques is to prove the authenticity of the documents as when challenged by comparing extracted watermark with the original watermark [7–9]. Watermarking is interesting to investigate due to several reasons such as invisibility and robustness to various signal and image processing attacks such as cropping, geometric rotations, noise attack, and deformations. Therefore watermarking technique necessarily requires handling of in image imperceptibility, resistant to attacks, large number of watermarks, and robustness. Combinations of spatial and transformed domains techniques have been leveraged by many engineers and researchers to take advantage of both the domains. Also various other mathematical and statistical models and many other mostly applied interdisciplinary approaches in digital image processing: such as chaotic theory, fractal image coding, and adaptive techniques are explored. Transformed domain techniques using DCT and DWT are also exploited to embed digital image watermark into host image [10]. In [11] maximum entropy subband of the carrier image was explored for digital image watermarking, also subbands of the logo were shuffled to provide security to watermarked image. In [12] combination of DWT with singular value decomposition (SVD) techniques were presented for watermarking. In this method singular value matrix is embedded into the carrier image, whereas the remaining two unitary matrices are left intact which contents significant information. Whereas during extraction process these two unitary matrices are employed resulting in better robustness. In [13] watermark image was encrypted into random noise signal before watermarked into host image through DWT to enhance its security. Whereas in [14] Arnold transform was employed to enhance the security of the watermark image. Thereafter the logo embedded into LL subband of the carrier image that contains the high magnitude coefficients. In [6] proposed to identify region in carrier image for watermarking, thereafter the statistical properties of the region were employed to obtain a fusion of digital image watermark and statistical image synthesis for invisible image watermarking. However it is desired to build an invisible image watermarking technique that employs previously known techniques but adopting adaptive techniques without sacrificing on the security of the watermark. In this paper, we present a watermarking technique for digital images through adaptive texturization, statistical parameters, and Bhattacharya distance. The primary idea is to segment the host image into four different regions based on frequency distribution using discrete wavelet transform. Subsequently important statistical parameters mostly applied in image processing techniques such as mean, standard deviation, skew, kurtosis, and variance are calculated from the wavelet transform coefficients of each regions of host and watermark image. These statistical parameters of segmented regions of host and watermark image are then applied to obtain Bhattacharya distance. Wavelet transform coefficients of watermark image are embedded into wavelet transform coefficients of one of the regions of carrier image with minimum Bhattacharya distance through embedding factor. Performance of the technique was tested on multiple host and watermark images under common image processing attacks, which yield better results.

The paper is systematically prepared as Sect. 2 illustrates the embedding and extraction procedure, Sect. 3 demonstrates simulation results and finally conclusion derived from the results is stated in Sect. 4.

2 Watermark Embedding and Extraction Procedure

Digital image watermark embedding algorithm comprises several stages as segmentation in frequency using DWT, statistical parameter extraction, computing Bhattacharya distance, and watermark embedding in frequency/wavelet domain. Robustness of the proposed algorithm is enhanced by embedding all frequency/wavelet coefficients of watermark image into frequency/wavelet coefficients of selected region of host image. Whereas computing Bhattacharya distance in this manner enhances imperceptibility through matching texture of watermark image with the region of host image. Figure 1 shows the several stages in digital image watermarking through block diagram and it's embedding and extraction procedure is explained.

2.1 Watermark Embedding Procedure

Let input host image (I) and watermark image (w) are 8-bit gray scale of the size $p \times q$. Watermark embedding algorithm is detailed point wise as follows:

1. Read the input host (I) and watermark (w) image.
2. Convert color host and logo image into gray scale image.
3. Host image was segmented into four regions to identify the best region for watermark embedding.
4. Each segmented region of host image (I) was decomposed into four (4) subbands LL, LH, HL, and HH using Debauches 2D DWT.
5. Similarly watermark image was decomposed into four (4) subbands LL, LH, HL, and HH using Debauches 2D DWT.
6. Determine statistical parameters such as mean (μ), standard deviation (σ), variance (ν), skew (α), and kurtosis (γ) for each region of the carrier image and logo.

$$\mu_{mn} = \frac{E_{mn}}{p \times q} \tag{1}$$

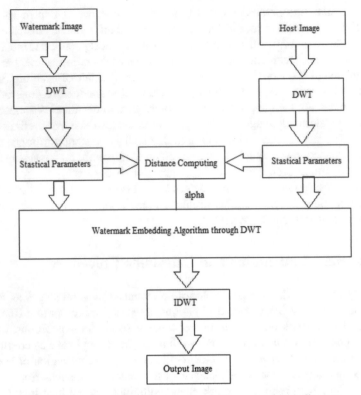

Fig. 1 Stages in digital image watermarking using block diagram

$$\sigma_{mn} = \frac{\sqrt{\sum_x \sum_y (|G_{mn}(x, y)| - \mu_{mn})^2}}{\sqrt{(p \times q) - 1}} \tag{2}$$

$$v_{mn} = \frac{\sum_x \sum_y (|G_{mn}(x, y)| - \mu_{mn})^2}{(p \times q) - 1} \tag{3}$$

$$\alpha_{mn} = \frac{\sum_x \sum_y (|G_{mn}(x, y)| - \mu_{mn})^3}{\sigma^3} \tag{4}$$

$$\gamma_{mn} = \frac{\sum_x \sum_y (|G_{mn}(x, y)| - \mu_{mn})^4}{\sigma^4} \tag{5}$$

7. Determine the Bhattacharya distance (d_{xy}) between each subband of the region of carrier image and logo.

$$d_{xy} = 0.125 * \left(\frac{(\mu_x - \mu_y)}{v_{xy}}\right) * \left(\frac{(\mu_x - \mu_y)}{v_{xy}}\right)^T + 0.5 * \log\left(\frac{v_{xy}}{\sqrt{v_x + v_y}}\right) \tag{6}$$

where

$$v_{xy} = \frac{(v_x + v_y)}{2} \tag{7}$$

8. Determine embedding factor alpha for each subbands through Bhattacharya distance and visibility controlling factor k.

$$\text{alpha} = k * \frac{1}{\left(1 + e^{d_{xy}}\right)} \tag{8}$$

9. Embed the wavelet coefficients of logo image into least Bhattacharya distance region of the carrier image through alpha an embedding factor.

$$Ie(x, y) = (\text{alpha} * w(x, y)) + I(x, y) \tag{9}$$

10. Two (2) dimensional Debauches (2D) IDWT is applied on the LL, LH, HL, and HH subband coefficients of the watermark embedded host image (Ie).

The complete process of watermark embedding into input host image through wavelet domain is described in the algorithm, whereas the process of watermark extraction is discussed in the next section.

2.2 Watermark Extraction Procedure

Let I_e and I represent watermark embedded and host image of the size $p \times q$, respectively. The steps involved in the complete process of extraction of watermark from the host image are illustrated as follows.

1. Read the original host and watermarked host image of the size $p \times q$.
2. Convert color host image into gray scale image.
3. Watermarked host and original host image were segmented into four regions.
4. Each region of watermark embedded host image was decomposed into four subbands LL, LH, HL, and HH using Debauches 2D DWT.
5. Similarly each region of the carrier image was decomposed into four (4) subbands LL, LH, HL, and HH using Debauches 2D DWT.
6. Extract watermark from each region of the watermarked host and original watermarked image through embedding factor alpha.

$$w(x, y) = \frac{(I_e(x, y) - I(x, y))}{\text{alpha}} \tag{10}$$

7. Two (2) dimensional Debauches (2D) IDWT is applied to the resultant of all subband coefficients of watermark image.
8. Determine the watermarking parameters between original and extracted watermark image with and without image processing attacks such as noise, compression, and geometric.

The complete process of embedding and extraction of watermark is discussed in this section, simulation results obtained after implementation of the above algorithm using computer programming language MATLAB is discussed in the next section.

3 Simulation Results

The comprehensive qualitative and measurable performance evaluation of the proposed adaptive technique using adaptive texturization, statistical parameters, and Bhattacharya distance is presented in this section. Figure 2 depicts the several host images and watermark image applied for demonstrations, which are classified based on varying texturization. Image Lena with ironic mix of directional texture which is distributed globally all over the entire image. Image garden has ironic mix of undirectional/random texture all over the host image. Whereas images fruits and scenery are poor in texture with fewer distribution all over the images and more color concentration in small areas. Performance evaluation of the proposed algorithm through peak signal to noise ratio (PSNR) and normalized correlation coefficient (NCC) was attempted. Robustness of the algorithm was measured by introducing geometrical attack, salt and pepper noise attack, and JPEG compression attack. Figures 3 and 4 show the respective extracted watermark images from Lena, garden, scenery, and fruits host images without and with the introduction of above attacks, respectively.

Lena   Garden   Scenery   Fruits

(a) Carrier images

(b) Watermark image

Fig. 2 Host and watermark images

Fig. 3 Retrieved watermark images from host image without attacks

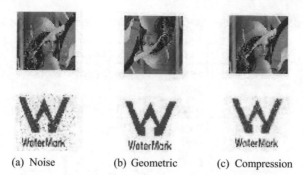

(a) Noise   (b) Geometric   (c) Compression

Fig. 4 Retrieved watermark images from host image after several attacks

The measured parameters PSNR and NCC are depicted in Tables 1 and 2 for the retrieved watermark images without and with introduction of above attacks, respectively. Comparison of the proposed method with similar method proposed in [8] is depicted in Table 3. The method in [8] does not segment host image into four regions for effective texture matching between host and watermark images.

Table 1 PSNR and CC parameters without geometric, noise, and compression attack

Parameters	Lena	Garden	Scenery	Fruits
PSNR (dB)	61.48	62.37	62.56	62.4
NCC	0.83	0.89	0.92	0.82

4 Conclusion

In this paper, we present a watermarking technique for digital images through adaptive texturization, statistical parameters, and Bhattacharya distance. The primary idea is to segment the host image into four different regions based on frequency distribution using discrete wavelet transform. Subsequently important statistical parameters mostly applied in image processing techniques such as mean, standard deviation, skew, kurtosis, and variance are calculated from the wavelet transform coefficients of each region of host and watermark image. These statistical parameters of segmented regions of host and watermark image are then applied to obtain Bhattacharya distance. Wavelet transform coefficients of watermark image are embedded into wavelet transform coefficients of one of the regions of carrier image with minimum Bhattacharya distance through embedding factor. Several host images and watermark image applied for demonstrations were classified based on varying texturization. Images with ironic mix of directional texture and undirectional/random texture all over the host image also image with poor in texture and more color concentration in small areas. Adaptive frequency domain approach suggests better watermarking technique in terms of robustness and imperceptibility. Further

Table 2 PSNR and CC parameters with geometric, noise, and compression attack

Attacks	Parameters	Lena	Garden	Fruits
Noise	PSNR (dB)	41.24	40.67	39.43
	NCC	0.51	0.52	0.49
Geometric	PSNR (dB)	38.95	37.23	36.45
	NCC	0.65	0.66	0.67
Compression	PSNR (dB)	34.54	35.66	36.73
	NCC	0.56	0.57	0.59

Table 3 Comparison of the proposed method through PSNR and CC values for Lena image

Parameters		Proposed method	[8]
PSNR (dB)	Without attack	61.48	47.79
CC		0.83	1.0
PSNR (dB)	With noise attack	41.24	32.15
CC		0.51	0.1458

security of the watermark image can be added to the existing algorithm to make it more secure through adopting encryption algorithms.

References

1. Natarajan, M., Makhdumi, G.: Safeguarding the digital contents: digital watermarking. DESIDOC J. Library Inf. Technol. **29**(3), 29–35 (2009)
2. Podilchuk, C.I., Delp, E.J.: Digital watermarking: algorithms and applications. IEEE Signal Process. Mag. **18**(4), 33–46 (2001)
3. Zhao, X., Ho, A.T.S.: An introduction to robust transform based image watermarking techniques. In: Intelligent Multimedia Analysis for Security Applications, vol. 282, pp. 337–364. Springer, Berlin (2010)
4. Halder, R., Pal, S., Cortesi, A.: Watermarking techniques for relational databases: survey, classification and comparison. J. Univers. Comput. Sci. **16**(21), 3164–3190 (2010)
5. Nema, A., Mohan, R.: Digital image watermarking: a review. Glob. J. Multidiscipl. Stud. **3**(6), 84–94 (2014)
6. Andalibi, M., Chandler, D.M.: Digital image watermarking via adaptive logo texturization. IEEE Trans. Image Process. **24**(12), 5060–5073 (2015)
7. Huang, X., Zhao, S.: An adaptive digital image watermarking algorithm based on morphological Haar wavelet transform. In: International Conference on Solid State Devices and Materials Science, vol. 25, pp 568–575. Published by Elsevier Physics Procedia (2012)

8. Vaidya, P., et al.: Adaptive digital watermarking for copyright protection of digital images in wavelet domain. In: 2nd International Symposium on Computer Vision & Internet, vol. 58, pp. 233–240. Published by Elsevier Procedia Computer Science (2015)
9. Chen, L., Zhao, J.: Adaptive digital watermarking using RDWT and SVD. In: IEEE International Symposium on Haptic, Audio and Visual Environments and Games (HAVE) (2015)
10. Roldan, L.R., Hernández, M.C., Chao, J., Miyatake, M.N., Meana, H.P.: Watermarking-based color image authentication with detection and recovery capability. IEEE Latin Am. Trans. **14**(2), 1050–1057 (2016)
11. Roy, A., Maiti, A.K., Ghosh, K.: A perception based color image adaptive watermarking scheme in YCbCr space. In: 2nd IEEE International Conference on Signal Processing and Integrated Networks (SPIN) (2015)
12. Yadav, N., Singh, K.: Transform domain robust image-adaptive watermarking: prevalent techniques and their evaluation. In: IEEE International Conference on Computing, Communication and Automation (2015)
13. Shukla, D., Tiwari, N., Dubey, D.: Survey on digital watermarking techniques. Int. J. Signal Process. Image Process. Pattern Recognit. **9**(1), 239–244 (2016)
14. Tao, H., et al.: Robust image watermarking theories and techniques: a review. J. Appl. Res. Technol. **12**, 122–138 (2014)

Singleton Wavelet Packet Matrix

Anita Devasthali and Pramod Kachare$^{(\boxtimes)}$

Department of Electronics and Telecommunication Engineering, RAIT, Nerul, Navi,
Mumbai, Maharashtra, India
anitadevasthali10@gmail.com, pramod_1991@yahoo.com

Abstract. Wavelet Packet Transform (WPT) is one of the generalized forms of the wavelet transform that has been a subject of interest for a variety of researchers. Owing to the multi-resolution characteristic of WPT, it has stepped into various research fields. The literature presents WPT as an iterative filtering and sub-sampling procedure, generally applied over a set of input feature vectors. In this work, we propose WPT as a linear encoder that packs the iterative process in the form of a transformation matrix. This results in a one-step matrix realization of complex and iterative WPT. The proposed matrix implementation is compared with state-of-the-art iterative DWT and Wavelet Packet Decomposition used in a variety of hardware- and software-defined languages. The proposed Wavelet Packet Matrix surpasses the baseline methods in terms of increased speed and reduction in process complexity. The generated sparse matrix can be a rapid transformation stage in real-time systems. It becomes handy in compressed sensing, harmonic analysis of signal and other applications involving iterative WPT analysis.

Keywords: Wavelet Packet Transform · Discrete wavelet transform · Refinement relation

1 Introduction of Wavelet

Wavelet Transform (WT) has been a topic of interest for researchers among various disciplines for more than three decades. The multi-resolution analysis is one of the salient features of WT. Applications like audio, image and video compression, time-frequency analysis, noise cancellation, image watermarking and voice conversion explore this basic property of WT [1]. Each of these systems represents input signal space in wavelet domain using iterative decomposition of smaller segments of input signal. Each level of decomposition views signal at resolution independent of the others. Hence, WT is also sometimes regarded as, signal processing microscope [2]. Depending on applications, some of the resulting sub-bands can be neglected as in noise cancellation and compression or modified as in watermarking and voice conversion.

© Springer Nature Singapore Pte Ltd. 2020
B. Iyer et al. (eds.), *Applied Computer Vision and Image Processing*,
Advances in Intelligent Systems and Computing 1155,
https://doi.org/10.1007/978-981-15-4029-5_19

WT is the result of incompetence of Fourier and short-time Fourier transforms in solving simultaneous time and frequency localization. The first Haar function, named after a Hungarian mathematician Alfred Haar, is the simplest form of wavelet transform introduced as functional decomposition of any real-time continuous domain signal. But this type of analysis function does not support lossless re-synthesis. Later, a British-Hungarian physicist, Denis Gabor, introduced short-time analysis of signals using Fourier transform called as Gabor transform. Such an analysis provided valuable directional information in solving image classification problems at the cost of increased computational complexity. The main disadvantage of Gabor transform is that it is not orthogonal and hence non-separable. In 1984 Meyer officially coined a word 'Wavelet' and introduced the concept of 'orthogonality' for designing of WT as a separable transform in analysis of seismic signal using finite duration signals of wave nature. Not later than a year or two, Mallat introduced the concept of multi-resolution analysis along with sequential pyramidal structure for WT and Wavelet Packet Transform (WPT). According to [1], vector representation of wavelet transform can be regarded as a choice of orthogonal based from an infinite set and it is only application concern. Similar results were delivered by Ingrid Daubechies while presenting new class of wavelets as digital filters (known as db1–db10). Meanwhile, many variants of similar transform such as morlet, coiflet, symlet, biorthogonal and reverse biorthogonal have been introduced in the context of variety of applications.

The theoretical institution suggests wavelet as an iterative implementation of filtering followed by sub-sampling [1]. But, the modern-day computational analysis demands fast and efficient representation of such an invaluable technique for real-time system integration. In [3], the author proposes linear procedure of wavelet tree decomposition as a matrix representation. In this work, we explore a similar approach to generate matrix representation for more generalized WPT decomposition.

2   Wavelet Analysis

Technically, WT analysis is a combination of conjugate mirror filtering followed by sub-sampling of signals. Mathematically, WT is defined as the cross-correlation between input signal $f(t)$ and Wavelet function (also called as Mother Wavelet) $\Psi(t)$, where

$$\Psi_{m,\mu}(t) = \frac{1}{\sqrt{m}} \Psi\left(\frac{t-\mu}{m}\right) \tag{1}$$

where (m, μ) are dilation and translation factors, respectively. Hence, wavelet transform is calculated as

$$W_{m,\mu}(t) = \langle f(t), \Psi(t) \rangle = \int_{-\infty}^{\infty} f(t) \, \Psi_{m,\mu}^{*}(t) \tag{2}$$

The admissible conditions for any function to be a Wavelet function are zeros mean (i.e. Wave) and finite energy (i.e. let).

The simplest class of WT is Orthogonal Dyadic Wavelets include filtering by two conjugate mirror filters followed by down-sampling by a factor of 2. Hence, Eq. (2) can be modified to represent discrete time signal $x \in \mathbb{R}$ which can be represented using dyadic sampled version of Eq. (1) as

$$W_{m,\mu}^k = \sum_{-\infty}^{\infty} x \, 2^{-k/2} \Psi \left(2^{-k} t - \mu^k \right) \tag{3}$$

where k is the level of decomposition. Here, translation factor μ^k has to be a discrete value. Discrete WT (DWT) from filters point of view is an iterative procedure, best defined using refinement relation,

$$S^k[n] = \sum_{j=m\mu}^{m\mu+N} h(j - m\mu) \, S^{k+1}[n] \tag{4}$$

$$D^k[n] = \sum_{j=m\mu}^{m\mu+N} g(j - m\mu) \, S^{k+1}[n] \tag{5}$$

where (h, g) are conjugate mirror filters and N is support of mirror filter (i.e. one less than filter length). S is the approximate part and D is the detail part of output DWT.

On the other hand, linear algebra defines DWT as simple matrix multiplication [3]. For First level of decomposition input signal $X \in \mathbb{R}^l$, we have

$$W^1 = \mathbf{\Phi}_l^1 X = \begin{bmatrix} S_{l/2}^1 \\ D_{l/2}^1 \end{bmatrix} X \tag{6}$$

where $\mathbf{\Phi}$ is $(l * l)$ Wavelet matrix generated using n column transformations of identity matrix [3]. DWT at each level can be calculated using Wavelet matrix at that level ($\mathbf{\Phi}^k : k \in \mathbb{Z}$). The wavelet matrices at consecutive level are related as

$$\mathbf{\Phi}_l^{k+1} = \begin{bmatrix} \mathbf{\Phi}_{l/2^k}^k & \mathbf{0} \\ \mathbf{0} & \mathbf{I}_{l-l/2^k}^k \end{bmatrix} \mathbf{\Phi}_l^k \tag{7}$$

One of the ways of generating initial matrix ($\mathbf{\Phi}_l^1$) has been proposed in [3]. Such type of decomposition works fine for generating wavelet tree, where only approximate part is analyzed at successive decomposition levels. In this work, we modify the Wavelet matrix to evaluate most general WPT decomposition at the output.

3   Wavelet Packet Matrix

Wavelet Packet analysis is a comprehensive WT, where not only approximate but also detail component of decomposition is further analyzed. The distinct characteristic between Wavelet Tree and Wavelet Packet Tree is depicted in

Fig. 1. It can be seen that third level Wavelet Tree is signal representation using one approximation component (H) and detail components (G) at third as well as all lower levels. On the other hand, Wavelet Packet Tree at third level is a set of orthogonal representations with *four* approximation components and *four* detail components. WPT, as shown in Fig. 1, is a conjugate mirror filtering where each component yields two new components at successive decompositions. Such a procedure can be replicated to generate higher decomposition levels. As discussed in section II, we propose a singleton Wavelet matrix representation to encase such an iterative process, which can be easily implemented at hardware as well as software levels.

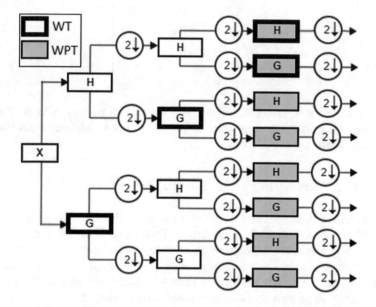

Fig. 1. Level 3 Wavelet Tree and Wavelet Packet Tree

The name 'singleton' itself suggests substitution of a complex but repetitive process in WT by a one-step matrix multiplication process. Such a matrix representation requests a closer look at refinement equations and deduces the similarity and differences between ordinary WT and WPT. Accordingly we must modify matrix equations as discussed in Sect. 2. The combination of Eqs. (6) and (7) presents the most important conclusion,

$$\mathbf{S}^k_{l/2^k} = \prod_{j=1}^{k-1} \mathbf{\Phi}^j_{l/2^j} S^j_{l/2^j} \tag{8}$$

$$\mathbf{D}^k_{l-l/2^k} = \prod_{j=1}^{k-1} I^j_{l-l/2^j} D^j_{l-l/2^j} \tag{9}$$

The equation intuitively supports the theoretical difference between WT and WPT, i.e. detail coefficients are left for higher levels of decomposition. Therefore, the preliminary modification we suggest is replacing identity matrix ($\mathbf{I}^j$) by sub-sampled Wavelet matrix at same level ($\mathbf{\Phi}_{l/2^j}^j$). The modified equation for WPT analysis becomes

$$\mathbf{W}_{l/2^k}^{k,s} = \prod_{j=1}^{k-1} \Phi_{l-l/2^j}^j \mathbf{W}_{l/2^j}^{j,s} \tag{10}$$

where $\{s \in \mathbb{Z}_{0 \le s < 2^k}\}$ denotes the sub-band index at each level of decomposition. The WPT matrix can be written as

$$\mathbf{\Phi}_l^{k+1} = \begin{bmatrix} \mathbf{\Phi}_{l/2^k}^k & 0 & \dots & 0 \\ 0 & \mathbf{\Phi}_{l/2^k}^k & \dots & 0 \\ \dots & \dots & \dots & \dots \\ \dots & \dots & \dots & \dots \\ 0 & 0 & \dots & \mathbf{\Phi}_{l/2^k}^k \end{bmatrix} \mathbf{\Phi}_l^k \tag{11}$$

Now that we have generalized the WPT analysis matrix, the only concern is obtaining the initial Wavelet matrix ($\mathbf{\Phi}$) at each level of decomposition using just the conjugate mirror filters.

4   Matrix Through Filter Coefficients

In [3], the author proposed some sort of sub-band alignment technique to generate a relation between impulse response of filters and Wavelet Tree Matrix. We can use exactly the same algorithm to generate initial decomposition matrices, as only the organization of the matrix has been changed while the processing remains the same. Though we have a former method, we would like to propose a much simpler approach to reach the same output matrix.

We recall refinement relation from Eq. (4), expanding wavelet coefficients in terms of conjugate mirror filters with impulse responses h (low pass) and g (high pass). Consider a wavelet with finite impulse response of length 'N', we have $\{h_i \in \mathbb{R} : i \in \mathbb{Z}_{0 \le i \le N}\}$ and $\{g_i \in \mathbb{R} : i \in \mathbb{Z}_{0 \le i \le N}\}$. The conjugate high-pass filter (g) can be easily calculated using low-pass impulse response as shown in Eq. (12).

$$g[n] = (-1)^n h[N - n] \tag{12}$$

Wavelet as a filter is a mapping defined as $\{\Psi : \mathbb{R}^l \mapsto \mathbb{R}^l\}$. Hence, we understand wavelet mapping as a simple circular convolution, preserving signal dimensionality for variable size of impulse responses [4].

The detailed procedure for generating initial Wavelet matrix is described in Table 1. As discussed in Table 1, we pad required number of zeros to maintain input–output dimensionality matching. We generate two separate matrices for low-pass (H_k) and high-pass (G_k) filtering. Finally, full Wavelet matrix (Φ) is a simple row concatenation of both the matrices.

Table 1. Steps to generate wavelet packet matrix

1. Rearrange low pass impulse response (h) coefficients as single row vector. • **Filter length < Signal dimension ($N < l$)** Pad ($l - N$) zeros at the end of impulse response to make dimensionality of both vectors to be same. $$H \equiv [h \ \ zeros(l - N)]$$ • **Filter length > Signal dimension ($N > l$)** Pad zeros at the end of impulse response to make length of impulse response, as integral multiple of signal dimension (i.e. $N = m * l$). Then, split the impulse response in smaller segments, each of dimension 'l'. $$hseg_{i,j} = \{h_i(j * i) : i \in \{0 : l - 1\}, j \in \{1 : m\}\}$$ where, i, j are indices of impulse response coefficient and segment, respectively. • **Filter length = Signal dimension ($N = l$)** Use original impulse response as it is for further processing. $$H \equiv h$$
2. **If ($N > l$)** Add coefficients of all segments element-by-element to generate a single filter response of the order of 'l'. This is same as generating circular convolution result using linear convolution. $$H \equiv \sum_{\forall j} hseg_{i,j}$$
3. Appoint the current impulse response vector (H) as, the first row of wavelet matrix (H_1).
4. Each successive row is circularly shifted version by factor of 2 (Dyadic) of former row. We can generate $l/2$ such distinct translations to get upper half-circulant Wavelet Matrix. $$H_k \equiv H_{k-1}(i+2)|_{modulo \ \ l}, \quad k = 2, 3, \ldots, l/2$$
5. Repeat steps (1– 4) to generate lower half-circulant Wavelet Matrix (G_k) using high pass impulse response (g).
6. Full Wavelet Matrix is concatenation of low pass and high pass half-circulant matrices.

Let's generate fourth order ($n = 4$) 'db3' Wavelet matrix for 1 level ($k = 1$) decomposition. The original low-pass impulse response will be $h = \{0.0352, -0.0854, -0.1350, 0.4599, 0.8069, 0.3327\}$. We now divide this response into segments each of dimension '4'and add both segments to generate desired low-pass filter response. This becomes the first row of Wavelet matrix, while successive rows are generated by circular shift of 2 (i.e. Dyadic) [1]. Along similar lines, we generate high-pass response with the help of Eq. (12). Finally, the Wavelet matrix is row concatenation of low- and high-pass filters.

$$\Phi_4^1 = \begin{bmatrix} H_1 \\ H_2 \\ G_1 \\ G_2 \end{bmatrix} = \begin{bmatrix} 0.8421 & 0.2473 & -0.1350 & 0.4599 \\ -0.1350 & 0.4599 & 0.8421 & 0.2473 \\ -0.2473 & 0.8421 & -0.4599 & -0.1350 \\ -0.4599 & -0.1350 & -0.2473 & 0.8421 \end{bmatrix}$$

Hence, similar results can be achieved with less complexity as presented in [3]. Higher level of decomposition, i.e. $k = 2$ can be obtained using relation defined in Eq.(11).

$$\Phi_4^2 = \begin{bmatrix} \Phi_2^1 & 0 \\ 0 & \Phi_2^1 \end{bmatrix} \Phi_4^1 = \begin{bmatrix} 0.5000 & 0.5000 & 0.5000 & 0.5000 \\ -0.6909 & 0.1504 & 0.6909 & -0.1504 \\ -0.5000 & 0.5000 & -0.5000 & 0.5000 \\ -0.1504 & -0.6909 & 0.1504 & 0.6909 \end{bmatrix}$$

We can obtain Φ_2^1, by replicating the above procedure with segments of dimension '2'. Another simpler alternative is to reformat Φ_4^1 to $[4 * 2]$ by circular addition and then sub-sample [rows:columns] in ratio $[1 : 2]$.

5 Matrix Analysis

The primary goal of this work is to design a simple and efficient predefined matrix for Wavelet Packet analysis. As seen in the above sections, we are able to generate WPT matrix for given signal dimension and desired decomposition level. Hence, 1D WPT analysis becomes a single-step matrix multiplication.

$$W = \Phi_n^k \, X \tag{13}$$

where W : Wavelet Packet Coefficient Matrix, Φ_n^k : k^{th} level WPT matrix of order n, X : $\{x_1, x_2, \ldots, x_m : x_i \in \mathbb{R}^n\}$ WTs focused in this work are orthogonal transforms. Hence, for higher dimensional signals we can calculate WPT coefficients by applying separable transform for each dimension. For example, 2D WPT coefficients for image $(I_{(n*n)})$ can be calculated as [1],

$$W = \Phi_n^k \, I \, (\Phi_n^k)^T \tag{14}$$

Similar procedure can be followed for higher dimensional WPT analysis. Such matrix multiplication act as rapid and efficient transformation alternative to iterative filtering and sub-sampling procedures. Having witnessed the implementation, we must comprehend the advantages and limitations of proposed Wavelet Packet Matrix.

5.1   Advantages

In this section, we compare the performance of the proposed matrix method with baseline Wavelet Packet Decomposition (WPDEC) technique used in MATLAB and state-of-the-art iterative DWT decomposition algorithms used in several other languages. The speed here refers to the process time taken to execute a particular operation. Lower the process time, faster the method and thus, better is the implementation. The process times required by three different methods at various levels and increasing number of decomposition samples are plotted on a semi-log scale as shown in Fig. 2.

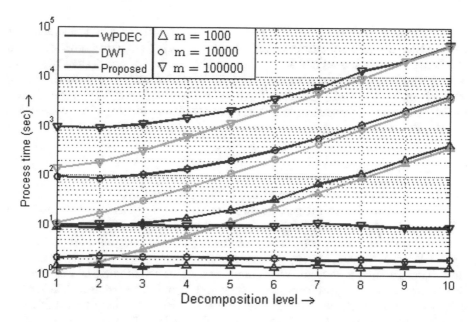

Fig. 2. Comparing process times for WPDEC, DWT and proposed matrix method

Process time for the proposed method is significantly reduced compared to rest of the approaches. It can also be observed that, with increased decomposition levels, both WPDEC and DWT-based implementations consume more time, whereas increase in time for the proposed method is insignificant. We have also examined the effect of increasing wavelet filter length on speed and observed no significant variation. This may be due to the fact that signal dimension is much greater than the filter length and thus dominates its effect on process time.

The complexity of the process is estimated as the number of multiplications and additions required to a complete particular operation, defined as process cost. Both WPDC and DWT algorithms are equally costly processes. Table 2 lists the process costs for wavelet filter of the order 'N' and input signal dimensionality 'l' and m decomposition samples.

Table 2. Process complexity for WPDEC, DWT and proposed matrix method

Method	Process cost
WPDEC/ DWT	Mul : $(N^2 + l * N) * m$ Add : $(l + N) * m$ Considering general statistics, $(N << l << m)$ Mul : $l * N * m$ Add : $l * m$
Proposed	Mul : $(N^2 + l * N) * m$ Add : $(l + N) * m$ Considering sparse representation, Mul : $N * N * m$ Add : $l * m$

As shown in Table 2 general cost for all the process remains the same. But considering the general statistics $(N << l << m)$, the sparse representation of Wavelet matrix has significantly reduced the number of multiplications. This is very much advantageous in case of DSP and Embedded processors, where multiplication is a costly process. Though we have not considered process cost for Wavelet matrix formation, it would be insignificant as most of the operations are simple shifting.

5.2   Limitations

Besides the various advantages put forth, the proposed matrix implementation also imposes certain limitations, as discussed below. For baseline WPDEC and DWT algorithm, we need to store only the filter coefficients for successful implementation. On the other hand, the proposed algorithm requires additional storage for Wavelet matrix, directly proportional to square of the input signal dimension. Though such a storage requirement can be easily fulfilled in modern-day processors, it will be problematic for very high signal dimensionality. Also, the proposed method will not be efficient on rare occasions, where the number of decomposition samples is small and comparable to signal dimensionality.

The maximum decomposition level is directly proportional to input signal dimensionality. In case of dyadic wavelets, it is defined as the '$\log_2 l$', where 'l' is input signal dimensionality. Hence, all decomposition matrices must have dimensions of the order of 2. Similarly, to follow the rules of matrix multiplication, we must make input dimensionality also of the order of 2. For perfect reconstruction, care must be taken to eliminate required zeros padding during analysis process.

6   Conclusion

Discrete implementation of Wavelet and Wavelet Packet Transforms present a repetitive architecture. The primary aim of this work is to generalize such procedures in the form of a transformation matrix. We use general refinement expressions to establish relation between wavelet coefficients and filter responses.

Taking the advantage of the orthogonal nature of transform a Wavelet matrix is formed. This reduces iterative filtering and sub-sampling process into a simple matrix multiplication. Along with reduced computational complexity, it also provides an efficient algorithm that can be employed by any hardware device. Sparse implementation of Wavelet matrix further reduces the processing time.

References

1. Mallat, S.: A Wavelet Tour of Signal Processing. Academic press (1999)
2. Mallat, S.G.: A theory for multiresolution signal decomposition: the wavelet representation. Pattern Anal. Mach. Intell., IEEE Trans. **11**(7), 674–693 (1989). https://doi.org/10.1109/34.192463
3. Yan, J.: Wavelet matrix. Department of Electrical and Computer Engineering, University of Victoria, Victoria, BC, Canada (2009)
4. Oppenheim, A.V., Schafer, R.W., Buck, J.R.: Discrete-Time Signal Processing, vol. 2. Prentice-hall, Englewood Cliffs (1989)

An Experimental Analysis of Gene Feature Selection and Classification Methods for Cancer Microarray

S. Akila$^{(\boxtimes)}$ and S. Allin Christe

Department of Electronics and Communication Engineering,
PSG College of Technology, Coimbatore, India
akilascs19@gmail.com, allinchriste@gmail.com

Abstract. DNA Microarray data analysis has emerged as a prominent tool for identifying and classifying the normal and abnormal genes related to cancer. The microarray data is comprised of samples and highly informative gene features. The dimension of the data is high and is difficult to handle during classification. Hence the dimension has to be reduced by obtaining the predictive gene features without affecting the classification accuracy. In this work, an experimental analysis of various feature selection methods along with the classifier for gene features selection is done and the analysis results are presented. Experiments have been conducted on the ALL-ML, Breast, Colon, Ovarian, Lung and Lymphoma datasets. With the results in terms of classification accuracy and number of gene features selected, the suitable algorithm for microarray dataset is suggested.

Keywords: Cancer classification · Gene selection and classification · Microarray classification · Feature selection · Classifier

1 Introduction

Classification and decision making plays a vital role in the healthcare system. This process has to be supported by classification algorithms. Cancer being one of the deadly diseases needs time-consuming diagnosis. The process is also error prone. Hence the need of machine learning algorithm in cancer screening and classification is high. This aids the process of diagnosis and prognosis which further aids in proper treatment and cure. In recent years, the microarray data for the cancer have gained importance in the cancer diagnosis process of classification. Handling the microarray data is challenging due to its large size [1]. The number of samples is less while the number of gene expressions is high. To overcome the challenges faced during the time of handling the microarray data, it is necessary to reduce the size of the data by obtaining the highly predictive gene features from the dataset. Gene feature selection is done to increase the recognition and improve the classification accuracy. The large dimension of gene features directly or indirectly is related to the classification accuracy and computational complexity [2–4]. For an efficient cancer classification model, the cancer microarray has to be reduced

© Springer Nature Singapore Pte Ltd. 2020
B. Iyer et al. (eds.), *Applied Computer Vision and Image Processing*,
Advances in Intelligent Systems and Computing 1155,
https://doi.org/10.1007/978-981-15-4029-5_20

to highly predictive gene features. To reduce the number genes present in the microarray, filter-based techniques are employed. These techniques are data-independent and work with the local optima. From the literature, it is clearly known that the filter-based techniques were widely used to reduce the microarray genes due to its ability of fast computation and easy implementation [5, 6]. The main aim of this work is to study and find the best-suited method to reduce the number of gene features without affecting the classification accuracy.

In the experimental study, six cancer microarray datasets were considered. Feature selection methods like Minimum Redundancy Maximum Relevance (mRMR) [7], and Fast Correlation Based Feature selection (FCBF) [8] and Relief (Re) [9] algorithm were experimented. The above feature selection methods were evaluated with the Decision Tree (DT) [10], Random Forest (RF) [11], Naïve Bayes (NB) [12], K Nearest Neighbor (KNN) [13], Artificial Neural Network (ANN) [14], and Sequential Minimal Optimization (SMO) [15, 16] classifiers, respectively.

The paper is organized as follows: Sect. 2 discusses the feature selection methods along with the various classifiers used for evaluation, Sect. 3 describes the cancer DNA microarray datasets that were used for the experimentation, and Sect. 4 contains the experimental results along with discussions and Sect. 5 contains the conclusion, respectively.

2 Methods for Gene Feature Selection and Classification

Gene feature selection is the process of obtaining the predictive gene features from the datasets. The dimension of the microarray data is high which leads to computation complexity and reduces the classification accuracy. The following are the feature selection techniques from gene feature selection discussed with respect to the cancer microarray dataset containing N samples and M gene feature expressions.

2.1 Feature Selection Algorithms

The various features selection algorithms chosen to experiment in this work are as follows.

Minimum Redundancy Maximum Relevance (mRMR)
To obtain the m subsets from the M gene feature expression using Minimum Redundancy Maximum Relevance (mRMR) algorithm, it is necessary to maximize the relevance between the gene-class interactions and minimize the redundancy between the gene-gene interaction. For a given m gene feature subset, the relevance is calculated by averaging the mutual information $I(x_i, C)$, where $x_i \in m$ gene feature subsets and class C. Next, the redundancy is calculated by averaging mutual information $I(x_i, x_j)$, where $x_i, x_j \in m$ gene feature subsets [7].

Fast Correlation Based Filter (FCBF)
To obtain the gene feature subsets that are most relevant to the target and less redundant to other gene is done using the Fast Correlation based Filter. Mutual Information I is used to implement the concept of FCBF. Consider X and Y as two random variables. $I(X, Y)$ refers to the mutual information of one variable obtained from another variable.

Once the mutual information is obtained, they are sorted in the decreasing order. The highly ranked gene features are obtained [8].

Relief (Re)

Using the relevance of the feature with the target class, weight is assigned to each feature. The weights are calculated using the nearest hit and nearest miss. The nearest neighbor of the same class and different classes are considered which is named as nearest hit and nearest miss, respectively [9]. High weights are assigned to the m gene features if they take different instances belonging to different classes or similar values of instances of the same class.

2.2 Classification Algorithms

The cancer classification models deployed in the microarray analysis had many statistical and machine learning methods in it. The classification model is further classified as unsupervised and supervised models. The clustering algorithms like hierarchical clustering and two ways clustering are the unsupervised models. The self-organizing maps SOM falls under the unsupervised model category. The widely used supervised classification model in the microarray cancer classification models is discussed in the following subsections.

Decision Tree (DT)

A Decision Tree (DT) is knowledge representation based on a tree structure for the process of classification. In a DT, each internal node represents the test features while the external nodes referred to as leaves give the possible results. The results as the classes to which the input sample belongs to after traversing from the root node to leave node through all internal nodes. C4.5 is a learning algorithm used to build a tree in top-down approach using information entropy. The dataset is given as an input to the tree, which is further split into small subsets. The depth of the tree is gradually increased with the splitting up of the data until the leaf node is reached. Leaf node is the termination point of any tree. The normalized entropy gain is used as the slitting criterion by the C4.5 algorithm with which it is possible to check the homogeneity of the m gene feature subset obtained with respect to the class [10].

Random Forest (RF)

Random forest is an ensemble of tree-based classifiers. Most of the literature works that referred random forest as the original method of forest RI were proposed by Breiman [11]. Considering our cancer microarray data with N sample and M gene features, the following steps are implemented to obtain the predictive gene features using forest-RI algorithm. For each tree, obtain M gene features randomly from the N samples which will be used as a training data to build the starting node of the tree. Set a number m < <M. Randomly select m gene features from the whole set at each node. For every m gene feature selected, using Gini index, select the best binary split values. Further, by choosing the best m gene feature split the samples associated into two new nodes. Until the maximum size of the tree is achieved, increase the set of the tree. Define the stopping criterion with the following: (i) when the number of samples to be spilt is less than the threshold. (ii) When all the samples fall under the same 6. The tree is not pruned. After the forest is built, according to the majority vote rule, the unknown sample is labeled with the frequent class in the ensemble trees.

Naïve Bayes (NB)
The algorithm is a statistical classification algorithm that works with the help of the Bayes theorem. According to the NB, a feature M in a class is independent of other features. To implement the algorithm the prior probability for class labels and likelihood probability for all features in the class is calculated [12].

K Nearest Neighbor (KNN)
KNN is a widely used classification algorithm known for its simplicity and performance. When an unknown sample N is present, it is assigned to the most frequent class that is present in the k nearest sample. To assign the sample to an unknown class, the algorithm uses the posteriori probability. $P(C_iN)$ is the probability calculated using the class label of the k nearest neighbor of N after performing experimental trials, the value of $K = 1$ which outperformed other values. The higher the value of K, it moved far away from the neighbors belonging to the same class was fixed [13].

Artificial Neural Networks (ANN)
An Artificial Neural Network [14] works with the phenomenon behind the biological nervous systems processing the information. Highly interconnected elements called neurons are used for processing. Each neuron contains a local memory. The information is processed locally and the output is transmitted to all neurons in a unidirectional manner. The neural networks are configured as application-specific through a learning process. As a result of the learning process, the neurons adjust themselves with the information. In this work, a completely connected feed-forward multilayer perceptron network is built. Training is done using the backpropagation algorithm.

Sequential Minimal Optimization (SMO)
The algorithm obtains the reduced feature by breaking the data into random small subsets. When there occur large SVM learning problems, the SMO is used to perform a series of small optimization tasks [15].

3 Datasets Used for Experimental Analysis

The cancer microarray datasets used for the experimental analysis of the various gene selection and classification algorithms were obtained from http://csse.szu.edu.cn/staff/zhuzx/Datasets.html [17, 18]. The datasets contained N samples with M gene expressions. The datasets contained samples ranging from 60 to 253 and the number if gene features between 2000 and 24481. The datasets belonged to binary classes and are shown in Table 1 along with its detailed description. The datasets obtained from the repository are preprocessed to remove the noise. Further, it is subjected to min-max normalization such that dataset $\in (0,1)$. The algorithms were implemented using the Weka 3.6 [19] and Matlab2016Ra and simulated in Intel i5 core processor.

4 Experimental Analysis and Discussion

The datasets after the preprocessing are given as the input to the feature selection algorithms. In this work, MRMR, FCBF, and Relief are the feature selection algorithms that were implemented. The predictive gene features were obtained using the feature selection algorithms. To evaluate the features selected by the feature selection algorithms,

Table 1 Cancer Microarray Dataset description

Datasets(Samples × Genes)	Sample composition
ALL-AML (72 × 7129)	Acute Lymphoblastic Leukemia (ALL) and Acute Myelogenous Leukemia (AML)
Breast (97 × 24481)	46 samples of distance metastases and 51 samples that remained healthy after primary diagnosis at an interval of minimum 5 years
Colon (62 × 2000)	biopsy sample containing 22 normal and biopsies 40 cancer biopsies
Ovarian (253 × 15154)	91 normal persons and 162 ovarian cancer patients
Lung (181 × 12533)	31 samples of malignant pleural mesothelioma and 150 adenocarcinoma
Lymphoma (62 × 4026)	A total count of 62 constitutes the Lymphoma microarray datasets

Table 2 Statistical results showing features selected using MRMR and evaluated using various classifiers

Datasets	Metric	DT	RF	NB	KNN	ANN	SMO
ALL-AML	Accuracy	83.47	86.32	98.61	85.14	95.97	97.22
	Features	45	45	45	45	45	45
Breast	Accuracy	88.16	90.35	55.67	90.75	89.60	78.35
	Features	10	10	10	10	10	10
Colon	Accuracy	81.53	85.89	87.09	86.45	83.95	87.09
	Features	50	50	50	50	50	50
Ovarian	Accuracy	85.05	88.19	98.41	87.92	92.89	100
	Features	100	100	100	100	100	100
Lung	Accuracy	96.99	99.45	95.07	99.17	100	96.05
	Features	10	10	10	10	10	10
Lymphoma	Accuracy	72.80	80.78	100	67.24	92.45	100
	Features	35	35	35	35	35	35

classifiers like Decision Tree (DT), Random Forest (RF), Naïve Bayes (NB), K Nearest Neighbor (KNN), Artificial Neural Network (ANN), and Sequential Minimal Optimization (SMO) are deployed. The evaluation metric used is the classification accuracy. All the experiments were conducted using the WEKA [17] tool for the evaluations using the classifiers.

The statistical results containing the classification accuracy and the number of predictive features selected using various feature selection algorithms are shown in Tables 2, 3, and 4, respectively. Also various classifiers were used to evaluate the obtained features with their classification accuracy.

From the experimental results obtained, analysis is performed. Different classification accuracy values were obtained for different classifier algorithms. The result of

Table 3 Statistical results showing features selected using FCBF and evaluated using various classifiers

Datasets	Metric	DT	RF	NB	KNN	ANN	SMO
ALL-AML	Accuracy	84.93	98.40	97.22	98.74	99.44	98.61
	Features	51	51	51	51	51	51
Breast	Accuracy	86.95	91.56	53.60	90.75	90.56	83.50
	Features	87	87	87	87	87	87
Colon	Accuracy	89.59	88.15	85.48	85.4	86.45	87.09
	Features	14	14	14	14	14	14
Ovarian	Accuracy	88.42	94.58	100	90.92	92.41	100
	Features	18	18	18	18	18	18
Lung	Accuracy	96.49	99.45	96.05	99.86	100	95.56
	Features	65	65	65	65	65	65
Lymphoma	Accuracy	78.43	90.78	100	89.68	97.92	100
	Features	50	50	50	50	50	50

Table 4 Statistical results showing features selected using Re and evaluated using various classifiers

Datasets	Metric	DT	RF	NB	KNN	ANN	SMO
ALL-AML	Accuracy	93.05	97.92	98.61	97.22	98.61	98.61
	Features	5	5	5	5	5	5
Breast	Accuracy	87.34	89.56	79.38	89.86	89.58	82.47
	Features	74	74	74	74	74	74
Colon	Accuracy	81.94	85.48	72.58	87.10	84.19	85.48
	Features	5	5	5	5	5	5
Ovarian	Accuracy	88.31	92.92	96.44	92.59	97.69	100
	Features	5	5	5	5	5	5
Lung	Accuracy	98.62	99.45	90.64	98.90	100	94.08
	Features	5	5	5	5	5	5
Lymphoma	Accuracy	82.55	92.29	100	87.50	95.21	100
	Features	10	10	10	10	10	10

analysis showing the best value of accuracy, and the features along with the corresponding feature selection algorithm and classifier are shown in Table 5. From the summary in Table 5, Re-based feature selection algorithm has relatively performed well with all the datasets. On considering the classifiers, the ANN and the SMO relatively performed well irrespective of the feature selection algorithm and the datasets. The SMO gave the best results for most of the datsets in all the features selection algorithms. Thus from the analysis, it is clearly shown that the Re-based SM algorithm works well for most of the datasets.

Table 5 Summary of best feature selection algorithm with the corresponding classifier

Datasets	Accuracy	Features	Feature selection algorithm	Classifier
ALL-AML	98.61	5	Re	ANN
			Re	SMO
Breast	90.75	10	MRMR	KNN
Colon	89.59	15	FCBF	DT
Ovarian	100	5	Re	SMO
Lung	100	5	Re	ANN
Lymphoma	100	10	Re	ANN
			Re	SMO

5 Conclusion

In this paper, an experimental investigation of feature selection methods and classifiers for cancer microarray data is implemented and analyzed. Feature selection algorithms like MRMR, FCBF, and Re were implemented and evaluated with six classifiers. With the experimental results obtained, the Re with SMO as classifier outperformed the other existing algorithms. The results were consistent with all the cancer microarray datasets that were subjected to the experiment. Finally, it arrived to a conclusion from the experimental results that the Re-based SMO algorithm is suitable on all the datasets.

References

1. Bayarri, M.J., Berger, J.O., Dawid, A.P., David, H.: Bayesian Factor Regression Models in the " Large p, Small n " Paradigm, Bayesian Statistics 7, Oxford University Press 723–732 (2003)
2. Wu, M.-Y., Dai, D.-Q., Shi, Y., Yan, H., Zhang, X.-F.: Biomarker identification and cancer classification based on microarray data using Laplace Naive Bayes model with mean shrinkage. IEEE/ACM Trans. Comput. Biol. Bioinform. **9**(6), 1649–1662 (2012)
3. Jain, A., Zongker, D.: Feature selection: Evaluation, application, and small sample performance. IEEE Trans. Pattern Anal. Mach. Intell. **19**(2), 153–158 (1997)
4. Saeys, Y., Inza, I., Larra naga, P.: A review of feature selection techniques in bioinformatics. Bioinformatics 23(19), 2507–2517 (2007)
5. Miao, J., Niu, L.: A survey on feature selection. Procedia Comput. Sci. **91**, 919–926 (2016)
6. Seijo-Pardo, B., Bolón-Canedo, V., Alonso-Betanzos, A.: Using a feature selection ensemble on DNA microarray datasets. In: Proceedings of the 24th European Symposium on Artificial Neural Networks, Computational Intelligence and Machine Learning (ESANN 2016), pp. 277–282, Bruges. Belgium (27–29 April 2016)
7. Ding, C., Peng, H.: Minimum redundancy feature selection from micro array gene expression data. J. Bioinform. Comput. Biol. **3**(02), 185–205 (2005)
8. Yu, L., Liu, H.: Feature selection for high-dimensional data: a fast correlation-based filter solution. ICML **3**, 856–863 (2003)

9. Wang, Y., Makedon, F.: Application of relief-F feature filtering algorithm to selecting informative genes for cancer classification using microarray data. In: Proceedings 2004 IEEE Computational Systems Bioinformatics Conference, CSB2004, pp. 497–498 (2004)
10. Chen, K.-II., Wang, K.-J., Wang, K.-M., Angelia, M.-A.: Applying particle swarm optimization-based decision tree classifier for cancer classification on gene expression data. Appl. Soft Comput. **24**, 773–780 (2014)
11. Breiman, L.: Random forests. Mach. Learn. **45**, 5–32 (2001)
12. Friedman, N. et al.: Using Bayesian Networks to Analyze Expression data. In: Fourth Annual International Conference on Computational Molecular Biology, vol. 7, pp.127–135 (2000)
13. Cho, S., Won, H.: Machine learning in dna microarray analysis for cancer classification. In: First Asia-Pacific Bioinformatics Conference on Bioinformatics, vol. 19, pp. 189–198 (2003)
14. Fernández-Navarro, F., Hervás-Martínez, C., Ruiz, R., Riquelme, J.C.: Evolutionary generalized radial basis function neural networks for improving prediction accuracy in gene classification using feature selection. Appl. Soft Comput. **12**(6), 1787–1800 (2012)
15. Bolon-Canedo, V., Sanchez-Marono, N., Alonso-Betanzos, A.: An ensemble of filters and classifiers for microarray data classification. Pattern Recogn. **45**(1), 531–539 (2012)
16. Al Snousy, M.B., El-Deeb, H.M., Badran, K., Al Khli, I.A.: Suite of decision tree-based classification algorithms on cancer gene expression data, Egypt. Inform. J. **12**(2), 73–82 (2011)
17. Zexuan, Z., Ong, Y.S., Dash, M., Markov Blanket-embedded genetic algorithm for gene selection. Pattern Recognit. **49**(11), 3236–3248 (2007). http://csse.szu.edu.cn/staff/zhuzx/Datasets.htmllast visited October 2019e
18. Hall, M., Frank, E., Holmes, G., Pfahringer, B., Reutemann, P., Witten, I.H.: The WEKA data mining software: an update. SIGKDD Explor. **11**, 10–18 (2009)

Deep Learning-Based Smart Colored Fabric Defect Detection System

Srikant Barua$^{(\boxtimes)}$, Hemprasad Patil, Parth Dharmeshkumar Desai, and Manoharan Arun

Vellore Institute of Technology, Vellore, Tamil Nadu, India
srikantbarua@gmail.com, hemprasadpatil@gmail.com,
parth.dharmeshkumar2019@vitstudent.ac.in, arunm@vit.ac.in

Abstract. Due to the huge increase in customers for different fabrics in this generation, the texture of the fabrics becomes an important issue thus bringing the requirement for correct and perfect detection of the fabric defects. In the existing semiautomated systems, a quality inspector takes 5 m/min with a defect and 15 m/min without defect to identify and rectify the defects with the resolution of 1 mm/pixel. This process results in the loss of the factory's overall throughput and efficiency. While manufacturing fabrics, there may be various defects like hole, missing yarn, broken yarn, stain, etc. These defects incur huge losses to the textile industry as they cause customer dissatisfaction. In order to reduce such losses, detection of defects beforehand is very important. Our project uses the concept of deep learning for the detection of colored fabric defects. The working and reliability of the fabric defect detection system presented is evaluated through vigorous experiments of real fabric samples with different defects.

Keywords: Fabric defect detection · Deep learning · CNN · LeNet · Inspection systems

1 Introduction

In the textile or yarn industry inspection of the quality of fabric plays a vital role, as any kind of defects on the surface of the fabrics can influence heavily on the garment qualities. In the world of automation, the need for making highly efficient, fast and reliable system is a must. The majority of fabric industries detect defects manually, which is inefficient and time consuming. The human eye can see only 60–70% of the defects when the system is in real time. There may be different distractions while testing the fabrics such as visual distraction (looking somewhere else), cognitive distraction (lost in thoughts), auditory distraction (interrupted by a co-worker) and fatigue. Thus, the human presence of mind is crucial as a distraction may result in a considerable loss. Also, the challenges have increased as per the industrial point of view. The industry should uphold the quality as well as quantity of production for retaining their reliability in the market. Human beings thus can't do this inspection task for a long time with the required accuracy. The automation of the inspection process is much needed by the

© Springer Nature Singapore Pte Ltd. 2020
B. Iyer et al. (eds.), *Applied Computer Vision and Image Processing*,
Advances in Intelligent Systems and Computing 1155,
https://doi.org/10.1007/978-981-15-4029-5_21

fabric manufacturing industries worldwide. This motivates us to work on this challenging problem as a research statement.

The digital image processing is applicable for the human interpretable pictorial information processing as well as storage and representation of data. Thus, computer vision as well as MATLAB, OpenCV and Keras play an important role and prove to be effective to solve the challenges thrown by the modern industrial environments. A part of the fabric that doesn't meet the requirements or an attribute of a fabric is defined as a defect which in turn results in customer dissatisfaction and can incur a huge loss for the textile industry. The different types of defects in a fabric are missing yarn, broken yarn, double yarn, stain, hole, etc. [1]. The normal colored fabric is indicated in Fig. 1a. The yarn may get entangled sometimes during manufacturing, which may result in holes, which is the most undesirable defect as shown in Fig. 1b. Also due to machine malfunctions, it can cause a defect like missing yarn as shown in Fig. 1c. Another major defect is stain marks as shown in Fig. 1d.

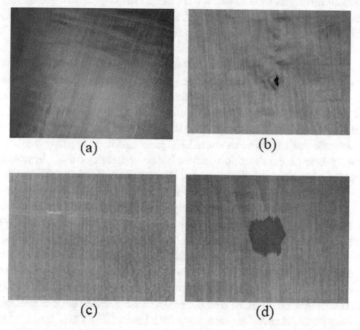

(a) (b)

(c) (d)

Fig. 1 Different classes **a** no defect **b** hole **c** missing yarn **d** stain

Deep learning is a state-of-the-art technology which can be used to detect the defects in the fabric as shown in Fig. 1 using the CNN algorithm. The system uses a high-resolution camera to capture the images and a stepper motor for conveyor arrangement. In addition, a proper illumination source is also important as less illumination may result in improper digital images. The proposed system uses a neural network to identify and classify the images. With the help of a neural network, the efficiency of the system can be increased. With the help of high-end GPU or CPU, better processing speed can be achieved, and computational time can be reduced.

This paper covers many sections as follows: Sect. 2 covers all the related works published in the last few years. The various techniques that are adopted are discussed in detail. In Sect. 3, system description of the proposed system, i.e., the various hardware modules used to drive the system are described. In Sect. 4, Convolution Neural Network is discussed followed by its implementation, Sect. 5 shows the results obtained with the method discussed and conclusive remarks are given in Sect. 6.

2 Related Works

A well-automated system enables lower manpower cost and shortest production time [2]. And, this problem attracts many researchers, that is why there are so many publications which designed a fabric defect detection system using different techniques. Wenbin et al. [3] have worked on detecting fabric defects using the Embedded Convolution Neural Network activation layer in their paper, using the fabric image autocorrelation for establishing the cloth motifs as well as using the same in the capacity of the eminent feature. They predicted high accuracy and proposed PPAL-CNN. Hanbay et al. [4] proposed the fabric defect detection system based on optics. The number of yarns as well as fibers acts as fundamental units. On the other side, potential fabric defects could be avoided by testing yarn and fibers before fabric development. Weninger et al. [5] also proposed a fault identification system for plain woven clothes with fully convolutional networks as well as yarn tracking. They have presented a method for cloth fault identification. Threads of fabric can be localized and tracked without any conditional settings. Later on, wave defects are detected. Previously, this problem has been solved with the help of neural network but that was restricted for gray fabric only. This paper proposes a smart colored fabric defect detection system based on deep learning, in which a camera has been placed above the moving fabric at the required distance with the illumination source. It will capture the image and feed it to the system for processing (testing). If the error occurs, the motor will immediately stop. Therefore, this system will be very helpful in small-scale as well as large-scale textile industries to reduce the labor cost and get better quality of fabric within the shortest period. Naleer and Senarathne [6] utilized an image analysis-oriented approach to classify the fabric images into ten distinct categories. The statistical variance as well as the image luminance variations were taken into consideration for efficient feature extraction. Hoang and Rebhi [7] explored the classical color spaces for analyzing the effect of LBP on the accuracy of fabric defect detection. They have achieved an accuracy of 92.1% in LUV color space when 10 class classifications were experimented.

3 System Description

Figure 2 shows the prototype of a fabric defect detection model. This system consists of an arrangement of lighting (illumination), a camera with proper adjustment, a rotor for conveyor and a computer with deep learning-related software. The illumination is an essential parameter while designing the system as less light may give inappropriate results. Therefore, a light source should be very intense and should cover the area of interest. While considering the lighting for illumination spectrum, geometry, size or area, the intensity of light is considered, also, the light should be uniform in the region of interest. A lens directs the light from a region of space to form an image onto the sensor of the camera. The lens defines the angle of view for the camera. It constrains and controls the amount of light reaching the camera. A high-resolution camera is used for this setup.

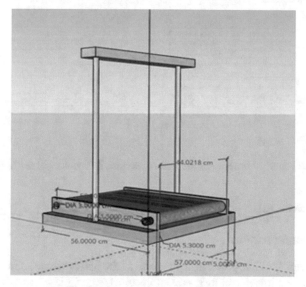

Fig. 2 Illustrative system prototype

In the proposed system, the rotor is rotating at the speed of 200 rpm which takes approximately 0.7615 s for reaching up to the next image. The speed of the rotor is adjusted using the PWM. With this system, the production of 1.5 m/min to process the images and the resolution of 0.0006 m/pixel can be obtained. Thus, at the cost of the resolution, the productivity can be increased.

The proposed system uses a convolution neural network for training and testing purposes. Training process flowchart is indicated in Fig. 3a. Dataset is needed for training the model, so raw images with defects are captured and augmented. The rows are divided into patches and are classified for making the dataset. Patches are resized into 512×512 sizes which is further scaled into [0, 1] range. The dataset is divided into two parts: 25% for testing after every epoch and 75% for training. Model is initialized in terms of batch size, epoch and initial learning rate. Testing process flowchart is indicated in Fig. 3b.

For the testing process, initially, a testing image is captured and divided into patches of size 512×512 pixels and passed to the trained model. The model gives output in terms of probability, whichever the maximum probability that has been concluded as a result (class).

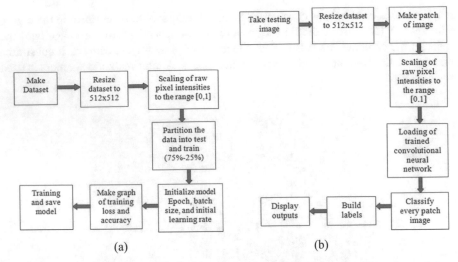

Fig. 3 **a** Flowchart of training of the neural network **b** flowchart of testing of a model

4 Implementation

4.1 System Configuration

Figure 4 illustrates the functional flow diagram of the hardware setup in the system. The camera acquires the image of fabric and subjects it to the CPU for inspection. The CPU has a pretrained model, through which the image is processed and gives the results. If the error arises, the system interprets it and maintains a log file for a certain amount of time. Some errors can be solvable under the given situation. So the drive can be stopped and the inspector can resolve the error. For this purpose, the motor driver arrangement is configured.

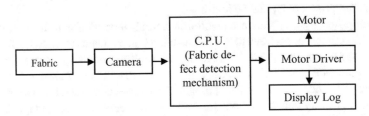

Fig. 4 Functional flow diagram

4.2 Tools

1. For computation, Intel Core i5 processor with a RAM of 8 GB and a 64-bit operating system is utilized.

2. Spyder environment: Spyder is an open-source cross-platform IDE for scientific data, written in Python language [8].
3. Google Colaboratory: Colaboratory provides Jupyter notebook environment which runs on the cloud [9].
4. OpenCV library: For real-time computer vision, OpenCV library is used. It is an open-source library [10].
5. TensorFlow: An open-source machine learning platform, developed by Google Brains [10].
6. Keras library: An open-source platform for deep learning. It is a TensorFlow API written in Python [10].
7. Arduino: An open-source prototyping platform for creating the interface between electronic objects [8].

5 Results

A dataset of 22000 images consisting of each class (hole, no error, stain and missing yarn) has been prepared. Further, the images are resized to 28 × 28 pixels. Subsequently, the data is partitioned into two categories namely training and testing, and the deep learning-based model is subjected to following initial configurations such as epoch = 30, batch size = 32 and initial learning rate = 1e-3. It requires 1000.85 s for training and 1.5910 s for testing an unknown image with an accuracy of 99.75%. The various published methods have utilized customized datasets and algorithms for fabric defect detection as indicated in Table 1.

Table 1 Published approaches with metrics and the proposed method

Sr. no.	Authors	No. of test images	#Classes	#Accuracy (%)
1	Naleer and Senarathne [6]	150	6	81.33
2	Hoang and Rebhi [7]	115	10	92.10
3	Proposed	6562	4	99.75

Figure 5 shows the graph of training loss and accuracy of the multi-label classification. It is observed that this classification has fewer values of losses, and training accuracy is very high. The system output is described subsequently.

The system with multi-label classification segregates among the different errors and defects in the dataset. Figure 6a illustrates the probability of 99.99% as missing yarn and Fig. 6b indicates a probability of 100% as a hole. The limitation of the system is, if more than one error occurs in a single image, then model may yield incorrect output. The basic layers of the LeNet model [11] are used for making this model.

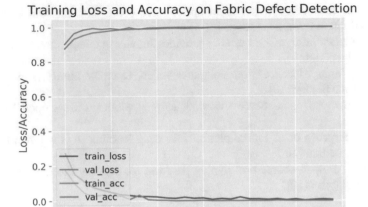

Fig. 5 Training accuracy and loss for multi-label classification

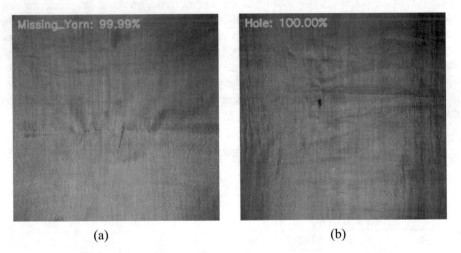

Fig. 6 Output images of multi-label classifiers **a** missing yarn, **b** hole

Output images:

```
[INFO] loading network...          [INFO] loading network...
Hole: 2.8416101e-08                Hole: 1.0
Missing_Yarn: 0.99994326           Missing_Yarn: 7.1914674e-10
No_defect: 1.798027e-14            No_defect: 9.885869e-09
Stain: 5.6775167e-05              Stain: 1.8016507e-11
----------------------            ----------------------
```

6 Conclusion

This paper presents a system which uses an algorithm that is capable of finding accurate defects in colored fabrics. The algorithm uses deep learning-based convolution neural networks to differentiate between different errors like stain, hole and missing yarn. Experimental results using different algorithms are described above. The trained model is accurate and provides reliable results.

7 Future Scope

In future, using this methodology, the throughput of the system can be improved by interfacing a frame grabber with the camera. In addition, an FPGA can be used to implement the system. Further, a GPU can be used for higher computational speed.

References

1. Patil, M., Verma, S., Wakode, J.: A review on fabric defect detection techniques. Int. Res. J. Eng. Technol. **4**(9), 131–136 (2017)
2. Kim, S.C., Kang, T.J.: Texture classification and segmentation using wavelet packet frame and gaussian mixture model. Pattern Recogn. **40**, 1207–1221 (2006)
3. Ouyang, W., et al.: Fabric defect detection using activation layer embedded convolutional neural network. IEEE Access **7**, 70130–70140 (2019)
4. Hanbay, K., Talu, M.F., Özgüven, Ö.F.: Fabric defect detection systems and methods—a systematic literature review. Optik **127**(24), 11960–11973 (2016)
5. Weninger, L., Kopaczka, M., Merhof, D.: Defect detection in plain weave fabrics by yarn tracking and fully convolutional networks. In: IEEE International Instrumentation and Measurement Technology Conference (I2MTC). IEEE (2018)
6. Naleer, H.M.M., Senarathne, D.N.: Fabric defect detection using variance and intensity analysis. In Proceedings of 8th International Symposium, Seusl (2018)
7. Hoang, V.T., Rebhi, A.: On comparing color spaces for fabric defect classification based on local binary patterns. In IEEE 3rd International Conference on Signal and Image Processing (ICSIP), pp. 297–300. IEEE (2018)
8. Desai, P.: Python Programming for Arduino, Packt Publishing Ltd, (2015)
9. Bisong, E.: Google colaboratory. In: Building Machine Learning and Deep Learning Models on Google Cloud Platform, 59–64. Apress, Berkeley, CA (2019)
10. Ketkar, N.: Deep learning with python. Apress (2017)
11. LeCun, Y., Bottou, L., Bengio, Y., Haffner, P.: Gradient-based learning applied to document recognition. Proc. IEEE **86**(11), 2278–2324 (1998)

Ship Image Classification Using Deep Learning Method

Pratik Patil[✉], Mohanasundaram Ranganathan, and Hemprasad Patil

Vellore Institute of Technology, Vellore, India
patilpratik699@gmail.com, mohanasundaramr@vit.ac.in,
hemprasadpatil@gmail.com

Abstract. Ship image classification in ocean background is of great significance for military and civilian domains which will further improve the automation in ship identification and naval domain perception. Deep stacked layers of neurons are extensively employed in recent years due to their ability to recognize the high-level features from an image in a hierarchal way. However, CNN lacks the capability of dealing with global rotation in an image of large size. This limits the accuracy of the CNN algorithm. Therefore, we have proposed a deep learning framework which uses a few layers of AlexNet for initial feature extraction and subsequently, KNN classifiers have been utilized for measuring the accuracy in classifying ships according to various categories.

Keywords: Ship classification · Transfer learning · CNN · VAIS dataset

1 Introduction

Ship classification has been the main importance in the security and safety of marine time environments. The enemy ship target tracking has got very high importance in defense-related operations. In fact, the maritime naval coastguards need this information for doing their duties. If the appearance of various ships from visible as well as infrared images is almost similar, then it becomes a very challenging task to decipher amongst the civilian ships which carry cargo and persons from the enemy ships which may contain ammunition. The resolution of the resultant ship image may be very low which complicated the task for its segregation between predefined classes. Hence this domain of ship classification has attracted many researchers and they have proposed the algorithms to classify the ship images.

The ship domain data has been collected from systems like a wide variety of sensors systems and they collect a large quantity of heterogeneous data. Hence, coping with this information and finding out insights from this data and effective recognition of ships from this is an attractive research domain. Henceforth, we need to develop a novel framework that can help to target recognition and identification which can be useful in various domains like security and commercial uses. The objective of this paper entails better ship identification accuracy upon prior information from metadata where raw data of images are labelled. The classical techniques for classification of images use features

© Springer Nature Singapore Pte Ltd. 2020
B. Iyer et al. (eds.), *Applied Computer Vision and Image Processing*,
Advances in Intelligent Systems and Computing 1155,
https://doi.org/10.1007/978-981-15-4029-5_22

which are handcrafted for specific tasks. Various natural factors such as weather and lightning conditions make it difficult to recognize ships. Also, wide inter-class variation makes it difficult to classify the ships. Multilinear principal component analysis (MPCA) [1] and multiple feature learning (MFL) [2]. Some methods have proposed algorithms which fuse low-level features extracted from Gabor filter and combine it with high-level features from CNN networks. In ship classification, the inter-class similarities are very small, hence in this the CNN layers may leave some important features as it does not capture low-level features. Thus, to compensate for this, the last few layers of the deep neural network are removed and features are extracted.

2 Literature Survey

Leclerc et al. [3] proposed an approach which focuses on already trained CNN networks which use the Inception and ResNet architectures for ship classification. In this approach rather than training CNN on random parameters, they finetuned already trained CNN to perform the classification of naval vessel images. They used Maritime Vessels dataset and they achieved significant improvements in accuracy on previous implementations of Maritime Vessels (MARVEL) dataset.

Wang et al. [4] used methodologies such as defogging the images containing cloud interference as well as SC-R-CNN also called scene classification network (SCN) to classify fog-containing images.

Yao et al. [5] researched and compared various ship recognition algorithms based on various CNNs. In this work, they specially created an image dataset of vessels and performed a variety of image preprocessing tests and recognition algorithms for particular vessel scenes.

Wang et al. [6] used the Italian COSMO-SkyMed SAR images for classifying cargo and non-cargo ships. It uses convolutional neural networks based on Google's TensorFlow environment. They do not classify subcategories of non-cargo ships due to variations in radar illumination directions and ship poses which require more data. To follow international naval traffic regulations, a vessel should have machine vision systems which recognize vessels throughout day and night. The authors in [6] addressed this challenge and they have experimented with the VAIS image set. For this, they used two algorithms Gnostic fields and Deep CNN.

Khellal et al. [7] used ELM (Extreme Machine Learning)-based approach to train the CNN system on discriminative features and to classify ships using those features.

Shi et al. [8] recognized ship images by exploiting both lower-higher hierarchical levelled features. They combined all the available feature-vectors acquired with deep convolutional neural network and this fusion of features is further fed to Support Vector Machine (SVM) classifier. For extraction of low-level features, it used Gabor filters and for high-level features it employed deep CNN.

Huang et al. [2] utilized Gabor multidirectional features, Fisher features and MS-CLBP-based features, BOW and pyramidal multilevel matching for ship image classification purposes. They have employed an SVM classifier for segregating the ship features into predefined classes.

Zhang et al. [9] have presented an algorithm for ship image segregation using a structural difference-based fusion approach. By employing a linear regression, the mapping between features has been accomplished for efficient feature vector formation.

3 Materials and Methods

3.1 CNN-Based Classification

CNNs are mostly utilized for image classification. CNN models are well known for the extraction of features from the images without any human intervention. Various layers like convolutional, Max pool and Flattening are eminent layers of any CNN. These types of networks are used with transfer learning for various image classification and recognition problems. The combination of transfer learning with CNN mostly offered better performances than the conventional approaches.

3.2 Transfer Learning

How is that in a real world when we perform some action and gain knowledge from it and then we apply that knowledge in finding a solution for some other problem? For example, knowledge acquired during learning to recognize cars can be applied to some extent to recognize trucks. Transfer learning is similar to a machine learning paradigm where a pretrained model that has been used for solving one problem is reused to address another problem as a starting point and finetuning it along the way according to the new problem. Usage and approach of Transfer Learning depend on both size of the new dataset and the similarity of the new dataset to the original dataset.

3.3 AlexNet

AlexNet is a neural network architecture which was designed in 2012 [10]. AlexNet is faster than older architectures while retaining a similar accuracy. AlexNet uses the rectified linear unit as an activation function instead of Tanh function. In this network, the overfitting challenge is dealt with using dropout instead of regularization but with doubling the training time and with a dropout rate of 0.5. The size of the network is reduced by overlap pooling which further reduces the error rates. The rectified linear unit is employed before the first and second FC as a part of this hierarchy. 227 * 227 is the input image size used in this architecture.

Our proposed solution uses 25-layer AlexNet architecture along with KNN as a classifier. In our architecture, we removed 5 layers from the bottom of the architecture and added the KNN classifier.

3.4 KNN Classifier

It is the most widely used classifier in the industry and mostly used for classification problems. This algorithm considers that similar things exist in close proximity, i.e. similar things are near each other.

4 Proposed Method

The proposed method based on transfer learning utilizes the pretrained model named AlexNet architecture with properly initialized weights. The features are then fed into the classifier which then predicts the suitable labels for the ships. Existing methods are used such as inception and ResNet which have lower accuracy than AlexNet hence the transfer learning combined with the features extracted from AlexNet network can be utilized to increase the accuracy. The block diagram for AlexNet-based proposed system is as depicted in Fig. 1.

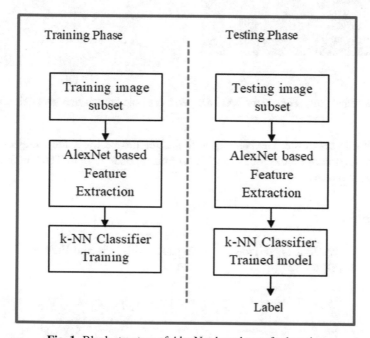

Fig. 1 Block structure of AlexNet-based transfer learning

The VAIS dataset also known as Visible and Infrared Ships is a well-known dataset for ship classification based research. VAIS dataset contains the images of maritime ships. It contains mainly 6 categories of images [11]. In this work, we will be working with the visible images.

VAIS dataset contains a total of 2865 images with a total of 1623 visible images and total of 1242 infrared images. The sample images are indicated in Fig. 2a.

The proposed framework works on the VAIS dataset which consists of total of 873 Training images and 750 Testing images.

The machine used for this is 9th Generation Intel Core i7-8750H, NVIDIA GeForce GTX 2050 Ti; 4 GB DDR5, 8/16/32 GB DDR4. The AlexNet used here consists of 25 layers out of which the layer named 'fc7' has been utilized to extract the eminent features. 4096*1 features are extracted per image. The input image of size $227 \times 227 \times 3$ is subjected to the network. All the layers are used for extracting the features except

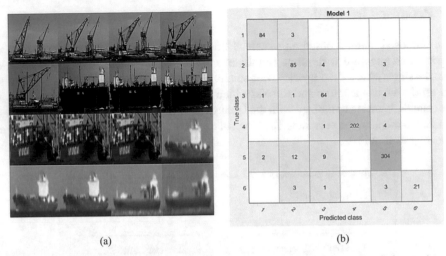

(a)   (b)

Fig. 2 a Sample images from the VAIS dataset, **b** confusion matrix and no. of observations

the last few layers of the AlexNet. Instead, this feature vector is then subjected to the k-Nearest Neighbour for classification. The confusion matrix, TPR, TNR, PPV and FDR are presented below.

5 Results

The True Positive Rates and True Negative Rates with PPR and FDR are indicated in Fig. 3a and b, respectively.

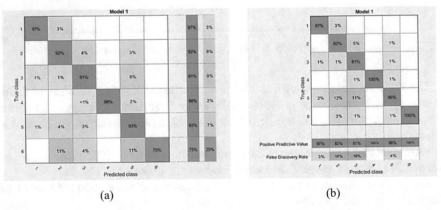

(a)   (b)

Fig. 3 a True positive and true negative rates, **b** positive predicate and false discovery rates

The AUC-ROC curve is the preformation measure for classification problems. Higher the AUC the better is the model performance. For this proposed deep learning-based

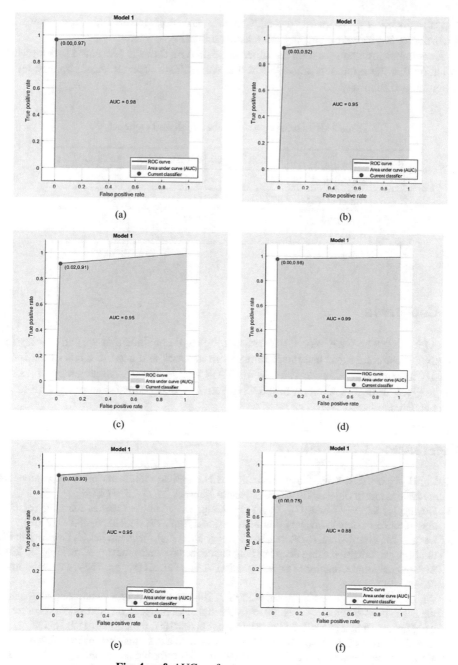

Fig. 4 a–f: AUC performance measure curves

ship classification experiment, the AUC curves are indicated in Fig. 4a–f where the performance is plotted by varying the class numbers.

Benchmarking analysis with the other published methods is presented in Table 1. It indicated that the approach presented in this work offers superior throughput in terms of percentage accuracy.

Table 1 Benchmarking with other published approaches

Sr. no.	Ref.	# Train images	# Test images	Accuracy (%)
1	Huang et al. [2]	873	907	87.60
2	Shi et al. [8]	873	907	88.00
3	Proposed method	873	907	93.70

6 Conclusion

Through this paper, we explored the methodology and conclude that we can employ the AlexNet architecture and use the features from network and use the KNN classifier to classify the various classes of ships in the VAIS visible dataset. An accuracy of 93.7% can be given from this AlexNet-based CNN network using transfer learning with KNN as a classifier.

References

1. Lu, H., Plataniotis, K.N., Venetsanopoulos, A.N.: MPCA: multilinear principal component analysis of tensor objects. IEEE Trans. Neural Network **19**(1), 1839 (2008)
2. Huang, L., Li, W., Chen, C., Zhang, F., Lang, H.: Multiple features learning for ship classification in optical imagery. Multimedia Tools Appl. **77**(11), 13363–13389 (2018)
3. Leclerc, M., Tharmarasa, R., Florea, M.C., Boury-Brisset, A., Kirubarajan, T., Duclos-Hindié, N.: Ship classification using deep learning techniques for maritime target tracking. In: 2018 21st International Conference on Information Fusion (FUSION), pp. 737–744. Cambridge, (2018)
4. Wang, R., You, Y., Zhang, Y., Zhou, W., Liu, J. Ship detection in foggy remote sensing image via scene classification R-CNN. 2018 International Conference on Network Infrastructure and Digital Content (IC-NIDC) (2018). https://doi.org/10.1109/icnidc.2018.8525532
5. Yao, B., Yang, J., Ren, Y., Zhang, Q. Guo, Z.: Research and comparison of ship classification algorithms based on variant CNNs. In: 2019 IEEE 3rd Information Technology, Networking, Electronic and Automation Control Conference (ITNEC), pp. 918–922. Chengdu, China (2019)
6. Wang, C., Zhang, H., Wu. F., Zhang, B., Tian, S.: Ship classification with deep learning using COSMO-SkyMed SAR data. In: IEEE International Geoscience and Remote Sensing Symposium (IGARSS), Fort Worth, TX, 2017, pp. 558–561 (2017)

7. Khellal, A., Ma, H., Fei, Q.: Convolutional neural network based on extreme learning machine for maritime ships recognition in infrared images. Sensors **18**(5), 1490 (2018). https://doi.org/10.3390/s18051490
8. Shi, Q., Li, W., Zhang, F., Hu, W., Sun, X., Gao, L.: Deep CNN with multi-scale rotation invariance features for ship classification. IEEE Access 1–1 (2018). https://doi.org/10.1109/access.2018.2853620
9. Zhang, E., Wang, K., lin, G.: Classification of marine vessels with multi-feature structure fusion. App. Sci. (2019). https://doi.org/10.3390/app9102153
10. Krizhevsky, A., Sutskever, I., Hinton, G.E.: Imagenet classification with deep convolutional neural networks. In: Advances in Neural Information Processing Systems, pp. 1097–1105 (2012)
11. Zhang, M.M., Choi, J., Daniilidis, K., Wolf, M.T., Kanan, VAIS, C.: A dataset for recognizing maritime imagery in the visible and infrared spectrums. In: Proceedings of the 11th IEEE Workshop on Perception Beyond the Visible Spectrum (2015)

A Selection Method for Computing the Ensemble Size of Base Classifier in Multiple Classifier System

Vikas Tomer[1], Simon Caton[2], Santosh Kumar[1(✉)], and Bhawnesh Kumar[1]

[1] Graphic Era Deemed to Be University, Dehradun, India
vikastomercse@geu.ac.in, amu.santosh@gmail.com
[2] National College of Ireland, Dublin, Ireland
simon.caton@ncirl.ie

Abstract. As a discipline, Machine Learning has been adopted and leveraged widely by researchers from several domains. There is a huge range of classifiers already available in machine learning and it has kept on growing with the advancement of this field. However, it is very hard to pick the best classifier among the several similar classifiers suitable for any problem. Recent advancement in this field for solving this issue is the Multiple Classifier System (MCS). It comes under the umbrella of ensemble learning and gives comparatively a better and definite result than a single classifier. MCS has two layers—(i) Base layer—contains a number of ML Classifiers appropriate for any specific task—and (ii) Meta Learner Layer—which aggregates the results from base layer classifiers by using techniques, such as Voting and Stacking. However, the job of selecting the appropriate classifiers from various classifiers or from a family of classifiers for a specific classification or prediction task on any dataset is still unraveling. This work emphasizes determining the characteristics of the selection method of base classifiers in the MCS. Moreover, which Meta Learner layer from Stacking and Voting aggregates the better result according to the different sizes of the base classifiers?

Keywords: Multiple Classifier System · Base classifier · Classifier's ensemble · Stacked generalization

1 Introduction

In the specialization of machine learning, sheer advancement was seen in the last two decades. To resolve the complex tasks of prediction, several machine learning algorithms have been proposed. Classification, Clustering, Regression with the numeric prediction are some of the complicated tasks of machine learning. To make the prediction task easier, the researchers have been discovering a huge list of novel ML algorithms into the family of the classifiers. As every family of these classifiers has their own capability for doing any specific classification task, several more classifiers are continuing to be accumulated to these classifiers list. Several areas have been adapting these novel classifiers daily for solving various issues from this huge list of existing classifiers. These new areas have

© Springer Nature Singapore Pte Ltd. 2020
B. Iyer et al. (eds.), *Applied Computer Vision and Image Processing*,
Advances in Intelligent Systems and Computing 1155,
https://doi.org/10.1007/978-981-15-4029-5_23

been utilizing rapidly growing machine learning applications, such as sentiment analysis, pattern recognition, text-based analysis, remote sensing of images, Email spam filtering, credit card fraud detection, and so on. Overall, machine learning can be categorized into two categories, i.c., supervised and unsupervised [1].

Multiple Classifier System is defined as multiple classifiers and is applied to complete any classification task in the base layer of the MCS and then the result of these multiple classifiers are aggregated through meta learner layer in the MCS. Thus, the entire procedure is completed through two layers: base and meta and provide better accuracy as compared to the independent algorithms [2, 3]. The selection criteria of the base classifiers in the base layer is the nucleus point of the MCS which actually delivers a better result than any other classifier [4]. In meta learner layer, stacking and voting are the algorithms applied in this work. However, it has been analyzed in related works that better outcome of prediction is dependent on the classifier selected in the base layer [5]. The selection of base classifiers along with the meta layer provides a better result for the prediction. Although its selection in the base layer is an enormous act and requires a long duration to solve this entire puzzle, this complex situation formulates the problem statement of this work—what are the criteria that simplify the selection method of the classifiers in the base layer of the Multiple Classifier System? This problem statement comprises some key questions that should be elaborated before unraveling the problem statement. These questions are: what is the base layer and base classifiers? Why is an appropriate selection of these base classifiers so significant?

First, it is necessary to investigate why a single classifier is insufficient? There might be several reasons for this investigation. For the entire dataset investigation, the selection of individual classifiers is cumbersome and a time-taking process and also it fails to provide the highest accuracy [6].

The second element is to consider diversity in the selection of classifiers. Now the question arises that how one can derive a disparity between these classifiers? Moreover, the disparity is not a single issue within an individual group of classifiers, but it is also a problem among the dissimilar groups of classifiers. So, the issue raises whether the disparity between the classifiers should exist within an individual group of classifiers or within the groups of classifiers?

2 Related Work

It is observed in the literature that many authors describe the benefits of MCS over individual classifiers and recommend it to use in data mining, as it provides higher accuracy in comparison to the individual classifier. Automatic detection of construction in the images promotes the use of MCS [7], whereas [8] exploring the credit score through base classifiers requires a comparison of various machine learning algorithms in the ensemble classifier. A novel microarray-based predicting model presented by [9] provides better results compared to the individual optimized model. Machine learning has been revolutionized by using the MCS in the domain of remote sensing of images (RSIs) by several novel approaches. A Hybrid MCS [10] has been introduced for sensing the medium-high remote images. This system is a fusion of the largest confidence algorithm, an ensemble method-Bagging, and an optimal set of sub-classifiers. The obtained results from this system show an improved accuracy [11].

RSIs prediction can be achieved by supervised and unsupervised learning [12]. Particularly, if laser segmentation fails to operate correctly, this domain studies the procedures for better optimization and recovery. Stack generalization is one more advanced development over the other ensemble methods. It resolved the various statistical issues in data science [13]. Presents three learning algorithms to implement stacking for providing the solution to the higher level learning issues [14]. Provides an introduction of the meta learner layer (MLL), and later on improved the quality of stacking. Stacking with MLL deals with the high dimensionality of the data for the higher number of classes.

More recently, the ensemble method has become more attractive by inculcating the new approaches [15]. Proposed a novel approach of Ant Colony Optimization (ACO) based on stacking ensemble configuration searching based on a pool of base classifiers. Stacked generalization of the ensemble method is one of the best methods and has better concordance between the layers of the MCS [16]. Proposed the procedure to lessen the size or number of base classifiers in the MCS. This procedure is called as one-vs-one using the Undirected Cyclic Graph. The outcome represents the cons and pros of using ML algorithms, such as Support Vector Machine, Ripper, and C 4.5 [17]. Raised an interesting adverse problem of classification, in which an individual classifier needs to tackle the intelligent adversary "who adaptively modifies the pattern to evade the system". One more noteworthy intention is deciding the benchmarks of picking features from the entire dataset as indicated by [18], which gave the automatic feature selection model to resolve this complex situation.

Overall, all the above observations may support to formulate the hypotheses for this work:

- Opting an individual classifier for bringing out the finest result for a multidimensional dataset is nearly infeasible [19].
- A fusion of base classifiers and a Meta learner classifier is more effective [19].
- The MCS does good as the final layer is Meta-Learner [14].

3   Methodology

This work contains some versatile datasets for the final experimentation. The requirement for selection of any dataset is that each one of the datasets needs to be different from the other. This work consist of three different datasets, each belongs to a different diversity in the context of their attributes and variables, such as Nominal and Numerical. The following is a brief detail of all datasets used:

i.   IRIS dataset contains the task of multiclass classification, because it has three classes for prediction. This dataset is extracted from the life science (Botany) domain. So, how can MCS apply to tackle some definite issues to classify this dataset?

ii. StressEcho dataset contains the task of binary classification, because it has only two classes for prediction. This dataset is extracted from the health science (Stress Echocardiography) domain. Through the evaluation of results from this dataset, we are provided with a lot of stringent information about the operational MCS.

iii. Credit Dataset again contains the task of binary classification, because it has only two classes for prediction. The data is extracted from a bank detail for loan aspirants. This dataset is a bit tricky, as it contains a blend of attributes, like continuous, and nominal and missing values.

Mathematical induction of this system is presented below.

Let $D \in d^n$ be multidimensional datasets, which contain multiple datasets d from several domains (in this work, three are used).

$$D(i) = \begin{bmatrix} F_{1,1}(di) \cdots F_{1,j}(di) \cdots F_{1,m}(di) \\ \vdots \qquad \vdots \qquad \vdots \\ F_{i,1}(di) \cdots F_{i,j}(di) \cdots F_{i,m}(di) \\ \vdots \qquad \vdots \qquad \vdots \\ F_{l,1}(di) \quad \cdots F_{l,j}(di) \cdots \quad F_{l,m}(di) \end{bmatrix}$$

$F_{l,1}(d1)$ is the lst feature in ith dataset.

Let $T = \{t_1, t_2, \ldots, t_m\}$ be the set of target class labels; and let $C = \{C_1, C_2, \ldots, C_l\}$ be a set of base classifiers in the base layer of MCS. Given the input pattern X in the dataset d_i, the output of the i th classifier is denoted as in

$$d_i(X) = \sum_{i=1}^{n} c_i(X), where i = 1, 2, \ldots, n. \tag{1}$$

represents the measure value of the possibility that classifier C_i considers that X belongs to class w_j.

The results from n classifiers at the base layer need to be aggregated, so that one consolidated can be determined. The final output of n classifiers from the base classifier is constructed in the meta learner layer as follows:

$$W_i(X) = \sum_{j=1}^{n} v_{ij} f_j(X), \max(b(X)) where i = 1, 2, \ldots, n. \tag{2}$$

where W_i is the final class which comes with the majority $(v_{ij} f_j(X))$ from all base classifiers $b(X)$.

4 Evaluation

This section contains a profound detail of each experimentation of this work in a spreadsheet. Each instance of this experimentation contains the size of the classifier, matrices of all available classifiers, meta learner layers, and their performance measures.

First, sparse matrices are prepared for defining which classifiers from 53 classifiers is or are used in each instance. The performance measurement has carried out as; Kappa statistic, Correctly and Incorrectly Classified Instances (Accuracy), Relative absolute error, Mean absolute error, Root relative squared error, and Root mean squared error. Throughout the experiments, parameters which were used in the models are tuned according to the subset of data assigned to them [20] and the compatibility of the base layer of the MCS to the meta learner layer is checked [2, 16]. Finally, According to the hypotheses, which was considered in prior, we check whether these are statistically significant or not [21, 22].

Figures 1, 2, and 3 represent the performance of base classifiers of the different sizes in terms of accuracy and kappa statistics. Figure 1 depicts that the accuracy is very much aligned with kappa statistics. Highest accuracy is observed as 95.45% with 0.9696 value of kappa statistics for base classifier size 1. Figure 2 again depicts similar information as given above for base classifier size 1, however, accuracy at the lowest point is aligned with even worst kappa statistics. Lowest accuracy is observed as 65.65% with 0.48 value of kappa statistics for base classifier size 2. Analogous information comes out from Fig. 3 for base classifier size 3.

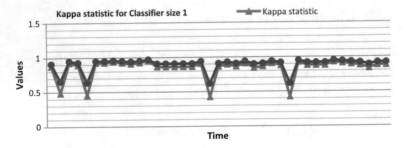

Fig. 1 Kappa statistic and accuracy for base classifier size 1

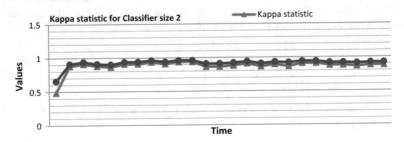

Fig. 2 Kappa statistic and accuracy for base classifier size 2

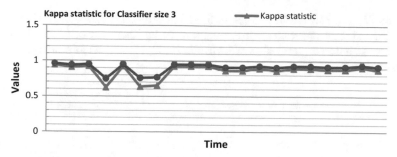

Fig. 3 Kappa statistic and accuracy for base classifier size 3

The overall result is summarized in Tables 1 and 2, which suggest that accuracy is aligned with kappa statistics for each and every size of base classifiers. It corroborates the fact that higher accuracy is not achieved by chance, as the kappa is on the peak when the accuracy is on its highest. Accuracy is highest with the size 3 of base classifier among all the sets of experimentation. Kappa statistics also corroborate the efficacy of accuracy in every instance.

Table 1 Instances when MCS perform best with different sizes/numbers of classifiers

Classifier sizes/numbers	RMSE	RAE	RRSE	Kappa	Accuracy
1	15.25	8.7	32.33	**0.9545**	**0.9696**
2	15.77	9.0023	33.44	**0.9393**	**0.9595**
3	16.22	10.92	34.38	**0.9242**	**0.9494**

Table 2 Instances when MCS perform worst with different sizes/numbers of classifiers

Classifier sizes/numbers	RMSE	RAE	RRSE	Kappa	Accuracy
1	38.85	69.87	82.36	0.4237	0.6161
2	28.28	56.08	59.95	0.4846	0.6565
3	35.88	62.088	76.07	**0.6235**	**0.7474**

Moreover, the extensive experimentation on the proposed MCS framework evaluates the conventional errors associated with different sizes of base classifiers. Figures 4, 5, and 6 depict the information about the conventional error losses in terms of Root Mean Squared Error, Root Absolute Error, and Root Relative Squared Error. Unlike, the first visualization on the performance of accuracy and kappa statistics, the errors prone to these base classifiers highly fluctuate. RRSE is the most susceptible among all the conventional errors for each of the base classifier sizes, as it depicts the value

82.36, 59.95, and 76.07% from Figs. 4, 5, and 6, respectively, whereas RMSE is most consistent and least prone to error throughout the instances for different sizes of base classifiers. As it depicts, the value of RMSE is 16.02, 15, and 10.92 from Figs. 4, 5, and 6, respectively.

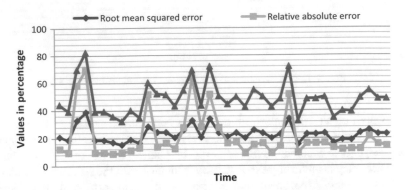

Fig. 4 RMSE, RAE, and RRSE for base classifier size 1

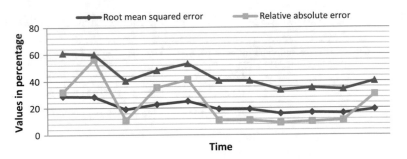

Fig. 5 RMSE, RAE, and RRSE for base classifier size 2

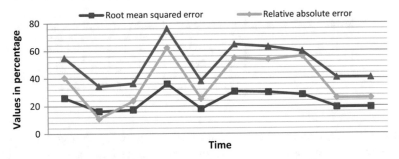

Fig. 6 RMSE, RAE, and RRSE for base classifier size 3

5 Conclusion

The pace of generating the data has been beating the earlier record at every second in almost every area. Hence it creates the pressure on data scientists to come up with the appropriate classifiers to help in the data science domain. MCS is one of the recent innovations in this direction with a few limitations. This work contributes to overcoming the insufficiency of picking the appropriate classifiers in the base layer of the MCS. This work effectively suggests the size or number of classifiers in the base layer of the MCS for any particular dataset. The procedure of picking up the appropriate size or number of base classifiers is finalized by conducting intensive experimentation.

This work also determines that Vote and Stack are the most compatible meta learner layer with any size or number of base classifiers. A number of intricate factors associated with this work, such as statistical learning are applied to choose the features at the beginning for the interpretation of results. Finally, this proposed framework advocates that the higher accuracy along with greater kappa value can be achieved by increasing the size or the number of base classifiers. Accuracy is not dependent on whether the classifiers belong to the same classifier's family or belong to different families. The intensive experimentation carried out for this work can be used on a larger scale by including several datasets from various dimensions in the future.

References

1. Ayodele, T.O.: Types of machine learning algorithms. New Adv. Mach. Learn. 19–49 (2010)
2. Arruti, A., Mendialdua, I., Sierra, B., Lazkano, E., Jauregi, E.: Expert systems with applications one method: NOV @. **41**, 6251–6260 (2014)
3. Cavalin, P.R., Sabourin, R., Suen, C.Y.: Dynamic selection approaches for multiple classiffier systems. Neural Comput. Appl. **22**(3–4), 673–688 (2013)
4. Tulyakov, S., Jaeger, S., Govindaraju, V., Doermann, D.: Review of classifier combination methods. Studies in Computational Intelligence 90 (Figure 1), 361–386 (2008)
5. Fernandez-Delgado, M., Cernadas, E., Barro, S., Amorim, D.: Do we need hundreds of classifiersto solve real world classification problems? J. Mach. Learn. Res. **15**, 3133–3181 (2014)
6. Valentini, G., Masulli, F.: Ensembles of learning machines, WIRN VIETRI 2002. In: Proceedings of the 13th Italian Workshop on Neural Nets-Revised Papers., pp. 3–22. ACM, London U.K (2002)
7. Son, H., Kim, C., Hwang, N., Kim, C., Kang, Y.: Classification of major construction materials in construction environments using ensemble classifiers. Adv. Eng. Inform. **28**(1), 1–10 (2014)
8. Marques, A.I., Garcia, V., Sanchez, J.S.: Exploring the behaviour of base classifiers in credit scoring ensembles. Expert Syst. Appl. **39**(11), 10244–10250 (2012)
9. Chen, M., Shi, L., Kelly, R., Perkins, R., Fang, H, Tong, W.: Selecting a single model or combining multiple models for microarray-based classifier development?-a comparative analysis based on largeand diverse datasets generated from the MAQC-II project. BMC Bioinformatics, vol. 12, (Suppl 10), p. S3 (2011)
10. Haibo, Yang., Hongling, Z., Zongmin, W.: Remote sensing classification based on hybrid multi-classifier combination algorithm. In: ICALIP 2010—2010 International Conference on Audio, Language and Image Processing, Proceedings (X), 1688–1692 (2010)
11. Yang, B., Cao, C., Xing, Y., Li, X.: Automatic classification of remote sensing images using multiple classifier systems. Math. Probl. Eng (2015)

12. Mukhopadhyay, A., Maulik, U., Bandyopadhyay, S., Coello, C.A.C.: A survey of multi objective evolutionary algorithms for data mining: Part i. IEEE Transac. Evol. Comput. **18**(1), 4–19 (2014)
13. Pari, R., Sandhya, M., Sankar, S.: A multi-tier stacked ensemble algorithm to reduce the regret of incremental learning for streaming data. IEEE Access **6**(8452944), 48726–48739 (2018)
14. Mohammed, M., Mwambi, H., Omolo, B., Elbashir, M.K.: Using stacking ensemble for microarray-based cancer classification. In: 2018 International Conference on Computer, Control, 12–14 (Aug. 2018)
15. Chen, Y., Wong, M.L., Li, H.: Applying ant colony optimization to configuring stacking ensembles for data mining. Expert Sys. Appl. **41**(6): 2688–2702 (2014). http://dx.doi.org/10.1016/j.eswa.2013.10.063
16. Beitia, M.: Contributions on Distance-Based algorithms, Multi Classifier Construction and Pairwise Classification, (April 2015)
17. Biggio, B., Fumera, G., Roli, F.: Multiple classifier systems for adversarial classification tasks. In: Multiple Classifier Systems, vol. 5519, pp. 132–141, (2009). http://dx.doi.org/10.1007/978-3-642-02326-2 14
18. Basu, T.: Effective text classification by a supervised feature selection approach. (2012)
19. Mendialdua, I., Arruti, A., Jauregi, E., Lazkano, E., Sierra, B.: Neuro computing classifier subset selection to construct multi-classifiers by means of estimation of distribution algorithms. **157**, 46–60 (2015)
20. Domingos, P.: A few useful things to know about machine learning. Commun. ACM **55**(10), 78 (2012)
21. Demsar, Janez: Statistical comparisons of classifiers over multiple data sets. J. Mach. Learn. Res. **7**, 1–30 (2006)
22. Ko, A.H.R., Sabourin, R.: Single classifier-based multiple classification Scheme for weak classifiers: An experimental comparison. Expert Syst. Appl. **40**(9): 3606–3622 (2013)

Segmentation of Knee Bone Using MRI

Anita Thengade$^{(\boxtimes)}$ and A. M. Rajurkar

MGM's College of Engineering, Nanded, India
thengadeanita@gmail.com, rajurkar_am@mgmcen.ac.in

Abstract. Segmentation of knee joint is an instrumental task in a medical fraternity during knee osteoarthritis (KOA) progression. In general, it is a primary concern to segment, detect and extract the defects from knee magnetic resonance images (MRI). Conventionally, this process is carried out manually in clinical practice but it is time consuming and observer dependent. It is a big challenge to segment cartilage manually from MRI, as cartilage structure has inadequate image contrast and complex tissue structure. Currently, semiautomatic and automatic methods are used to overcome this limitation. In the proposed method, we present a segmentation approach for the extraction of bone and cartilage from MRI. Here, in this paper we perform preprocessing for image enhancement and noise removal of knee MR image using Gaussian blur and block matching 3D method. Larger bones in the knee joint are segmented first to perform reliable cartilage segmentation. Hence distance regularization level set evolution (DRLSE) is used for bone extraction successively followed by cartilage segmentation. Further, the performance of the proposed method is analyzed on a variety of knee MR images and experimental results demonstrated the improvement in accuracy compared to existing approaches. In conclusion, it is observed that the proposed technique improves significant performance with consistency and robustness during the segmentation process.

Keywords: Magnetic resonance imaging · Bone segmentation

1 Introduction

Anatomical structure can be deteriorated slowly and progressively over time in the human body. Musculoskeletal part as knee is complex joint and frequently injured component of the human body. Common causes of knee pain are osteoarthritis and rheumatoid arthritis. Progressively degradation of articular cartilage in the joint is called osteoarthritis (OA) disease. OA may be found mostly in the synovial joint of the human body. Commonly affected joints of human beings are lower back, neck, hand, hip, knee, foot and knee joint in case of OA disease. According to the World Health Organization (WHO), knee osteoarthritis (KOA) is an extremely prevalent joint disease with a high prevalence rate in men as well as women aged over 60 years [1–3]. According to statistics presented on the Indian population [1], nearly about 12 million people suffer from KOA disease that needs to be treated. As per the report released by the Australian Institute of Health and Welfare (AIHW) government organization, one in ten people are affected by OA.

© Springer Nature Singapore Pte Ltd. 2020
B. Iyer et al. (eds.), *Applied Computer Vision and Image Processing*,
Advances in Intelligent Systems and Computing 1155,
https://doi.org/10.1007/978-981-15-4029-5_24

Monitoring the progression of KOA requires the morphology assessment of the knee by measuring structures such as bone interface, cartilage thickness and cartilage volume [4]. Bone and cartilage morphology can be delineated manually by a radiologist or expert in clinical practice in the assessment of KOA. Manual delineation could take approximately 4 h to analyze a single scan making the diagnosis excessively time consuming and laborious. Well-known diagnostic imaging methods suggested by physicians are Computed Tomography (CT) scan and MR images. These medical images are used to provide valuable information about size, shape and location. Nevertheless, in the early stages of KOA, structural changes of bone such as shape and size can be visualized well by the MR imaging technique.

Due to the increasing usability of MR imaging in clinical routine, accurate segmentation of bone and cartilage from MRI is required for quantitative assessment of KOA and also to detect the problem in the early stage. Before applying any treatment, it is essential to segment bone and measure cartilage thickness, volume, area and surface roughness. In current clinical practice, a radiologist or expert uses his knowledge to perform manual segmentation slice-by-slice and could take 3–4 h for a single scan. In addition to this, it has consistency issues, is observer dependent and excessively time consuming [5, 6].

To address the limitations of manual delineation, it is advantageous to automate the segmentation of the cartilage. Hence numerous segmentation models have been introduced by research communities in recent years. Researchers have developed semiautomated and automated methods to segment articular cartilage. The main challenges in developing these methods are the thin structure of cartilage and low contrast between cartilage and surrounding tissues. In a semiautomatic method using MR images, it needs user intervention in the form of seed points or scribbles to limit the object boundaries. Semiautomatic segmentation of cartilage could be achieved using various algorithms such as graph cut, random walk and active contour. Even though the semiautomatic method seems to be the best method to provide efficient measurement and performance, they require user intervention, which may result in reliability issues. Several researchers are focusing on developing a fully automatic segmentation method. The main focus is to provide a detailed literature survey of existing methods and also perform bone segmentation followed by cartilage extraction using MR images.

The framework of the paper is given as follows: Sect. 2 summarizes existing approaches for bone and cartilage extraction. Section 3 discusses challenges in semiautomatic and fully automatic segmentation. Then Sect. 4 presents segmentation frameworks for bone and cartilage extraction. Section 5 describes experimental results and the paper closes with the conclusion.

2 Literature Survey

The most crucial step in KOA assessment is the segmentation of the knee joint. Over the last few decades, image segmentation has been adopted by researchers to give a quantitative measure of KOA such as cartilage thickness, volume or bone deformation. Tables 1 and 2 summarize existing segmentation techniques in two categories as semiautomatic and automatic, respectively. These techniques were developed in recent decades

Table 1 Existing bone/cartilage semiautomatic segmentation techniques

Segmentation method	Author/year	Region of interest	Performance measure
Thresholding	Dalvi et al. [4]	Femur, tibia and patella	Sens. (patella)—92.69 Sens. (tibia)—96.95 Sens. (femur)—97.05 Spec. (patella)—94.82 Spec. (tibia)—98.33 Spec. (femur)—98.79
	Jiann-Shu Lee	Femur and patella	Not specified
Region growing	Peterfy et al. [6]	Cartilage	Reproducibility—3.6–6.4%
	Piplani et al. [21]	Cartilage	Percentage of error—6.53%
Deformable model	Zohara Cohen et al. [7]	Cartilage	Root mean square Cartilage extraction—0.22 mm Cartilage thickness—0.31 mm
Graph-based	Hackjoon Shim et al. [8]	Cartilage	DSC (94.3 $\pm$ 1.3)
Clustering	Hong-Seng Gan et al. [9]	Cartilage	DSC—0.83, sensitivity—0.84 and specificity—0.99
	Hong-Seng Gan et al. [10]	Cartilage	Reproducibility—0.92 $\pm$ 0.051, sensitivity—0.96 $\pm$ 0.051 specificity—0.997 $\pm$ 0.0019
	Mutha et al. [20]	Cartilage	Not specified
Random walk	Hong-Seng Gan et al. [11]	Cartilage	DSC (tibial cartilage)—0.80 DSC (patella cartilage)—0.77

to overcome the difficulties of manual segmentation. Human intervention is provided in semiautomatic methods such as thresholding, region growing, deformable model, graph-based model and clustering. Semiautomatic segmentation techniques have been presented recently to segment bone and cartilage as summarized in Table 1.

As shown in Tables 1 and 2, it can be observed that most of the semi/fully automatic methods have achieved better accuracy compared to the ground truth. Thus to address these issues and improve the performance of the existing segmentation approach, a fully automatic segmentation method for bone extraction is presented in this paper. Proposed work uses distance regularization level set evolution (DRLSE) for an automatic delineation of bones such as fumer and tibia from MR images.

Table 2 Existing bone/cartilage fully automatic segmentation techniques

Segmentation method	Author/year	Region of interest	Performance measure
kNN/voxel classification	Folkesson et al. [12]	Cartilage	DSC: 0.80, sensitivity: 0.80 and specificity 0.99
	Folkesson et al. [13]	Cartilage	Cartilage extraction—40–60 min
Multi-atlas segmentation	Tamez-Pena et al. [14]	Femur/tibia/cartilage	Sens. (femur)—88 Spec. (femur)—99 Sens. (tibia)—88 Spec. (tibia)—99 DSC. (femur)—88 DSC (tibia)—84
	Shan et al. [15]	Femur/tibia	DSC. (femur)—85 DSC (tibia)—85
Deep learning	Ambellan et al. [16]	femur/tibia	DSC medial femoral cartilage 86.1 ± 5.3% and lateral femoral cartilage 90.4 ± 2.4%
	Liu et al. [17]	Femur/tibia	DSC femur bone—95.5% femur cartilage—90.2% tibial bone—94.3% tibial cartilage—89.5%
	Norman et al. [18]	Cartilage	DSC for T1—weighted cartilage—0.742 (95% confidence) DSC for DESS cartilage—0.867 (95% confidence)

3 Proposed Methods

The segmentation framework of KOA disease is presented in Fig. 1. The proposed framework consists of three phases: Phase I: Preprocessing of input image, Phase II: Segmentation of bone and cartilage and Phase 3: Classification of the image as normal knee or KOA. In this paper, we are mainly focusing on preprocessing and bone segmentation followed by cartilage extraction.

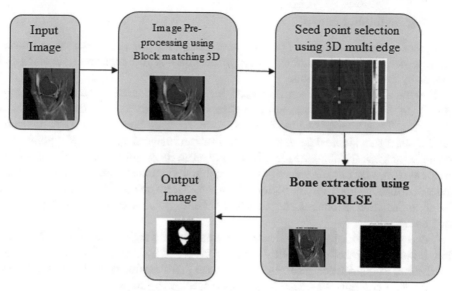

Fig. 1 Proposed segmentation approach

3.1 Preprocessing

Low contrast between cartilage and surrounding tissues calls the need for preprocessing before we proceed for segmentation. As mentioned in the introduction, inadequate brightness, noisiness and low tissue contrast are the challenges for bone segmentation. In this work, the development framework initially takes input as Knee MR image with a fast spin protein density sequence. The preprocessing phase resolves the image noise by applying the BM3D filtering technique [5] and contrast is enhanced by Gaussian filtering. We observe that the proposed deformable model is sensitive to noise and brightness and thus preprocessing can be applied to provide sharp edges between the tissues.

The general steps involved in preprocessing are as follows:

1. Upload noised MR image.
2. Apply block matching 3D method: in this step, the block matching 3D method is used to filter the speckle type of noise by preserving image features.
3. Apply Gaussian blur kernel: Gaussian blur is applied to enhance the contrast of the denoised image. Gaussian image can be obtained at three different scales such as 0.65, 1.1 and 2.5 mm using Eq. 1

$$G(a, b) = \frac{1}{2\Pi\sigma^2} e^{-\left(\frac{a^2 + b^2}{2\sigma^2}\right)} \tag{1}$$

4. High contrast denoised image as output image

3.2 Segmentation Method

In segmentation framework, the preprocessed images are passed on to the segmentation phase to carry out the extraction of bone corresponding to femur and tibia using MR images by the DRLSE method.

Several methods have been developed for the segmentation of bone and cartilage [3–22]. In this work, the concept of seed selection is derived from Gandhamal et al. [23] using 3D multi-edge overlapping technique. To select bone position automatically, middle slices of a single dataset are chosen from an image stack. Two seed points (one for fumer and one for tibia) are identified based on an image profile as peaks and valleys. In this system, x coordinate values of both the points are obtained from the left histogram and y coordinate value from the bottom histogram. These points are used as initialization for edge detection of bone during the segmentation process explained in detail in the next paragraph.

Seed point locations obtained in the previous section are used to extract femur and tibia bone region based on the DRLSE model proposed by Li et al. This model is based on distance regularization term and energy term that describes the desired location of zero level set contours which are defined as the following steps.

Segmentation of bone extraction is as follows:

1. Input image: noised and denoised MR image.
2. Define the initial level set function as follows.

$$\frac{\partial \phi}{\partial t} = F|\nabla \phi| + A \cdot \nabla \phi \tag{2}$$

3. Calculate the external energy function for LSE given as follows:

$$\varepsilon(\phi) = \mu R_p(\phi) + \varepsilon_{\text{ext}}(\phi) \tag{3}$$

4. Compute distance regularization term by Eq. 4

$$\frac{\partial \phi}{\partial t} = \mu \text{div}(\text{dp}(|\Delta \phi|))\Delta \phi + F|\Delta \phi| + A \cdot \Delta \phi \tag{4}$$

5. Stop the iteration if we reached the desired shape, otherwise go to step 3.
6. Cartilage extraction is performed based on Bone masking of fumer and tibia bone.

4   Results

A. R-time data analysis

Knee MR images were collected from the local multifaceted hospital of 150 patients (80 females and 70 male) and thus we have 1000 MR images in total. In this experiment, we have considered normal MR images and KOA MR images of the left knee as well as right knee images. Image acquisition was obtained by MRI 3.0 T (GE Medical System Discovery MR 750 w). For this research, sagittal plane proton density fat suppression MRI sequences were selected for bone/cartilage extraction. We have acquired 500 images in all for experimental purposes.

B. Qualitative analysis

In qualitative analysis, the performance of the preprocessing step and segmentation method are visually examined and interpreted by a human being. Figure 2 shows the result of contrast-enhanced image and noise removed image using Gaussian blur and block matching 3D and compared with histogram equalization. In the segmentation process, contour at various iteration and final contour of both bones (fumer and tibia) are shown in Figs. 3 and 4, respectively.

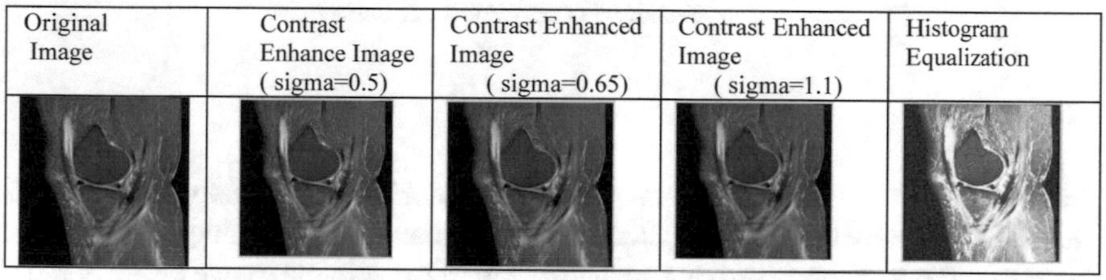

Original Image	Contrast Enhance Image (sigma=0.5)	Contrast Enhanced Image (sigma=0.65)	Contrast Enhanced Image (sigma=1.1)	Histogram Equalization

Fig. 2 Result of contrast-enhanced image and image

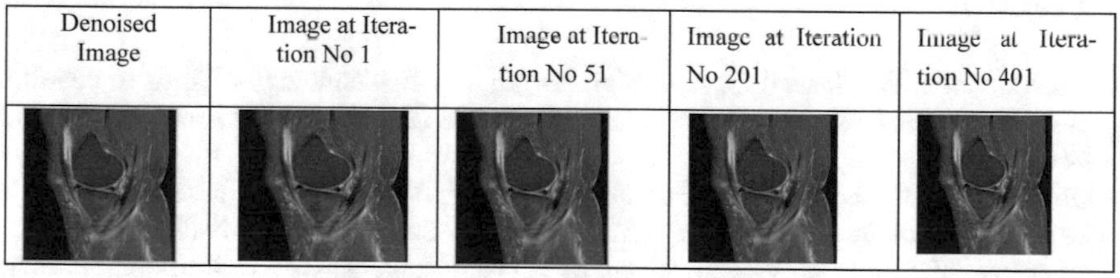

Denoised Image	Image at Iteration No 1	Image at Iteration No 51	Image at Iteration No 201	Image at Iteration No 401

Fig. 3 Phases of bone region extraction

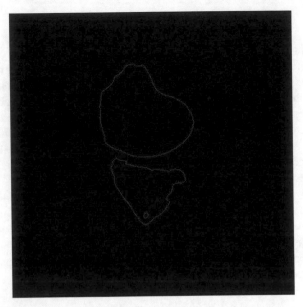

Fig. 4 Result of bone region extraction

5   Conclusion

In this paper, bone segmentation framework using MR image is developed. The performance of the developed method was compared with manual segmentation and obtained a better result. We have observed that segmentation of the original image takes more time compared to the denoised image. In future, segmentation of the patella will be included in this model along with fumer and tibia bone.

References

1. Pal, C., Singh, P., Chaturvedi, S., Pruthi, K., Vij, A.: Epidemiology of knee osteoarthritis in India and related factors. Indian J. Orthop. **50**, 518 (2016). https://doi.org/10.4103/0019-5413.189608
2. Hafezi-Nejad, N., Demehri, S., Guermazi, A., Carrino, J.A.: Osteoarthritis year in review 2017: updates on imaging advancements. Osteoarthr. Cartil. **26**, 341–349 (2018)
3. Nieminen, M.T., Casula, V., Nevalainen, M.T., Saarakkala, S.: Osteoarthritis year in review 2018: imaging. Osteoarthr. Cartil. **27**, 401–411 (2019). https://doi.org/10.1016/J.JOCA.2018.12.009
4. Dalvi, R., Abugharbieh, R., Wilson, D., Wilson, D.R.: Multi-contrast MR for enhanced bone imaging and segmentation. In: 2007 29th Annual International Conference of the IEEE Engineering in Medicine and Biology Society, pp. 5620–5623 (2007)
5. Qvist, P., Bay-Jensen, A.-C., Christiansen, C., Dam, E.B., Pastoureau, P., Karsdal, M.A.: The disease modifying osteoarthritis drug (DMOAD): Is it in the horizon? Pharmacol. Res. **58**, 1–7 (2008). https://doi.org/10.1016/j.phrs.2008.06.001
6. Peterfy, C.G., Genant, H.K.: Emerging applications of magnetic resonance imaging in the evaluation of articular cartilage. Radiol. Clin. North Am. **34**, 195–213 (1996)

7. Cohen, Z.A., McCarthy, D.M., Kwak, S.D., Legrand, P., Fogarasi, F., Ciaccio, E.J., et al.: Knee cartilage topography, thickness, and contact areas from MRI: in vitro calibration and in-vivo measurements. Osteoarthr. Cartil. **7**, 95–109 (1999). https://doi.org/10.1053/joca.1998.0165

8. Shim, H., Chang, S., Tao, C., Wang, J.-H., Kwoh, C.K., Bae, K.T.: Knee cartilage: efficient and reproducible segmentation on high-spatial-resolution MR images with the semiautomated graph-cut algorithm method. Radiology (2009). https://doi.org/10.1148/radiol.2512081332

9. Gan, H.S., Mohd Rosidi, R.A., Hamidur, H., Sayuti, K.A., Ramlee, M.H., Abdul Karim, A.H. et al.: Binary seeds auto generation model for knee cartilage segmentation. Int. Conf. Intell. Adv. Syst. ICIAS (2018). https://doi.org/10.1109/icias.2018.8540570

10. Gan, H.S., Tan, T.S., Karim, A.H.A., Sayuti, K.A., Kadir, M.R.A.: Interactive medical image segmentation with seed precomputation system: data from the osteoarthritis initiative. In: 2014 IEEE Conference on Biomedical Engineering and Sciences (IECBES) "Miri, Where Eng. Med. Biol. Humanit. Meet (2014). https://doi.org/10.1109/iecbes.2014.7047510

11. Khaizi, A.S.A., Rosidi, R.A.M., Gan, H.S., Sayuti, K.A.: A mini review on the design of interactive tool for medical image segmentation. In: 2017 International Conference on Engineering and Technology Technopreneurship, ICE2T (2017). https://doi.org/10.1109/ice2t.2017.8215985

12. Folkesson, J., Dam, E., Olsen, O.F., Pettersen, P., Christiansen, C.: Automatic segmentation of the articular cartilage in knee MRI using a hierarchical multi-class classification scheme. Lecture Notes on Computer Science (including Subseries Lecture Notes in Artificial Intelligence Lecture Notes in Bioinformatics), (2005). https://doi.org/10.1007/11566465_41

13. Folkesson, J., Dam, E.B., Olsen, O.F., Pettersen, P.C., Christiansen, C.: Segmenting articular cartilage automatically using a voxel classification approach. IEEE Trans. Med. Imaging (2007). https://doi.org/10.1109/TMI.2006.886808

14. Tamez-Pena, J.G., Barbu-McInnis, M., Totterman, S.: Knee cartilage extraction and bone-cartilage interface analysis from 3D MRI data sets. Proc SPIE **5370** (2004)

15. Huang, C., Shan, L., Charles, H.C., Wirth, W., Niethammer, M., Zhu, H.: Diseased region detection of longitudinal knee magnetic resonance imaging data. IEEE Trans. Med. Imaging (2015). https://doi.org/10.1109/TMI.2015.2415675

16. Ambellan, F., Tack, A., Ehlke, M., Zachow, S.: Automated segmentation of knee bone and cartilage combining statistical shape knowledge and convolutional neural networks: data from the Osteoarthritis Initiative. Med. Image Anal. (2019). https://doi.org/10.1016/j.media.2018.11.009

17. Liu, F., Zhou, Z., Jang, H., Samsonov, A., Zhao, G., Kijowski, R.: Deep convolutional neural network and 3D deformable approach for tissue segmentation in musculoskeletal magnetic resonance imaging. Magn. Reson. Med. (2018). https://doi.org/10.1002/mrm.26841

18. Norman, B., Pedoia, V., Majumdar, S.: Use of 2D u-net convolutional neural networks for automated cartilage and meniscus segmentation of knee MR imaging data to determine relaxometry and morphometry. Radiology (2018). https://doi.org/10.1148/radiol.2018172322

19. Danielyan, A., Katkovnik, V., Egiazarian, K.: IEEE Trans. Image Process. **21**(4), 1715–1728 (2012)

20. Thengade, A., Mutha, B.H.: Image segmentation for detection of knee cartilage. In: 2018 Fourth International Conference on Computing Communication Control and Automation (ICCUBEA), Pune, India, pp. 1–5 (2018). https://doi.org/10.1109/iccubea.2018.8697658s

21. Piplani, M.A. et al.: Articular cartilage volume in the knee: Semi automated determination from three dimensional reformations of MR images. Radiology **198**(3), 855–859 (1996)

22. Wilson, D., Paul, P.K., Roberts, E.D., Blancuzzi, V., Gronlund-Jacob, J., Vosbeck, K. et al.: Magnetic resonance imaging and morphometric quantitation of cartilage histology after chronic infusion of interleukin 1 in rabbit knees. Proc. Soc. Exp. Biol. Med. (1993)
23. Gandhamal, A., Talbar, S., Gajre, S., Razak, R., Hani, A.F.M., Kumar, D.: Fully automated subchondral bone segmentation from knee MR images: data from the Osteoarthritis Initiative. Compute. Biol. Med. **88**, 110–125 (2017)

Machine Learning Techniques for Homomorphically Encrypted Data

Hemant Ramdas Kumbhar$^{(\boxtimes)}$ and S. Srinivasa Rao

CSE Department, Koneru Lakshmaiah Education Foundation,
Vaddeswaram, Guntur, Vijayawada, India
hemant.kumbhar@gmail.com, srinu1479cse@kluniversity.in

Abstract. The application of cloud-oriented services has increased from personal use such as mailing systems to the medical field. Typically, a cloud service provider offers encryption services of sensitive data. However, decrypting this data in the cloud for data analysis using machine learning can lead to privacy concerns. To this end, learning over encrypted data without decrypting it is the need of the future. In this article, we present a holistic review and summarize the literature on homomorphic encryption, related issues, applications in machine learning over encrypted data, and potential future research directions.

Keywords: Homomorphic encryption · Machine learning · Privacy preservation

1 Introduction

Nowadays, huge data is generated by different software applications used in industries, government sectors, schools, colleges, medical field, and in research field also. Social networking sites like Facebook, YouTube, Twitter, etc., are generating large data per second. The banking sector, share market, and cryptocurrency are not behind too. In almost all of the mentioned areas, users prefer to store their data on the cloud.

Data owner, in this scenario, is very much concerned about the privacy of their data. Many encryption techniques are used for making data private.

As per current trends, the concept of learning from own historical data like finding trends, interests, recommendations, future predictions, and flaws detection is increasing rapidly. Here, data mining comes into the picture. Industries, banking or medical sectors want to use their historical data, analyze this data for knowledge mining. This data mining can be extended to machine learning as learning through experience [1]. But bankers, industrialists, or doctors are not professional about data analytics/machine learning implementations. They are dependent on professional software programmers who can do the analytics task and produce knowledge to learn new things which can help data owners for the betterment of their business/profession. Hence, data owners need learning from their data without sacrificing the privacy of their data to third-party professionals. Figure 1 shows how nowadays data owners encrypt their data and then store it on cloud space [2, 3].

B. Iyer et al. (eds.), *Applied Computer Vision and Image Processing*,
Advances in Intelligent Systems and Computing 1155,
https://doi.org/10.1007/978-981-15-4029-5_25

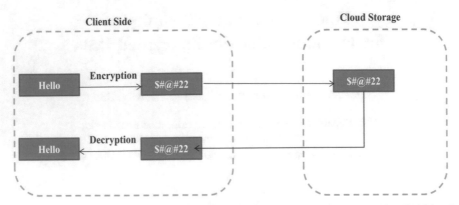

Fig. 1 A recent trend in data encryption and the subsequent data storage on the cloud by the owners

Let's consider a scenario about doctors from different hospitals who want to gain knowledge from their historic patient data. But as they don't have knowledge of data mining algorithms, neural networks, and deep learning; they will not able to gain knowledge on their own. Solutions they may use are listed next. For gaining this knowledge, the doctor has to go to some data mining expert. Give the historical patient data and ask experts to find all knowledge helpful for them for their future treatments. Here is the risk that doctors have to give sensitive data about their patient and treatments. No doctor will agree for it. Another solution is we may think of encrypting the data using different encryption/decryption algorithms. Encrypted data will be sent to data mining experts. Then data mining expert will decrypt it and then will apply the learning algorithms. Here is the possibility that other than the mining expert, cloud owners may misuse data.

As shown in Fig. 2 the third solution is, doctors have to depend on some Third Party Agent. The third party agent may generate public and private keys. The private key will be given to the doctor to encrypt their data. And the public key will be used by data mining experts to decrypt the data. It may be possible that the third party agent is also compromised.

The next solution many researchers used in privacy-preserving is to add some fake record to original data and then encrypt it and hand it over to data mining experts [4]. But we can easily understand that by adding fake data about patient and treatment and then learning on that modified data may give seriously bad knowledge after applying data mining techniques. Protecting privacy requires both preventing leakages of the training data and ensuring that the final model does not reveal private information. Preventing any and all privacy leaks is a significant challenge; existing systems developed for deep learning named Caffe, TensorFlow, DeepMind and many more were not designed with learning securely in mind.

Currently, other angles of privacy issues are identified. Applications like WhatsApp, Facebook, Picasa, YouTube, Gmail, Google Drive, Google forms, etc., are collecting videos, images, and lots of text data. They have centralized storage for this huge data, which they are providing for free to users. Actually, this is the private data of each user but the above applications have direct access and they use this data to

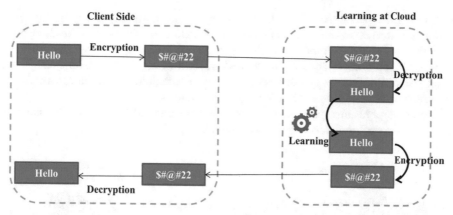

Fig. 2 Learning by decrypting data on a cloud system

train on strong applications like NLP, image-based search, personal assistance, language translators, traveling assistant, and many more. In the current scenario, they have a monopoly over artificial intelligence or machine learning model. Here we need the privacy of user data [5].

Researchers in Medical Science Institute have hunger of patient and treatment data. With need to large and variety of data for more accurate research, researchers expect data from another Medical Science Institutes, but due to privacy concerns these institutes don't share data to each other [5].

How to discover knowledge without decrypting the data? The need of the hour is that data owner must get knowledge from his own data at the same time not allow anyone to decrypt the data. Does learning from data happen over encrypted data without decrypting it? Fortunately yes! In this paper, we review the efforts taken by researchers in the context of machine learning techniques for encrypted data. The remaining part of this paper is organized as Sect. 2: related work; Sect. 3: tools and challenges. It also covers how to use homomorphic encryption, milestones in that Homomorphic Encryption, success, failure, future opportunities.

2 Related Work

2.1 Homomorphic Encryption

Ronald Rivest et al. [6] very firstly raised the possibility of learning over encrypted data. They raised limitations regarding encryption/decryption techniques used for preserving the privacy of user data. They described four possible ways to preserve the data. The first one is, instead of giving sensitive data to computer professionals outside the home organization, purchase organization's own computer system to store data. But it is expensive. Second, hire computer for storage of encrypted data but do the operations in-house by using some terminal to connect to a rented computer for learning. In this solution, a large communication cost is an issue. Third, design the new computer such that decryption will happen in CPU only and decrypted data cannot be accessed outside the CPU. This special hardware will face many issues. Difficulty in key management. Slow execution occurs due to more number of load/store for each encryption-decryption. Also simultaneously encryption and doing an operation on data is impossible. In all the above three solutions, decryption has to be done for learning from data. It means privacy is not preserved. Fourth, use different encryption techniques such that operations on encrypted data without decrypting it may be possible.

Homomorphic encryption means we can do some calculation, computation on encrypted data without decrypting it as shown in Fig. 3.

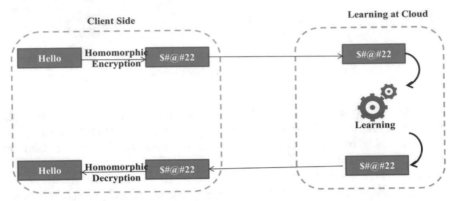

Fig. 3 Learning over encrypted data using homomorphic encryption

Pascal Paillier [7] presented probabilistic homomorphic encryption comparable to RSA. A new trapdoor mechanism based on residues is proposed which is more secure. No proof of security is proposed by him, so we still need to find a defense against different possible attacks on this scheme. Also, they proposed the need for distributed solutions.

Craig Gentry [8] raised the issue that if we store our all data on the cloud without encryption, it is illegal for medical-like fields and also there is a high risk to data privacy. Cloud service providers may misuse or take benefits of the unencrypted data over the cloud. So he raised the need of finding a new way of dealing with easy to use and privacy preservation of user data over the cloud. As without decryption, processing data to gain some knowledge was seen to be very impossible at fight glance. To understand this is possible, he presented a very good analogy of a jewelry shop. In this shop, the owner wants that their work should design very good jewelry but should not steal the gold. For this, the shop owner prepared a transparent box in which gold is stored and the worker is given special gloves. Using these gloves user can put his hands into the box and work for design. As this box is locked by the owner, the worker cannot take gold out and still design beautiful jewelry. In this analogy, a locked box with gold is encrypted data, gloves are homomorphism and designed jewelry is encryption function. He told that homomorphic encryption can be applied which can be symmetric or asymmetric. He proposed the need for semantic security and probabilistic encryption scheme for semantic security. In his Ph.D. work, he found that more malleable encryption is easy to break, which is not the problem with semantic security. He proposed a somewhat homomorphic encryption scheme. In a comparison of addition or subtraction operations, as multiplication adds noise faster-bootstrapping scheme is presented by him which uses bootstrap next level of operation when the current level reaches to max noise.

2.2 Machine Learning Techniques for Encrypted Data

Nowadays, machine learning is coming up as a popular service of the cloud. For commercial platforms like Google, Azure, etc., this platform provides an infrastructure to store user data and also do training of prediction models on that user data [9]. Hence, three parties, cloud, data owner, and prediction model provider are benefited by this system. But the issue in this system is the privacy of data owners. Prediction over encrypted data is the need. As discussed above, previous work proposes the linear prediction models. But many of the problems and data in real life are a nonlinear one. Also, the accuracy obtained by the linear prediction model is limited for nonlinear data like speech data or image data from which we have to recognize the object. So the machine learning model which is nonlinear will be beneficial. Pengtao Xie et al. [10] presented the neural network as a nonlinear machine learning model to learn over encrypted data. As discussed above, homomorphic encryption is used to do the encryption and over this encrypted data, we can perform the different operations without decrypting it. With the neural network, issues are the activation functions used in NN are not represented as a polynomial. So, nonlinear activation functions are required to be transformed into polynomial form. Practical issues raised by Pengtao are first, computations over encrypted data are slower, especially for high degree polynomials. Therefore, care has to be taken that polynomials should be kept with a low degree for the approximated neural network. Secondly, for an l-layered neural network and for each layer, degree of the polynomial is d, total neural network polynomial degree will be dl. It means to reduce the polynomial degree, the number of layers should be as less as possible which is the contract to deep learning. So learning is difficult and feasible for limited datasets only.

If the neural network is to be built for data from different parties, it needs a different way to learn. Reza Shokri et al. [5] presented a learning model which is built using neural network over input from multiple parties. They introduced jointly learning using a neural network. In their work, they contributed algorithms which run in parallel manner resulting in optimization algorithms with privacy preservation. Reza proposed a model in which different data owners run a learning algorithm at their own storage. Sharing of selective stochastic gradient descent values is done from local neural networks trained parameter to other parties who jointly use this learning model. This parameter sharing is actually used to train other parties' neural network for more accurate results. It means, in this secure multiparty computation, neural network parameters which are trained in a different environment are gathered together. Now two major issues are there in this method. First, will they give a good output? Second, random gradient values selection makes convergence slow (Table 1).

Table 1 Summarised evolution of homomorphic encryption

Paper title	Findings
Data banks and privacy homomorphism [6]	Very firstly proposed the possibility of homomorphism
Public-key cryptosystems based on composite degree residuosity classes [7]	Presented probabilistic homomorphic encryption is comparable to RSA. A new trapdoor mechanism based on residues is proposed which is more secure
A brief review of machine learning and its applications [2]	Summarized 7 different machine learning methods with application
Computing arbitrary functions on encrypted data [8]	Presented good analogy of jewelry shop to understand homomorphic encryption for cloud storage
Private predictive analysis of encrypted data [11]	Presented heart attack prediction system over encrypted data using Microsoft Windows Azure cloud by using leveled homomorphic encryption
Crypto-nets: neural networks over encrypted data [12]	The neural network as a nonlinear machine learning model to learn over encrypted data. Secure multiparty computation
Privacy-preserving deep learning [5]	Multiparty jointly learning. optimization algorithms in Deep NN with parallelism. No encryption is needed
A review of homomorphic encryption and software tools for encrypted statistical machine learning [13]	Presented summarized issues in homomorphic encryption. Listed out machine learning tools available for encrypted data learning
The modification of the RSA algorithm to adapt fully homomorphic encryption algorithm in cloud computing [10]	Addition operation is made homomorphic by modifying the RSA formula
Machine learning and its applications: a review [14]	Presented how ML is better than rule-based systems. Stated DM and ML are totally different
Privacy-preserving extreme learning machine using additively homomorphic encryption [15]	Additive homomorphism only. Less communication overhead as one-way communication from participants to ELM. ELM can learn over multiple sources of input data. Due to the simple network structure and no iterative calculation, learning is very fast
Deep encrypted text categorization [16]	Optimal learning semantics of words in Natural language processing using RNN and LSTM

(continued)

Table 1 (*continued*)

Paper title	Findings
AHEad privacy-preserving online behavioral advertising using homomorphic encryption [17]	Provide solution over add-blocking software. Preserves privacy in OBA tasks
Comparison of selected homomorphic encryption techniques [18]	Presented circuit diagrams for operations like OR, integer addition. Also how polynomials are written for these different circuits and the operations performed by them
A homomorphic encryption-based system for securely managing personal health metrics data [19]	Learning of patient lifestyle and suggesting physical activities for the betterment of patients using HE. For FHE operations new operations like rotate (slot) and select (mask) along with addition and multiplication are proposed
Toward dynamic end-to-end privacy-preserving data classification [20]	Data classification over encrypted data. Proposed decision tree learning algorithm which is dynamically updated
Private machine learning classification based on fully homomorphic encryption [21]	Privacy-preserving decision-based classification and Naive Bayes classification using fully homomorphic encryption. FHE with SMID, Noise minimization by Modulus switching
Accelerating ElGamal partial homomorphic encryption with GPU platform for industrial Internet of Things [22]	Fast PHE technique using GPU for storing IoT applications sensor data on cloud
Learning from privacy preserved encrypted data on cloud through supervised and unsupervised machine learning [23]	Neural network for supervised and unsupervised learning is implemented on homomorphically encrypted data
Improved homomorphic discrete Fourier transforms and FHE bootstrapping [24]	Fast homomorphic DFT using structured linear transformation

3   Tools and Challenges

See Table 2.

Table 2 Tools, methods, and challenges

Tools/methods	Challenges
Presented algebraic system for homomorphic encryption [6]	Not fully secure; slow down encryption. Multiplication operation adds more noise
Probabilistic encryption scheme [7]	No proof of security is proposed, so still we need to find a defense against different possible attacks on this scheme. Also, they proposed the need for distributed solutions
Rote learning, learning by teaching, inductive learning, analog learning, explained learning, neural network, knowledge discovery [2]	Collecting data is a big issue for the training of learning algorithms due to its large size. More time required for collecting data. The problem of sampling. Application-specific learning algorithm design is rough, hard and limited
Somewhat homomorphic encryption (SWHE) and bootstrapable encryption [8]	Bootstrapable SWHE scheme is computationally expensive. The possibility of more noise addition in decrypted ciphertext
Logistic and linear regression over encrypted data [11]	Need for scalable systems for performance improvement for huge medical data
A polynomial approximation to neural networks [12]	Computations over encrypted data are slower, especially for high degree polynomials. A number of layers need to keep minimum for keeping polynomial degree small. So learning is difficult and feasible for limited datasets only
Supervised learning, Distributed selective stochastic gradient descent (DSSGD) [5]	Neural network parameter leakage. Averaging NN parameters by aggregating differently trained NN does not give good performance
C library by Gentry for HE, Scarab—C library for HE. HELib—C++ library for FHE, SIMD parallelism, HE-R package [13]	Semantic encryption operations on ciphertext may add noise. For Linear Classification—no division and comparison is possible. Linear Regression is possible up to 5 dimensions only
Extreme learning machine (ELM) [15]	Data analyst and data owner are assumed as an honest one. Data size should be greater than hidden layers for ELM
A recurrent neural network, long short-term memory, Tensorflow, Keras [19]	For complex RNN computation cost is more
Online behavioral advertising (OBA), real-time bidding (RTB), threshold homomorphic encryption [17]	Need performance improvement for increased profile data size

(continued)

Table 2 (*continued*)

Tools/methods	Challenges
Basic SWHE, FHE over integers [18]	No good speed of execution for deeper circuits. These techniques use mathematical structures and hence need optimizations [25]
Safe BioMetric, tools—Apache Spark, IBM Bluemix, IBM OpenWhisk programming [17]	FHE operations must be optimized to reduce data transfer, data processing, and data storage
Decision tree for classification [20]	Computation over encrypted data; there is very high overhead. Randomization of data causes an issue of accuracy. Not both training and prediction phases focused simultaneously
Relinearization technique. Modulus switching [21]	No thought on how to store large data

4   Conclusion

To conclude, the issue which is open in learning over encrypted data is slow evaluation of operations on encrypted data. The set of functions which can be computed on ciphertext space is very restricted, e.g., addition and multiplication operations only. Limited number of times addition and multiplication operation can be performed because multiplication operation multiplies noise too. Arithmetic operations over encrypted data are more complex than standard arithmetic operations. On encrypted data: addition operation can be done with good speed as compared to multiplication operation. Divisions and comparisons operations cannot be performed on encrypted data. Large space is required for homomorphically encrypted data. Space is Complex. The depth of operations is limited. Semantic encryption operations on ciphertext may add noise. For Linear Classification—no division and comparison is possible. Linear Regression is possible up to 5 dimensions only.

References

1. Roberts, J.C., Al-Hamdani, W.: Who can you trust in the cloud?, pp. 15–19 (2011)
2. Hua, W., Learning, A.R.: A brief review of machine learning and its application. In: 2009 International Conference on Information Engineering and Computer Science (ICIECS 2009), p. 1 (2009)
3. Rangasami, K., Vagdevi, S.: Comparative study of homomorphic encryption methods for secured data operations in cloud computing. In: 2017 International Conference on Electrical, Electronics, Communication, Computer, and Optimization Techniques (ICEECCOT), vol. 2018, pp. 551–556 (2018)
4. Mendes, R., Ao, J.O.: Privacy-Preserving Data Mining : Methods, Metrics, and Applications, vol. 5 (2018)
5. Shokri, R., Shmatikov, V.: Privacy-Preserving Deep Learning *, pp. 909–910 (2015)
6. Rivest, R.L., Adleman, L., Dertouzos, M.L.: On data banks and privacy homomorphism. Found. Secur. Comput. 4(11), 169–178 (1978)

7. Paillier, P.: Public-key cryptosystems based on composite degree residuosity classes. Eurocrypt, 223–238 (1999)
8. Gentry, C.: Computing arbitrary functions of encrypted data. Commun. ACM **53**(3), 97 (2010)
9. Halevi, V., Shai and Shoup: GitHub—shaih/HElib: An Implementation of homomorphic encryption (2014). Available: https://github.com/shaih/HElib. Accessed 12 Nov 2018
10. Sha, P., Zhu, Z.: The modification of RSA algorithm to adapt fully homomorphic encryption algorithm in cloud computing. In: 2016 4th International Conference on Cloud Computing and Intelligence Systems (CCIS), vol. 1, pp. 388–392 (2016)
11. Bos, J.W., Lauter, K., Naehrig, M.: Private predictive analysis on encrypted medical data. J. Biomed. Inform. **50**, 234–243 (2014)
12. Ata, C.D., Bilenko, M., Finley, T., Gilad-bachrach, R., Lauter, K., Naehrig, M.: Crypto-Nets: Neural Networks over Encrypted Data, pp. 1–9 (2015)
13. Aslett, L.J.M., Esperança, P.M., Holmes, C.C.: A review of homomorphic encryption and software tools for encrypted statistical machine learning, pp. 1–21 (2015)
14. Angra, S., Ahuja, S.: Machine learning and its applications: a review. In: Proceedings of the 2017 International Conference On Big Data Analytics and Computational Intelligence, ICBDACI, pp. 57–60 (2017)
15. Kuri, S., Hayashi, T., Omori, T., Ozawa, S.: Privacy preserving extreme learning machine using additively homomorphic encryption (2017)
16. Vinayakumar, R., Kp, S., Poornachandran, P.: Deep Encrypted Text Categorization, pp. 364–370 (2017)
17. Helsloot, L.J., Tillem, G., Erkin, Z.: AHEad : Privacy-preserving Online Behavioural Advertising using Homomorphic Encryption, pp. 4–7 (2017)
18. Ogiela, M.R., Oczko, M.: Comparison of selected homomorphic encryption techniques. In: 2018 IEEE 32nd International Conference on Advanced Information Networking and Applications (AINA), pp. 1110–1114 (2018)
19. Bocu, R., Costache, C.: A homomorphic encryption-based system for securely managing personal health metrics data. IBM J. Res. Dev. **62**(1), 1–10 (2018)
20. Talbi, R., Bouchenak, S., Chen, L.Y.: Towards dynamic end-to-end privacy preserving data classification. In: 2018 48th Annual IEEE/IFIP International Conference on Dependable Systems and Networks Workshops (DSN-W), pp. 73–74 (2018)
21. Sun, X., Zhang, P., Liu, J.K., Yu, J., Xie, W.: Private machine learning classification based on fully homomorphic encryption, vol. 6750 (2018)
22. Chong, R.J., Lee, W.K.: Accelerating ElGamal partial homomorphic encryption with GPU platform for industrial internet of things. In: 2019 7th International Conference on Green and Human Information Technology (ICGHIT), pp. 108–112 (2019)
23. Khan, A.N., Fan, M.Y.: Learning from privacy preserved encrypted data machine learning. In: 2019 2nd International Conference on Computing and Mathematical Engineering and Technology, pp. 1–5 (2019)
24. Han, K., Hhan, M., Cheon, J.H.: Improved homomorphic discrete fourier transforms and FHE bootstrapping. IEEE Access **7**, 57361–57370 (2019)
25. van Dijk, M., Gentry, C., Halevi, S., Vaikuntanathan, V.: Fully Homomorphic Encryption Over the Integers, pp. 24–43. Springer, Berlin, Heidelberg (2010)

Improving Leaf Disease Detection and Localization Accuracy Using Bio-Inspired Machine Learning

Bhavana Nerkar[1](✉) and Sanjay Talbar[2]

[1] NIELIT Aurangabad, Aurangabad, India
bhavananerkar7@gmail.com
[2] Department of E&TC Engineering, Shri Guru Gobind Singhji
Institute of Engineering and Technology, Nanded, India
sanjaytalbar@yahoo.com

Abstract. Disease detection and localization from leaf imagery have found its way into various applications of leaf-based image processing. These applications include, but are not limited to, yield improvement, re-fertilization, disease spread detection, etc. This process requires careful selection of segmentation, feature extraction, feature selection, classification, and post-processing algorithms which should work in tandem for improved system efficiency. In this paper, we propose a bio-inspired machine learning-based classification algorithm, which adaptively learns to improve the accuracy via active feature selection. This process is followed by an adaptive post-processing unit which indicates the level of infection, and suggests the depth of infection in the field. The system was tested against multiple datasets which included both real-time and static data, and it was observed that the classification accuracy is more than 98%, and the spread detection accuracy is more than 94%. These results are also compared with state-of-the-art classifiers like random forest, Naïve Bayes, and convolutional neural networks. The comparisons prove that the proposed algorithm is not only superior in terms of classification accuracy, but also improves the overall system performance in terms of disease localization.

Keywords: Leaf · Disease · Detection · Machine learning · Bio-inspired · Localization

1 Introduction

Detecting and locating disease and infected regions in leaf-based imagery requires processing at multiple levels. For example, in order to find out the location of disease-infected regions, we first need to evaluate the kind of disease the leaf possesses. For that, we need to classify the leaf-image into one of many categories. Classification requires training data, which must contain pre-labeled leaf images. Pre-labelng requires manual evaluation of the images, which requires proper image capture. Thus, in order to evaluate the disease type and the localization of disease, there are multiple stages of processing the leaf image. These stages can be categorized as follows:

© Springer Nature Singapore Pte Ltd. 2020
B. Iyer et al. (eds.), *Applied Computer Vision and Image Processing*,
Advances in Intelligent Systems and Computing 1155,
https://doi.org/10.1007/978-981-15-4029-5_26

- Capture stage,
- Preprocessing stage,
- Feature processing stage,
- Classification stage, and
- Post-processing stage.

The capture stage requires in-depth knowledge about the kind of leaf images which will be processed. Capturing these images with proper angle, proper lighting, and a proper field of view will result in a highly accurate system. Any misalignment in the capturing phase will exponentially affect the performance of the system. Thus, it is recommended that a field professional be consulted for this purpose.

Post the capturing stage, the images pass through various preprocessing stages, which include but are not limited to

- Image denoising to obtain clearer images.
- Scale-based registration for proper sizing of the image.
- Angle and sheer-based registration for properly aligning the images.
- Removal of any kind of watermark, or prints from the image.
- Segmentation of the leaf region from the image via clustering mechanism.

Generally, segmentation is a separate process in classification systems. But, for leaf-based processing systems, there is no need to complicate the segmentation process. A simple clustering mechanism is sufficient to extract the leaf region from the input image. The segmented image is then processed by the feature processing stage. This stage includes feature extraction and selection. Extracting features from leaf images requires careful selection of algorithms, which can describe one disease distinctly from another. Generally, algorithms like gray-level co-occurrence matrix (GLCM), wavelet features, Fourier features, color-based maps, texture-based maps, and edge-based maps are used. These features need proper selection in order to obtain the best features suited for classification using the given training set. Generally, once the training set is changed, the feature selection algorithm needs to be re-run in order to re-evaluate the most useful features for optimum classification.

Once the features are selected, and an optimal feature-training set is obtained, then the process of classification comes into the picture. Classification algorithms vary from simple linear algorithms like k-Nearest Neighbors (kNN), to complex algorithms like convolutional neural networks (CNN). The computational complexity of these algorithms is generally directly proportional to the accuracy obtained from the algorithm. For leaf-disease processing, CNN, support vector machines (SVM), random forest (RF), and Naïve Bayes (NB) algorithms are proven to give better results. In fact, this text also uses the abovementioned algorithms for performance comparison with the proposed classifier. The classification process is followed by the post-processing stage. This stage is very specific to the kind of disease class the image belongs. For example, bacterial blight images require segmentation of reddish regions in order to find out the region of infection, while alternaria blight requires segmentation of dark regions. Therefore, researchers must develop disease-specific post-processing algorithms in order to segment the regions-of-infection, which will indicate the percentage spread of the disease. In this text, we have worked on each stage of processing and developed a novel bio-inspired machine learning algorithm for classification and post-processing of diseases; this algorithm is mentioned in the section next to the literature survey section. The selection of bio-inspired machine learning algorithm has the following benefits during classification:

- The algorithm is adaptive, therefore there is no need to retrain the system from time to time.
- As the number of entries in the training set increases, the algorithm retrains itself internally to manage the changes, thus guaranteeing high level of accuracy.
- Feature selection unit works in tandem with the classification layer, thereby further optimizing the system performance.
- The training set is grown automatically with the confidence score obtained from the testing set, thereby reducing the error in classification.

These advantages of the classification process are due to the bio-inspired learning and thus are the choice of selection for this research. The next section is a literature survey of various classification and post-processing techniques for leaf-based disease detection and other baseline approaches for the same purpose.

2   Literature Review

Throughout the years, numerous procedures have been exhibited to identify plant leaf illnesses. In [1], bolster vector machines (SVM) has been utilized, and it has analyzed more than five various types of maladies. In our exploration, we found that SVMs as a rule have a testing precision of 75%, however, the paper asserts that the exactness of SVM is around 91%, which may be on the preparation set itself. An AI approach is referenced in [2], where specialists have used arbitrary timberlands (RF) in the mix with histogram of slope (HoG) highlights for grouping. The testing set precision of this strategy is professed to be 70%, which is a sensible measure, as arbitrary timberlands contrast the sets and randomized highlights, and consequently the exactness is by and large around 70–80%. Another learning-based approach which uses move learning is proposed in [3], wherein scientists have built up an easy to utilize interface for plant leaf infection location, however, no remarks are made for the exactness of the framework. They have asserted it to be genuinely exact, however, the precision level isn't referenced in the content, so it's on the specialists to execute it and check for its exhibition. A second SVM-based approach is referenced in [4], wherein analysts have built up a technique utilizing a dark level co-event framework (GLCM) highlights. These highlights are genuinely adequate for separation between different sorts of plant illnesses. The unclear correlation in the paper shows a 90% preparing level exactness, which is very low, as SVM execution depends on a grouping procedure, which probably won't be precise. Comparative MATLAB-based execution is finished utilizing [5], where specialists are utilizing RF strategy for the grouping process. In this technique, the analysts have discovered that the framework is genuinely precise for the identification of illnesses, and furthermore restriction of the measure of malady in the given information picture. No remark is done on the degree of exactness; however, the yield pictures appear to be genuinely broke down by the specialists.

A short examination of discovery calculations is done in [6], where specialists have looked at SVM, RF, k-Nearest Neighbors (kNN), and neural systems. They guarantee that fake neural systems have a more significant level of precision when contrasted with different calculations, and along these lines they should be utilized for constant grouping process. Another exploration in [7] additionally demonstrates that repetitive neural systems (RNNs) have better arrangement exactness, for the plant illness location process, when contrasted with different strategies like Decision Tree Classifier (DTC), Support Vector Machine (SVM), Naive Bayes Classifier, k-Nearest Neighbor (kNN), K-implies grouping, Genetic calculation, and so forth. It additionally guarantees that RNN requires less postponement when contrasted with different strategies, and along these lines ought to be the technique for decision for some applications. One more audit in [8] makes reference to that the back engendering neural organize (BPNN) has better execution when contrasted with different techniques, which demonstrates that neural system ought to be the strategy for decision for the plant malady identification application. Chipping away at this line, the examination referenced in [9] utilizes OTSU thresholding joined with neural systems so as to assess that the general exactness of the neural system is higher than SVMs; there is no notice of any numerical qualities, however from their investigation and scientific models, it is very characteristic that neural systems are a superior decision for the characterization procedure. While in [10], the analysts have utilized different strategies like kNN, SVM, and ANN for arrangement, yet they additionally guarantee that neural systems are a decent decision for the order procedure. Two additional surveys distributed in [11, 12] likewise demonstrate this point; neural systems are the calculation of decision for the procedure of plant ailment characterization. A comparative survey is distributed in [13], wherein specialists have professed to analyze in excess of 8 distinct techniques for plant malady identification, and they see that their proposed novel division calculation gives 95% precision when joined with an SVM classifier. The correlation is done on minor 10–15 pictures, and hence their case doesn't appear to be legitimized; it is encouraged to perform steadiness on the examination calculation before utilizing it for any constant frameworks. Another examination in [14] utilizes direct relapse and kNN in order to accomplish a precision of 95%, which is like [13], however in [14], there is no depiction of the quantity of pictures utilized, and consequently this exploration likewise requires due persistence from the specialists before usage. The next section describes the proposed bio-inspired machine learning algorithm for the detection and localization of leaf diseases. The section is split into two more sections, one for classification and the other for post-processing.

3   Bio-Inspired Machine Learning Disease Classification

The proposed bio-inspired-based machine learning algorithm has the following processing blocks.

- Preprocessing for reducing Gaussian noise.
- Adaptive clustering for leaf segmentation.
- Feature extraction using multilevel features.
- Genetic algorithm-based training layer.

- Machine learning evaluation and adaptive training layer.

Initially the image under test is given to a Weiner filter, which checks for the presence of any kind of Gaussian noise in the image. Weiner filter is used because Gaussian noise is present in the images taken from digital cameras. Once the denoising process is completed, then an adaptive clustering algorithm based on bisecting k-means is used. The following steps are used in the algorithm design:

- Divide the image into 2 clusters using bisecting k-means algorithm.
- Obtain images I1 and I2, where each image represents one particular cluster.
- Evaluate the number of green pixels in each of the clusters, let the number of green pixels in each cluster be G1 and G2.
- Use Eq. 1 to find the cluster with leaf as follows:

$$I_{\text{leaf}} = I1, \text{ if } G1 > G2$$
$$\text{else, } I_{\text{leaf}} = I2, \text{ if } G2 \geq G1 \tag{1}$$

- Here, I_{leaf} is the segmented image which contains all the leaf pixels.
- Fill the holes in the I_{leaf} image, and use this hole-filled mask on the original input image.
- Hole filling is needed in order to include any disease pixels which might have been mis-clustered.

Using the above algorithm, we obtain the segmented leaf image, which contains the leaf and the disease regions from the input image. The segmented image is given to a feature extraction layer, wherein the following features are evaluated:

- Color map
  Color map or extended histogram map is obtained by plotting the quantized color levels on the X-axis and the number of pixels matching the quantized color level on the Y-axis. The obtained graph describes the color variation of the image and thus is used to describe the image during the classification stage. The color map resembles the gray level histogram of the image with one minor difference that the color map quantizes the R, G, and B components of the image before counting them, while the histogram directly counts the pixels belonging to a particular gray level and plots them. This ensures that all the color components of the image are taken into consideration by the descriptor.
- Edge map
  The extended edge map a.k.a. edge map describes the edge variation in the image. To find the edge map, the image is first converted into binary, and then Canny edge detector is applied to it. The original RGB image is quantized the same as in the color map. The locations of the edges are observed, and the probability of occurrence edge on a particular quantized image level is plotted against the quantized pixels in order to evaluate the edge map of the image. The edge map is used to define the shape variation in the image and is a very useful and distinctive feature for any image classification system.

- Gray-level co-occurrence matrix (GLCM)
  The GLCM calculation method uses the information about adjacent pixels and creates a histogram kind of matrix out of it. It is a statistical method of examining texture that considers the spatial relationship of pixels is the gray-level co-occurrence matrix (GLCM), also known as the gray-level spatial dependence matrix. The GLCM functions characterize the texture of an image by calculating how often pairs of pixels with specific values and in a specified spatial relationship occur in an image, creating a GLCM, and then extracting statistical measures from this matrix.

Upon extraction of these combined features, the feature set is prepared. This feature set is a combination of the following:

- The feature vectors of the color map, edge map, and GLCM.
- Tagged images with their disease names.
- The total number of input images is divided into training and testing feature sets in a ratio of 70:30, respectively. Once the training and testing feature sets are created, then the following genetic algorithm-based classifier is applied on the testing set:

Inputs.

- Number of iterations $= Ni$
- Number of solutions $= Ns$
- Minimum and maximum number of features to be used for classification process $= F_{min}$ and F_{max}
- Learning rate $= Lr$

Initially mark each of the solutions as 'to be changed'

- For each iteration in 1 to Ni
- For each solution in 1 to N_s

If the solution doesn't need to be changed, then move to next solution
Else,
find the number of features to be used, using Eq. 2,

$$F_{num} = RAND(F_{min}, F_{max}) \qquad (2)$$

Select F_{num} random features from the training and testing feature set for each image. Mark the indices of these selected features as Fidx.
Apply a convolutional neural network-based classifier for the F_{num} selected features, and evaluate the testing accuracy for these features; let the accuracy be A_{si}, where 'i' is the solution number.

- For all the solutions, obtain the values of $A_{s1}, A_{s2}, A_{s3} \ldots A_{sNs}$.

- Evaluate the mean accuracy using Eq. 3,

$$A_{\text{smean}} = \frac{\sum_{i=1}^{Ns} A_{\text{si}}}{Ns} \tag{3}$$

.
- Now evaluate the learning threshold using Eq. 4,

$$A_{\text{sth}} = A_{\text{smean}} * Lr \tag{4}$$

.
- Mark all solutions as 'to be changed' which satisfy Eq. 5,

$$A_{\text{Si}} < A_{\text{sth}} \tag{5}$$

.
- Repeat the process of Ni iterations, and obtain Table 1.

Table 1 Machine learning table

Solution	F_{num}	Feature indices (Fidx)	Accuracy	A_{sth}

Once the table is created, we then select the features with index Fidx, having the highest accuracy. These features are used for classification, and continuous accuracy evaluation is performed. Upon continuous accuracy evaluation, if we find out that the overall accuracy is reducing, then this new accuracy is updated in Table 1. After the update process, the table is re-scanned and again the highest accuracy feature index is used for the classification process.

This process is repeated for each of the images in the test set, and for any new images which might need evaluation. Once the images are classified into a particular disease type, then a post-processing layer is applied on each of the images. This layer has different algorithms for different kind of diseases. Each of these algorithms is described in the next section.

4 Post-processing for Evaluation of Disease Percentage

For the purpose of this research, we considered the following leaf diseases:

Reddening, Fig. 1, causes the edges of the leaf to become reddish in color, due to which there is a lack of photosynthesis in the image, and therefore the plant growth is affected. Moreover, due to reddening the flower quality of the plant is also affected. These red spots on the leaf are permanent, and there is only one solution, i.e., removal of all kinds of red leaf from the plant. The process is very time consuming, and due to the contagious nature of this disease, even a single reddening leaf can cause a lot of damage to the plant over a short period of time.

Fig. 1 Reddening

Fig. 2 Gray mildew

Gray mildew, Fig. 2, also called as fungi-like disease, occurs due to treatment of plants with chemical fertilizers, and has many disastrous effects on both the stem and the leaf of the plant. This disease spreads very quickly, and reduces the life span of the plant. It can be removed by medicinal treatment.

Alterneria, Fig. 3, is the most common plant-leaf disease. It causes bright spots on the leaf, and is very hard to remove. It is treated same as reddening, but it is not contagious. This disease is carried to the plant by insects or other plant-eating species.

Bacterial blight, Fig. 4, is similar to alternaria but spreads throughout the leaf. It reduces the strength of the leaf and makes it easily breakable. As the name suggests, it occurs due to bacteria formation on the leaf. Eliminating leaf affected by bacterial blight is the only option for plant treatment.

Each of these disease images is needed to be post-processed with a different kind of post-processing algorithm. Each of these algorithms is described as follows:

Fig. 3 Alternaria

Fig. 4 Bacterial blight

Reddening

- Divide the image into two clusters using the bisecting k-means algorithm.
- Find the cluster in which the intensity of red pixels is highest than green and blue pixels, and mark it as the reddening cluster.
- Count the number of pixels in the reddening cluster, and divide it by the total number of leaf pixels. This will evaluate the percentage of reddening in the leaf.

Gray Mildew

- Divide the image into two clusters using the bisecting k-means algorithm.
- Find the cluster in which intensity of red, green, and blue pixels is more than 128 (for an 8-bit image), and mark it as the mildew cluster.
- Count the number of pixels in the mildew cluster, and divide it by the total number of leaf pixels. This will evaluate the percentage of gray mildew in the leaf.

Alternaria and Bacterial blight

- Divide the image into two clusters using the bisecting k-means algorithm.
- Find the cluster in which the intensity of green pixels is highest than red and blue pixels, and mark it as the non-disease cluster.
- Subtract the input image from the non-disease cluster to obtain the disease cluster.
- Count the number of pixels in the disease cluster, and divide it by the total number of leaf pixels. This will evaluate the percentage of alternaria/bacterial blight in the leaf.

Using these specific algorithms, post-classification will help in finding out the percentage of infection in the given leaf image. The next section describes the results of the proposed classification and post-processing algorithm.

5 Results and Analysis

We tested the proposed system under various dataset sizes, and evaluated the accuracy for ANN, CNN, and the proposed bio-inspired methods. The following table demonstrates the statistical result evaluation for ANN.

From Table 2, we can observe when the number of training images is low, then the accuracy of ANN reduces linearly with the number of evaluations for a given number of training records, but as the number of training images increase, we observe a steady increase in the number of correctly classified images, and thus the overall accuracy of the ANN increases. For real-time datasets, the accuracy is between 75 and 80%, and it saturates around 80%. In contrast, as we can observe from Table 2 as follows, the overall accuracy of CNN is much higher than that of ANN.

From Table 3, we can observe that the accuracy of CNN also reduces linearly for a low number of records, but increases steadily for a larger dataset. This is due to the fact that the proposed convolutional neural network performs pattern analysis at each layer, and thus even a small chance of matching with the provided input features is taken into consideration by the network under test. A similar analysis is done for the proposed classifier, and the results can be observed in Table 4 as follows.

The accuracy of the proposed classifier is very high due to the continuous learning process. The average accuracy for ANN is around 76%, while for CNN is around 94%, but for the proposed classifier the accuracy is more than 97% across multiple disease

Table 2 Accuracy analysis for ANN

Number of images in database	Number of images tested	Correct outputs ANN	Accuracy (%) ANN
10	10	9	90.00
20	15	13	86.67
20	20	16	80.00
30	25	20	80.00
30	30	22	73.33
50	35	26	74.29
50	40	30	75.00
50	50	36	72.00
80	60	42	70.00
80	70	49	70.00
80	75	53	70.67
80	80	57	71.25
100	85	61	71.76
100	90	65	72.22
100	95	70	73.68
100	100	75	75.00
125	105	80	76.19
125	110	85	77.27
125	115	90	78.26
125	120	94	78.33
125	125	99	79.20

types. Thereby, we can comment that the proposed classifier is superior in terms of accuracy when compared with simple ANN and CNN. This superiority is majorly due to the active selection of features, and continuous network evaluation. Similarly, the outputs of the post-processing unit are shown in the following figures.

From Fig. 5 it is inherent that the infected regions from the gray mildew disease are clearly segmented out, and we can observe the percentage of infection more clearly, as compared to the original image. Similar results were obtained for the remaining diseases, which are shown in Figs. 6 and 7.

Thus, the proposed post-processing layer is able to detect and obtain the infected disease regions from the input leaf images.

Table 3 Accuracy comparison of CNN

Number of images in database	Number of images tested	Correct outputs CNN	Accuracy (%) ANN
10	10	10	100.00
20	15	14	93.33
20	20	18	90.00
30	25	24	96.00
30	30	28	93.33
50	35	33	94.29
50	40	37	92.50
50	50	47	94.00
80	60	56	93.33
80	70	66	94.29
80	75	70	93.33
80	80	75	93.75
100	85	80	94.12
100	90	85	94.44
100	95	90	94.74
100	100	94	94.00
125	105	99	94.29
125	110	103	93.64
125	115	108	93.91
125	120	113	94.17

Table 4 Accuracy observation of the proposed algorithm

Number of images in database	Number of images tested	Correct outputs proposed	Accuracy (%) proposed	Number of images in database	Number of images tested	Correct outputs proposed	Accuracy (%) proposed
10	10	10	100.00	80	80	78	97.50
20	15	15	100.00	100	85	83	97.65
20	20	19	95.00	100	90	88	97.78
30	25	24	96.00	100	95	93	97.89
30	30	29	96.67	100	100	97	97.00
50	35	34	97.14	125	105	102	97.14
50	40	39	97.50	125	110	107	97.27
50	50	49	98.00	125	115	112	97.39
80	60	58	96.67	125	120	117	97.50
80	70	68	97.14	125	125	122	97.60
80	75	73	97.33				

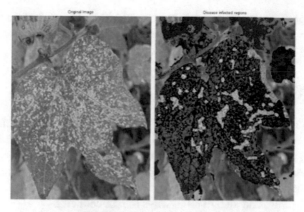

Fig. 5 Detected gray mildew

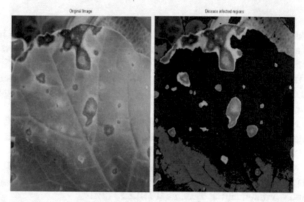

Fig. 6 Detection of alternaria

Fig. 7 Detection of bacterial blight

6 Conclusion

From the result section, we can observe that the proposed algorithm is able to identify diseases with a very high level of accuracy. It outperforms industry-leading methods like artificial neural networks and convolutional neural networks. The proposed bio-inspired machine learning algorithm produces an accuracy of more than 97% on the testing dataset. This accuracy is about 3% higher than the accuracy produced by CNN.

Also, the post-processing layer is able to correctly identify and segment out the region of diseases from the input image. This layer is specific for each disease, and thus must be continuously updated in order to include more disease types.

7 Future Work

Generally, after machine learning there is a very little future scope for this work in terms of classification accuracy. But, researchers can always work on ways to improve the post-processing segmentation technique and make it generic. This way, in spite of any kind of input image, the algorithm will adaptively select the segmentation process, and produce an accurate segmentation output. It will help farmers to detect new kind of diseases in their yield, and also find out ways to reduce them.

References

1. Pooja, V., Das, R., Kanchan, V.: Identification of plant leaf diseases using image processing techniques. In: IEEE International Conference on Technological and Innovations in ICT For Agriculture and Rural Development, pp.130–133 (2017)
2. Ramesh, S., Hebbar, R., Vinod, P.V. et al: Plant disease detection using machine learning. In: International Conference on Design Innovations for 3Cs Compute Communicate Control, pp 41–45 (2018)
3. Boikobo, Muhammad: Design of plant disease detection system: a transfer learning approach work in progress. In: Proceedings of IEEE International Conference on Applied System Innovation 2018, pp 158–161. IEEE ICASI (2018)
4. Meena, R., Saraswathy, G.P., Ramalakshmi, G.: Detection of leaf diseases and classification using digital image processing. INn: International Conference on Innovations in Information, Embedded and Communication Systems (ICIIECS) (2017)
5. Devaraj, A., Sarvepalli, K.: Identification of plant disease using image processing technique. In: International Conference on Communication and Signal Processing, pp. 749–753 India April 4–6 (2019)
6. Dhaware, C.G., Wanjale, K.H.: A modern approach for plant leaf disease classification which depends on leaf image processing. In: International Conference on Computer Communication and Informatics (ICCCI-2017), Coimbatore, India, Jan 05–07 (2017)
7. Mishra, B., Lambert, M., Nema, S. et al.: Recent technology of leaf disease detection using image processing approach: a review. In: International Conference on Innovations in Information, Embedded and Communication Systems (ICIIECS) (2017)
8. Santhosh Kumar, S., Raghavendra, B.K.: Diseases detection of various plant leaf using image processing techniques: a review. In: 5th International Conference on Advanced Computing and Communication Systems (ICACCS) (2019)
9. Khirade, S.D., Patil, A.B.: Plant disease detection using image processing. In: International Conference on Computing Communication Control and Automation, pp. 768–771 (2015)
10. Dhaware, C.G., Wanjale, K.: A modern approach for plant leaf disease classification which depends on leaf image processing. In: 2017 International Conference on Computer Communication and Informatics (ICCCI-2017), Coimbatore, India, Jan 05–07 (2017)
11. Singh, J., Kaur, H.: A review on: various techniques of plant leaf disease detection. In: Proceedings of the Second International Conference on Inventive Systems and Control (ICISC 2018), pp. 232–237 (2018)

12. Mishra, B., Lambe, M., Nema, S.: Recent technologies of leaf disease detection using image processing approach: a review. In: International Conference on Innovations in Information, Embedded and Communication Systems (ICIIECS) (2017)
13. Singh, V., Misra, A.K.: Detection of plant leaf diseases using image segmentation and soft computing techniques. Inf. Process. Agric. 4(1), 41–49 (2016)
14. Prasad, S., Peddoju, S.K., Ghosh, D.: Multi-resolution mobile vision system for plant leaf disease diagnosis. Springer, London, pp. 379–388 (2015)

Identification of Medicinal Plant Using Image Processing and Machine Learning

Abhishek Gokhale, Sayali Babar, Srushti Gawade$^{(\boxtimes)}$, and Shubhankar Jadhav

Department of Computer Engineering, Pimpri Chinchwad College of Engineering, Pune, India
abhishekgokhale1599@gmail.com, sayali9779@gmail.com,
srushtigawade1999@gmail.com, shubhankarjadhav33@gmail.com

Abstract. Medicinal plants are the backbone of the system of medicines; they are the richest bioresource of drugs of traditional systems of medicine, modern medicines, nutraceuticals, food supplements, folk medicines, pharmaceutical intermediates, and chemical entities for synthetic drugs. These plants are classified according to their medicinal values. Classification of medicinal plants is acknowledged as a significant activity in the production of medicines along with the knowledge of its use in the medicinal industry. Medicinal plant classification based on parts such as leaves has shown significant results. An automated system for the identification of medicinal plants from leaves using Image processing and Machine Learning techniques has been presented. This paper provides knowledge of the process of identification of medicinal plants from features extracted from the images of leaves and different preprocessing techniques used for feature extraction from a leaf. Many features were extracted from each leaf such as its length, width, perimeter, area, color, rectangularity, and circularity. It is expected that for the automatic identification of medicinal plants, a web-based or mobile computer system will help the community people to develop their knowledge on medicinal plants, help taxonomists to develop more efficient species identification techniques and also participate significantly in the pharmaceutical drug manufacturing.

Keywords: Leaf recognition · Medicinal plants · Feature extraction · Image processing · Machine learning · OpenCV

1 Introduction

Medicinal plants have long been utilized in traditional medicine. Identification of medicinal plants is a very challenging task without external resources or assistance. Identification of the right medicinal plants that are used for the preparation of medicines is important in the medicinal industry. In various countries, there is a trend toward using traditional plant-based medicines alongside pharmaceutical drugs. Therefore, there seems to be immense potential in this field. Various kinds of algorithms are integrated into the application software. Image analysis is one important method that helps segment image into objects and background [1]. One of the key steps in image analysis is feature detection. Transforming the input data into the set of features is called feature extraction. The

© Springer Nature Singapore Pte Ltd. 2020
B. Iyer et al. (eds.), *Applied Computer Vision and Image Processing*,
Advances in Intelligent Systems and Computing 1155,
https://doi.org/10.1007/978-981-15-4029-5_27

image processing nowadays have become the key technique for the diagnosis of various features of the plant [1].

The non-automatic method is based on morphological characteristics. Thus, classification here is based on the core knowledge of botanists. However, this non-automatic identification is tedious. Hence many researchers support this automated classification system and identification. There are a few systems developed so far where most of the processes are the same.

Following are the steps involved:

Step 1: Preparing the dataset.
Step 2: Preprocessing.
Step 3: Once the preprocessing is done, attributes have to be identified.
Step 4: Training.
Step 5: Classification of the leaves.
Step 6: Result evaluation (Fig. 1).

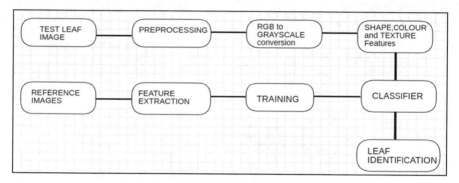

Fig. 1 Design of the proposed solution

This is an effective and efficient automated system (Fig. 1) that can be used by any student, pharmacist, or anyone from the non-botanical background. Motivation to undertake this project was given to us by an incident that happened with the head of the 'National Social Service' cell. He was trying to figure out a way to identify the medicinal plants correctly so that the villagers could make use of them for their pharmaceutical purposes. Seeing his difficulty gave us an idea of building this system.

2 Literature Review

See Table 1.

3 Methodology

The system will work in four stages:

Table 1 Literature review

Sr. No.	Research paper title	Year of publication	Accuracy
1.	Plant leaf recognition using a convolution neural network [2]	2019	94%
2.	Identification of Indian medicinal plant by using artificial neural network [1]	2018	75%
3.	Automatic recognition of medicinal plants using machine learning techniques [3]	2017	90.1%
4.	Plant identification system using its leaf features [4]	2015	More than 85%

A.   Obtaining dataset.
B.   Image segmentation/preprocessing.
C.   Feature extraction.
D.   Classification algorithm (Fig. 2).

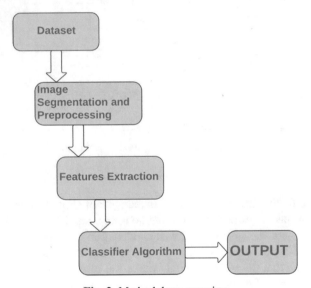

Fig. 2 Methodology overview

A. Obtaining dataset
Leaves are a feasible means to identify plants [5]. The image dataset used in this paper is Flavia leaves dataset which is obtained from http://flavia.sourceforge.net/. This image dataset consists of approximately 1900 image instances of leaves of 32 different species of plants. Sample images from one class are shown (Fig. 3).

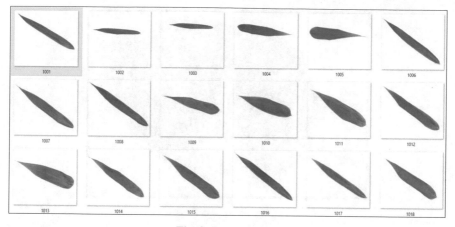

Fig. 3 Image dataset

For training and testing the model, a dataset was created using the extracted features of the leaf. The dataset was divided into two sets namely Train set (70%) and Test set (30%).

B. Image segmentation/preprocessing

Pixel values play a very important role in image analysis. They can be used to segment distinct objects. If there's a significant difference in the contrast values of the object and the image's background, then the pixel values will also differ. In this case, a threshold value can be set. Accordingly, an object or the background can be classified on the basis of the pixel values being less than or greater than a threshold value. This method is also known as Threshold Segmentation. It converts original image (Fig. 4) to grayscale (Fig. 5). If the image has to be divided into two regions, i.e., object and background, a single threshold value is defined. This is known as the global threshold (Fig. 6). If there are multiple objects along with the background, multiple thresholds need to be calculated. These thresholds are collectively known as the local threshold. This technique is preferred when there is a high contrast between object and background.

Fig. 4 Original leaf image

Fig. 5 Grayscale image

Two adjacent regions with different grayscale values are always differentiated based on the edge present between them. The discontinuous local features of an image can be considered as the edges. This discontinuity may prove to be helpful in defining a boundary of the object. This helps in discovering multiple objects present in an image along with their shapes. Filter and Convolutions are used in Edge detection. Edge detection is fit for images having better contrast between objects. When there are too many edges in the image and if there is less contrast between objects, it should not be used.

Digital image processing techniques are used for the classification of medicinal plants in the Plant Leaf Identification system. Firstly, all the images are preprocessed, for removing background area [6]. Then their features based on color, texture, and shape [7] are extracted from the processed image. The subsequent steps were followed for preprocessing the image.

(1) In this technique, we convert RGB to a grayscale image.
(2) After conversion, we smoothen the image using a Gaussian filter.
(3) Then Otsu's thresholding method is used for adaptive image thresholding (Fig. 7).
(4) Morphological Transformation is used for the closing of the holes.
(5) The last step for preprocessing is that the boundary extraction is done using contours.

C. Feature extraction

The major problem in image analysis arises due to the number of variables involved. These variables require a large amount of memory and computation. If the dataset is used as it is, it becomes less instructive and more redundant for doing analysis. When an algorithm has to process large datasets, then by applying this method, the dataset will be reduced to minimum dimensions. Extracting useful features from images in the dataset is the feature extraction process. Various types of leaf features were extracted (Fig. 8) from the preprocessed image which are listed as follows:

Fig. 6 Global threshold

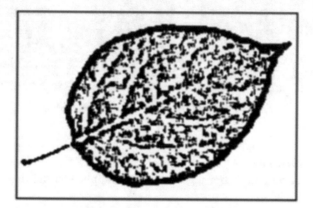

Fig. 7 Adaptive mean threshold

1. Features related to shape:

   - Length.
   - Width.
   - Total area.
   - Perimeter.
   - Proportional relationship between width and length (aspect ratio).
   - Rectangularity.
   - Circularity.

2. Features related to color:

- The sum of channels divided by the number of channels of R, G, and B (mean).
- Amount of variation of a set of values of R, G, and B channels (standard deviations).

3. Features related to texture:

- The difference between the textures (contrast).
- The similarities between the textures (correlation).
- Inverse difference.
- Entropy.

	A	B	C	D	E	F	G	H	I	J	K	L	M	N	O	P	Q
1	area	perimeter	physiologi	physiologi	aspect_rat	rectangulɛ	circularity	mean_r	mean_g	mean_b	stddev_r	stddev_g	stddev_b	contrast	correlatioɪ	inverse_di	entropy
2	197484	3479.036	1416	759	1.865613	5.442183	61.28948	6.395667	13.64341	4.388007	24.02533	40.20093	21.44841	12.63969	0.997666	0.911738	1.688689
3	101248	2490.382	1190	130	9.153846	1.527931	61.25555	7.049316	9.232018	10.87607	33.81621	37.38222	46.47923	8.137424	0.997191	0.944818	1.193795
4	86570.5	2290.683	1095	119	9.201681	1.505189	60.61222	3.434303	6.371511	2.644757	19.9757	29.05737	19.27505	8.553729	0.99661	0.959023	0.848758
5	190214	2856.479	1318	254	5.188976	1.759976	42.89629	7.670415	13.3036	6.049157	28.82289	40.22185	26.9486	8.440064	0.998419	0.914331	1.673914
6	227727	2917.249	1324	286	4.629371	1.662798	37.3708	8.992028	16.67117	6.294281	30.96716	45.0402	28.59533	8.641447	0.998568	0.898644	1.968081
7	233724	3689.81	1434	953	1.504722	5.847076	58.25117	7.31924	15.73062	4.432931	24.81753	42.22132	21.53427	14.56146	0.997708	0.895654	1.957151
8	258395	3543.678	1396	874	1.597254	4.721856	48.59866	9.674849	18.43227	6.657215	28.70796	46.43166	24.42757	11.79534	0.998101	0.884639	2.126015
9	244401	3732.957	1479	912	1.621711	5.518995	57.01682	9.033226	16.53425	6.947844	29.5701	43.37798	28.11403	13.62784	0.997945	0.889018	2.070195
10	223690	3142.317	1404	388	3.618557	2.435299	44.14214	8.594646	16.13478	5.935715	28.13373	43.99787	24.5651	12.59348	0.997737	0.896627	1.891695
11	288344.5	3083.27	1329	450	2.953333	2.074082	32.96944	9.552538	19.34629	6.716408	27.72883	45.99022	24.64435	10.30296	0.998595	0.874376	2.286608
12	28444	726.8183	224	183	1.224044	1.441148	18.5721	73.99541	94.84775	37.14786	75.08527	88.4658	66.17187	499.0942	0.940857	0.455425	8.665403

Fig. 8 Feature set of different leaf samples

D. Classifier algorithms

Four machine learning classifier algorithms were applied to the data, which are as follows:

1. KNN (k-Nearest Neighbor) Algorithm.
2. Logistic Regression.
3. Naïve Bayes Algorithm.
4. SVM (Support Vector Machine).

These classifier algorithms were applied to the preprocessed data. The results are shown in Table 2. The Logistic Regression classifier achieves the best performance with an accuracy of 83.04% (Table 2).

However, due to resource constraints, for finding the highest accuracy the important parameters of every classifier were varied. The k-Nearest Neighbor (KNN) classifier gave the best accuracy of 79.49% (Table 3).

Apart from the accuracy, the performance was also assessed on a class proportion of leaves, for each class, that was accurately chosen from the entire set [3]. Precision here is the proportion of precisely identified leaves out of the total leaves that are predicted to be a specific plant while F-measure here is considered as the average of these two values [3].

Table 4 shown gives useful knowledge which can be used to test the strong aspects of the system and address its weaknesses. Plants that have low precision and recall

Table 2 Performance of machine learning classifiers

Sr. no.	Classifier	Accuracy
1.	SVM	82.69
2.	Logistic Regression	83.04
3.	Naïve Bayes	72.90
4.	KNN	81.99

Table 3 Performance of machine learning classifiers after cross-validation

Sr. no.	Classifier	Accuracy
1.	SVM	78.74
2.	Logistic Regression	78.85
3.	Naïve Bayes	71.23
4.	KNN	79.49

must be reassessed. For example, new features must be designed and extracted that give uniqueness in such leaves and are determinative of their species [3].

In Fig. 9, the confusion matrix is shown which is obtained when using the k-Nearest Neighbor (KNN) classifier with specified attributes in each iteration. The identification was successful which is indicated by the highest values in the diagonal line. Classes ranging from 0 to 31 represent the different 32 species of plants.

4 Conclusion

The main aim of this paper is to identify the medicinal plant from a given sample of a leaf. For this, we proposed an automated system for the identification of species of plants from leaves on the basis of their Color, Shape, and Texture features by using image processing techniques. Accordingly, the features were extracted from the Flavia image dataset, which consists of a total of 1907 images, and machine learning algorithms like SVM, Logistic Regression, Naïve Bayes, and KNN were applied. Accuracies of 82.69%, 83.04%, 72.90%, and 82.99% were observed, respectively. After cross-validation of the extracted features, the accuracies changed to 78.74%, 78.85%, 71.23%, and 79.49%, respectively. As a result, an inference was deduced from the observed accuracies that KNN would be best suited to the proposed solution. This system takes less processing time with increased accuracy for identification .

Table 4 Performance assessment of species using KNN classifiers

Class	Precision	Recall	F1-score	Support
0	0.92	0.61	0.73	18
1	0.76	0.94	0.84	17
2	0.92	1.00	0.96	22
3	0.93	1.00	0.97	28
4	0.89	0.96	0.92	25
5	0.88	1.00	0.94	15
6	0.71	0.85	0.77	20
7	0.85	0.73	0.79	15
8	0.60	0.50	0.55	12
9	0.78	0.93	0.85	15
10	0.79	0.83	0.81	18
11	0.72	0.76	0.74	17
12	0.77	0.71	0.74	14
13	0.67	0.67	0.67	21
14	0.93	0.88	0.90	16
15	0.25	0.06	0.10	17
16	1.00	1.00	1.00	26
17	0.88	1.00	0.94	22
18	0.95	0.90	0.92	20
19	0.85	0.94	0.89	18
20	0.74	0.93	0.82	15
21	0.89	0.57	0.70	14
22	1.00	0.93	0.97	15
23	0.71	1.00	0.83	17
24	0.90	0.56	0.69	16
25	0.71	0.67	0.69	15
26	0.88	0.88	0.88	16
27	0.82	0.90	0.86	20
28	0.83	0.88	0.86	17
29	0.87	0.91	0.89	22
30	0.73	0.79	0.76	14
31	0.92	0.73	0.81	15
Average	0.814	0.813	0.805	17.85

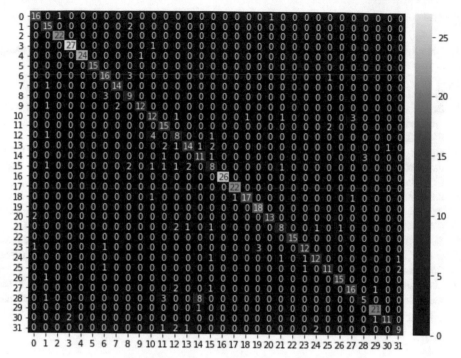

Fig. 9 Confusion matrix for KNN classifier

Acknowledgements. The authors would like to extend gratitude toward the faculty guide Dr. Anuradha Thakare and H.O.D Department of Computer Engineering Dr. K. Rajeswari for their constant support and guidance.

References

1. Aitwadkar, P.P, Deshpande, S.C, Savant, A.V.: Identification of Indian medicinal plant by using artificial neural network. Int. Res. J. Eng. Technol (IRJET) **5**(4), 1669–1671 (2018)
2. Jeon, W.-S., Rhee, S.-Y.: Plant leaf recognition using a convolution neural network. Int. J. Fuzzy Logic Intell. Syst. **17**(1), 26–34 (2017)
3. Begue, A., Kowlessur, V., Mahomoodally, F., Singh, U., Pudaruth, S.: Automatic recognition of medicinal plants using machine learning techniques. Int. J. Adv. Comput. Sci. Appl. (IJACSA) **8**(4) (2017)
4. Nijalingappa, P., Madhumathi, V.J.: Plant identification system using its leaf features. In: International Conference on Applied and Theoretical Computing and Communication Technology (iCATccT) (2015)
5. Khmag, A., Al-Haddad, S.A.R., Kamarudin, N.: Recognition system for leaf images based on its leaf contour and centroid. In: IEEE 15th student conference on research and development (SCOReD) (2017)
6. Sabu, A., Sreekumar, K., Nair, R.R.: Recognition of ayurvedic medicinal plants from leaves: a computer vision approach. In: Fourth International Conference on Image Information Processing (ICIIP) (2017)
7. Venkataraman, D., Mangayarkarasi, N.: Computer vision based feature extraction of leaves for identification of medicinal values of plants. IEEE International Conference on Computational Intelligence and Computing Research (2016)

Human Gait Analysis Based on Decision Tree, Random Forest and KNN Algorithms

Ayushi Gupta[1]($\boxtimes$), Apoorva Jadhav[1], Sanika Jadhav[1], and Anita Thengade[2]

[1] Department of Computer Engineering, MITCOE, Pune, India
ayushimg9@gmail.com, apoorvajadhav98@gmail.com,
jadhavsanika13@gmail.com
[2] School of CET, MITWPU, Pune, India
anita.thengade@mitwpu.edu.in

Abstract. Human Gait refers to motion accomplished through the movement of hand limbs. Gait analysis is a precise investigation of the human walking pattern using sensors attached to the body for recording body movements during activities like walking on a flat surface, treadmill and running. This paper addresses the analysis of human gait based on performance using various classification techniques involving Decision Tree, Random Forest and KNN algorithms. The paper highlights the comparison of these classification models based on various parameters using the RapidMiner Studio tool. The comparison is based on performance metrics and Receiver Operating Characteristic (ROC). The results show that the Random Forest algorithm performs better in classifying normal and abnormal gait.

Keywords: Human gait · Gait analysis · Decision tree · Random forest · KNN · Classification models

1 Introduction

Human Gait describes the walking pattern of a human. Human gait is characterized as a bipedal, biphasic front drive of human body gravity and exchange of development of various parts or fragments of the human body. Gait analysis is useful to determine the normal and abnormal walk of a person. Strength, coordination and sensation work together to allow a person to have a normal gait walking pattern. The data for gait analysis is recorded using sensors attached to the body and activities such as walking on a treadmill, flat surface, running, jogging, etc. are performed to classify normal and abnormal gait. Nowadays gait analysis research is trending worldwide due to its various applications. Gait analysis is required for athletes. Gait analysis helps athletes to run efficiently and distinguish pose related or development related issues in individuals with wounds [1]. In the medical field, gait analysis is used to determine neurologic, muscular or skeletal problems of a person depending on the gait [2]. Recently the medical field is advancing using Machine Learning techniques such as prediction, classification, regression using various algorithms. This paper provides the classification of normal

© Springer Nature Singapore Pte Ltd. 2020
B. Iyer et al. (eds.), *Applied Computer Vision and Image Processing*,
Advances in Intelligent Systems and Computing 1155,
https://doi.org/10.1007/978-981-15-4029-5_28

and abnormal gait using Decision Tree, Random Forest and KNN classifiers. The classification is performed on the 93 Human Gait (walking) database using the RapidMiner tool.

2 Literature Survey

Over the last few decades, several researchers have made the attempt on gait recognition and various applications of gait. Recently, Singh et al. have presented a detailed survey on Vision-Based Gait Recognition [3]. Extensive research efforts have been made by Singh et al. to present the review of existing work on gait recognition.

It is observed that gait recognition is performed for two different applications where the first reason is gait-based recognition of humans having various examples such as biometric identification, gender identification, crime department or surveillance purpose. The second most important aspect of gait study is in clinical practice for diagnosis of orthopedic as well as Parkinson patients [4]. Li et al. [5] have proposed the SVM classification approach to identify the gender of humans based on their gait. To overcome the drawback of Li et al. [5], gender identification was also challenged by Lu et al. [6] and Sudha et al. [7]. In clinical practice, human gait analysis plays an important role to identify gait patterns and which will further be used to diagnose walking disorder of patients. During the review of existing work, it has been observed that very few researchers have contributed to this area [8–10].

Various studies published on human gait are useful for biomedical engineering, physical medicine, physical therapy and orthopedics fields. Machine Learning techniques have been found to be useful in biomedical science for analysis and calculations on multidimensional set of examples. In [11], the authors examined if it is feasible to use sensors of inertial measurement and supervised learning techniques to differentiate normal walking gait patterns and foot drop gait disease. The data was collected from 56 adults having foot drop and 30 adults having normal gait with many walking trials. Random forest, Naive Bayes and SVM classifiers gave an accuracy of 88.45%, 86.08% and 86.87%, respectively. This study would be helpful in clinical predictions.

3 Methodology

3.1 System Architecture

In the first phase Human Gait data with its various attributes like age, height, weight, gender, etc. are considered (Fig. 1). Further the dataset is processed. The records are labeled as Y (normal walk) and N (abnormal walk). To split the data for testing and training, phase cross-validation is performed. Next the three classifiers Decision tree, Random forest and KNN are applied to the dataset and the results are obtained.

3.2 Dataset

The dataset comprises data on 72 people measured with 73 attributes. Two classes "Y" and "N" are used to classify the abnormality in gait metrics of the particular individual.

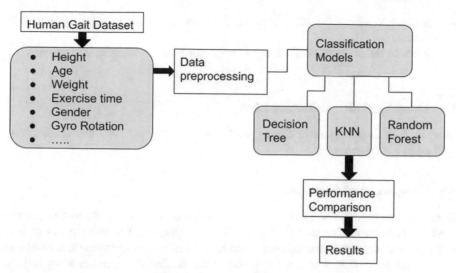

Fig. 1 System architecture

Table 1 Dataset layout with ranges of attributes

Sr. no.	Weight	Age	Height (cm)	Exercise time (Min)	Gender	Gyro RotationX.rad. s.	Alti meter relative altitude	Df1.1$Label
1 to 72	44–137	17–53	150–195	0–180	M/F	−4.346 to 3.95	−25.7 to 6.52	Y/N

The data has 70% normal and 30% abnormal records. Following are the predominant attributes influencing the classification models (Table 1):

Various other attribute readings like altimeter pressure, pedometer, motionrotation (*x*, *y*, *z*), location speed, MotionUser application, etc. are also recorded in the dataset.

3.3 Data Preprocessing

The dataset consists of a record of 72 people on 73 attributes for gait analysis.

- Removed all the rows with less attribute data or missing record of a person.
- Attributes having "0" values have been removed.
- "Location_floor" with value "-9999" through the dataset is removed.
- "Motion Altitude Reference Frame.txt" column with value "X Arbitrary Corrected Z Vertical" is removed.
- "Location Horizontal Acurracy.m" with value "5" throughout is removed.

- "Identifier Vendor.txt" with vendor ids does not contribute in classification so it is removed.
- "AltiMeter Reset Bool" and "State_N" with value "0" throughout is removed.

3.4 Cross-Validation Technique Used

Internal Split Data present in Cross-Validation Operator with automatic Sampling is applied to the dataset.

3.5 Comparison of Algorithms

Decision tree algorithm. Decision Tree is a supervised learning algorithm. It can be used for both regression as well as classification purpose. For solving the problem, decision tree uses tree representation. Each leaf of the tree represents a class label while the internal node of the tree represents the attributes. Decision tree classifies instances from root node to leaf node [12]. The challenging part of the decision tree is the selection of an attribute of a root node for each level. The two popular attribute selection methods are (i) information gain (ii) Gini index.

Information Gain

Change in entropy measures information gain. Entropy represents uncertainty of a random variable.

Mathematically information gain is represented as follows [12]:

$$\text{Gain}(C,\ A) = \text{Entropy}(C) - C((|Cv|/|C|) * \text{Entropy}(Cv)) \tag{1}$$

where

C: set A: attribute Cv: subset of C where attribute A has value v.

$|Cv|$: number of elements in subset Cv $|C|$: number of elements in set C.

Gini index.

Gini index estimates the degree or likelihood of a specific variable being wrongly ordered when it is randomly picked.

Mathematically Gini index is represented as follows [13]:

$$\text{Gini index} = 1 - \sum_{i=1}^{n} (p_i)^2 \tag{2}$$

where p signifies the probability of an item being arranged to a specific class (Table 2).

Random forest algorithm. Random forest categorizer algorithm put forward by Breiman et al. [14] is a machine learning method using numerous learning calculations together while figuring and creating anticipated outcome. Random forest combines bootstrap aggregation (bagging) [15] and random feature choosing [16] to develop an assortment of decision trees. It generates collectively multiple decision trees instead of predicting the class of human gait using a single unique decision tree and the data is collected using sensors. Multiple trees help to reduce the overfitting problem in the decision tree (Table 3).

Table 2 Performance matrix generated by decision tree algorithm

	True Y	True N	Class precision (%)
Pred Y	49	10	83.05
Pred N	4	9	69.23
Class recall	92.45%	47.37%	

Table 3 Performance matrix generated by random forest algorithm

	True Y	True N	Class precision (%)
Pred Y	58	13	81.69
Pred N	0	1	100
Class recall	100%	7.14%	

KNN algorithm. KNN, a supervised learning algorithm, is used to elucidate regression as well as classification problems. This algorithm presumes that analogous entities are prevalent in the near vicinity. It is a non-parametric and torpid learning algorithm which denotes that it does not make any suppositions on the prevalent data. Principally, it functions on feature resemblance. It collects the first set of categorized entries and then returns the labels of K-chosen entries. Presuming it is regression, it remits the mean of the K labels. If it is classification, it remits the mode of the K labels (Table 4).

Table 4 Performance matrix generated by KNN algorithm

	True Y	True N	Class precision (%)
Pred Y	54	14	79.41
Pred N	4	0	0.00
Class recall	93.10%	0.00%	

4 Result Discussion

4.1 Receiver Operating Characteristic (ROC) Curve

ROC is a common tool used for the evaluation of classifiers. ROC False Positive Rate (FPR) is plotted against True Positive Rate (TPR) [17] at different thresholds.

- False Positive Rate = (False Positive)/(False Positive + True Negative)
- True Positive Rate = (True Positive)/(True Positive + False Negative)

ROC describes the performance of the classification models.

From the given comparison curve, it is evident that Random Forest provides the highest accuracy (performance) among all the other models compared here. Figure 2 describes the ROC curve plot.

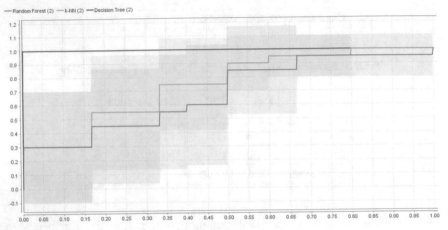

Fig. 2 ROC curve

Qualitative Comparison is described in Table 5.

Table 5 Quality comparison of algorithms

Algorithm name	Accuracy (%)	Precision (%)	Recall (%)
Random forest	82.14	95	30
Decision tree	80.72	69.23	54
KNN	75.36	60	28

5   Conclusion and Future Work

Gait Analysis is an important domain of study. It has various applications in the field of sports, orthopedics, physiotherapy, etc. It helps in improving body posture, deciding the right sports equipment and relieving the patients with postural defects due to the above. Through the above study, we have thus compared this data across three classification models—Decision Tree, KNN, Random Forest Algorithms. Random Forest gives the highest accuracy of 80.54%. In future, this analysis could be further applied for early detection of postural defects in infants and cure them with proper measures. This dataset was recorded using a smartphone's accelerometer which gave the above statistics. The accuracy and precision can be improved by using actual sensors mounted on different parts of the body.

References

1. Bakchy, S.C., Mondal, M.N.I., Ali, M.M., Hoque Sathi, A., Ray, K.C., Jannatul Ferdous, M.: Limbs and muscle movement detection using gait analysis. In: 2018 International Conference on Computer, Communication, Chemical, Material and Electronic Engineering (IC4ME2), Rajshahi, pp. 1–4 (2018). https://doi.org/10.1109/ic4me2.2018.8465598
2. Esquenazi, A., Talaty, M.: Gait analysis, technology and clinical applications. Phys. Med. Rehabil. 99–116 (2011)
3. Singh, J.P., Jain, S., Arora, S., Singh, U.P.: Vision-based gait recognition: a survey. IEEE Access **6**, 70497–70527 (2018). https://doi.org/10.1109/ACCESS.2018.2879896
4. Tian, Y., Wei, L., Lu, S., Huang, T.: Free-view gait recognition. PLoS ONE **14**(4), e0214389 (2019). https://doi.org/10.1371/journal.pone.0214389
5. Li, X., Maybank, S.J., Yan, S., Tao, D., Xu, D.: 'Gait components and their application to gender recognition. IEEE Trans. Syst. Man Cybern. C Appl. Rev. **38**(2), 145–155 (2008). https://doi.org/10.1109/tsmcc.2007.913886
6. Lu, J., Wang, G., Moulin, P.: 'Human identity and gender recognition from gait sequences with arbitrary walking directions'. IEEE Trans. Inf. Forensics Secur. **9**(1), 51–61 (2014). https://doi.org/10.1109/TIFS.2013.2291969
7. Sudha, L.R., Bhavani, R.: An efficient spatio-temporal gait representation for gender classification. Appl. Artif. Intell. **27**(1), 62–75 (2013). https://doi.org/10.1080/08839514.2013.747373
8. Weiss, R.J., Wretenberg, P., Stark, A., Palmblad, K., Larsson, P., Grondal, L., Brostrom, E.: Gait pattern in rheumatoid arthritis. Gait Posture **28**(2), 229–234 (2008)
9. Saad, A., Zaarour, I., Guerin, F., Bejjani, P., Ayache, M., Lefebvre, D.: Detection of freezing of gait for Parkinson's disease patients with multisensor device and Gaussian neural networks. Int. J. Mach. Learn. **8**(3), 941–954 (2017)
10. Ťupa, O., et al.: Motion tracking and gait feature estimation for recognising Parkinson's disease using MS Kinect. Biomed. Eng. Online **14**(1), 97 (2015)
11. Bidabadi, S.S., Murray, I., Lee, G.Y.F., Morris, S., Tan, T.: Classification of foot drop gait characteristic due to lumbar radiculopathy using machine learning algorithms. Gait and Posture **71**, 234–240 (2019). ISSN 0966-6362, https://doi.org/10.1016/j.gaitpost.2019.05.010
12. Prajwala, T.R.: A comparative study on decision tree and random forest using R tool. Int. J. Adv. Res. Comput. Commun. Eng. **4**(1), (2015)
13. Tahsildar, S.: Gini Index For Decision Trees. Quantinsti (2019)
14. Breiman, L.: Random forests. Mach. Learn. **45**(1), 5–32 (2001)
15. Breiman, L.: Bagging predictors. Mach. Learn. **24**(2), 123–140 (1996)
16. Amit, Y., Geman, D.: Shape quantization and recognition with randomized trees. Neural Comput. **9**(7), 1545–1588 (1997)
17. Brownlee, J.: How to use ROC curves and precision-recall curves for classification in Python. Machine Learning Mastery (2018)

Adaptive Mean Filter Technique for Removal of High Density Salt and Pepper (Impulse) Noise

Swati Rane[1]($\boxtimes$) and L. K. Ragha[2]

[1] Department of Electronics & Telecommunication Engineering, SIES Graduate School of Technology, Nerul, Navi Mumbai, Maharashtra, India
swati.r05@gmail.com
[2] Terna College of Engineering, Nerul, Navi Mumbai, Maharashtra, India

Abstract. In this paper, the adaptive mean filter technique for removal of high density salt and pepper also termed as impulse noise is presented. It is desired to remove high density impulse noise from the images due to hot pixels produced or generated because of current leakage in the image sensors placed inside the camera. Adaptive mean filter technique processes the images affected with noise using variable filter size that results in better removal of high density impulse noise as compared with fixed filter size. Variable filter size comparatively takes more processing time but results in better evaluation parameters such as signal to noise ratio (SNR), mean absolute error (MBE) and image enhancement factor (IEF). Various experiments were performed on images having different entropy to measure the performance of the presented filter technique. The adaptive mean filter with variable size filters effectively removes high density impulse noise up to density 0.5. The presented adaptive mean filter technique requires simple arithmetic operations that achieve faster processing with less computational overheads. The presented adaptive mean filter technique for removal of high density salt and pepper or impulse noise can be adapted in image acquisition systems.

Keywords: Adaptive mean filter · Variable filter size · Salt and pepper (impulse) noise · Noise removal

1 Introduction

Remarkable advancement in wireless communication technology and personal assistant devices has resulted in the production of large sophisticated mobile phone devices. Many mobile devices have inbuilt cameras that require mechanical shutters to avoid salt and pepper noise [1]. Due to the absence of these mechanical shutters, most of the images acquired through mobile phone cameras are affected by high density impulse noise. Therefore, it is required to have a simple, less computationally complex and effective impulse noise removal technique.

Standard median filter is mostly applied for the removal of impulse noise. These median filters are implemented using techniques such as "standard median filter" [2], "switching median filters" [3, 4], "adaptive median filter" [5] and "adaptive center

© Springer Nature Singapore Pte Ltd. 2020
B. Iyer et al. (eds.), *Applied Computer Vision and Image Processing*,
Advances in Intelligent Systems and Computing 1155,
https://doi.org/10.1007/978-981-15-4029-5_29

weighted median filter" [6]. These filters have demonstrated effective performance in removing salt and pepper (impulse) noise at lower densities. However, it is desired to remove salt and pepper noise at higher densities due to the absence of mechanical shutters in cameras. Several filters have been proposed using different methodologies for removing high density salt and pepper (impulse) noise. The use of fuzzy logic and weighted mean filtering [7], "impulse noise removal method based on non-uniform sampling and supervised piecewise autoregressive modeling" [8], "vector filter based on geometric information" [9], mean filter based on morphological image processing [10], "local mean and variance" to detect noisy pixels [11] and Newton–Thiele filter for removing of salt and pepper noise have been demonstrated. All these techniques suggest that median filter is untenable to remove high density salt and pepper (impulse) noise [12–14].

The performance of the standard median filter is harshly damaged in very high density impulse noise. The innovative, simple and effective algorithm for cancelation of noise densities from 10 to 90% without degrading the quality of the image is desired. The key success of such performance is mainly due to highly accurate noise detection and removal methodology employed in the techniques. Also, these techniques are computationally complex and require more processing time and power. Therefore, they are untenable in the personal assistant devices and mobile phones which requires simple, less computationally complex and effective salt and pepper (impulse) noise removal techniques.

The adaptive filtering technique that increases the filter size during the processes of filtering of noisy images has been proposed for the removal of high density salt and pepper noise. Mostly this adaptive filter starts filtering with a minimum filter size of 3×3. It periodically increases filter size by two if the condition specific to the image quality (noise removal) is not satisfied. Thus, the entire process of filtering is repeated with an adaptive filter of size 5×5, then 7×7 and so on. The filtration process is repeated either until the image quality (noise removal) condition is met or until a predefined maximum filter size is reached. Moreover, it is observed that adaptive filters are better at removing higher noise densities, but require larger processing time and are difficult to implement.

In this paper, an adaptive mean filter technique for removal of high density salt and pepper (impulse) noise is presented. It is desired to remove high density salt and pepper (impulse) noise from the images due to hot pixels generated or produced because of current leakage in the image sensors placed inside the camera. The adaptive mean filter technique processes the noisy image with variable filter size that results in better removal of salt and pepper noise as compared with fixed filter size. Variable filter size comparatively takes more processing time but results in better evaluation parameters such as signal to noise ratio (SNR), mean absolute error (MBE) and image enhancement factor (IEF). Various experiments were performed on images having different entropy to measure the performance of the presented adaptive mean filter technique.

The organization of the paper is as follows Sect. 1 introduces various filtering techniques employed for salt and pepper (impulse) noise removal, Sect. 2 describes the algorithm adopted for removal of salt and pepper (impulse) noise using adaptive mean filtering technique in detail. Results are discussed in Sect. 3 and finally concluded in Sect. 4.

2 Adaptive Mean Filtering Technique

The adaptive mean filtering technique is described in the section along with the algorithm. The salt and pepper noise removal technique is accomplished in two phases, the first phase detects the pixel corrupted with salt and pepper noise and the second phase restores the original image without affecting its quality. The overall block diagram of the adaptive high density salt and pepper noise removal technique is depicted in Fig. 1.

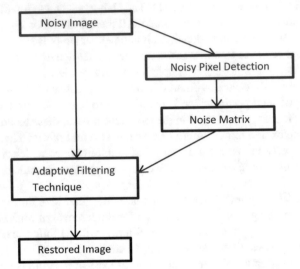

Fig. 1 Block diagram of the adaptive salt and pepper noise removal technique

Let input noisy image be represented as $I(x, y)$ corrupted with the salt and pepper; its gray level distribution can be expressed as

$$
\text{In}(x, y) = \begin{cases} n_{\min} \\ n_{\max} \\ \text{orginial image} \end{cases}
\tag{1}
$$

where $n_{\min}$ is with probability of $p_{\min}$, $n_{\max}$ is with probability of $p_{\max}$ and original image has probability of $1 - (p_{\min} + p_{\max})$. Also $n_{\min}$ and $n_{\max}$ are the minimum and maximum gray levels respectively adopted by salt and pepper (impulse) noise. These gray levels depend upon the number of bits used to represent pixel values, thus an 8-bit gray-level image has $n_{\min} = 0$ and $n_{\max} = 255$. The noise detection stage of the presented adaptive noise removal technique marks the pixels as noisy if the values $n_{\min}$ and $n_{\max}$ are detected. The output of the noise detection stage represented as noise matrix can be expressed as

$$
N(x, y) = \begin{cases} 1, \text{ if } n_{\min} \text{ and } n_{\max} \\ \quad = \text{In}(x, y) \\ 0, \text{ otherwise} \end{cases}
\tag{2}
$$

In this stage, the noisy image matrix obtained from the noisy image in the previous stage is processed further. It is learned from the literature that, mean filtering performance is better than median filtering and does not degrade the quality of the input image. Also, adaptive filters with variable filter size accomplish better experimental results than filters with a fixed window size. Therefore, to reinstate the original image or pixels data, our technique employs a selective mean filter with variable filter size. The size of the filter is increased during the processing of the noisy image. The adaptive mean filter selects noise-free pixels for calculating the mean. Also if the numbers of noisy pixels in 8/24/35/48 neighbor for filter size 3×3, 5×5, 7×7 are less than two, mean is calculated and its value is assigned to the noisy pixel, whereas, if the number of noise-free pixels is greater than two, mean filtering skips the averaging operation. The requisite operation will be performed in the next stage with an increase in the filter size by 2. The adaptive filtering operation is demonstrated in Fig. 2.

23	25	27
0	29	30
23	**255**	36

Number of noisy pixels = 2

Replace noisy pixels with mean of noise free pixels

Mean = 27

23	25	27
27	29	30
23	27	36

(a)Total number of noise free pixels more than 6 in the given filter size

23	255	27
0	29	0
23	255	36

Numbers of noise free pixels less than 7, increase the size of the filter by 2 and repeat the process.

(b) Total number of noise free pixels less than 7 in the given filter size

Fig. 2 Noisy image restoration stage

23	25	27
0	29	30
23	255	36

Number of noisy pixels = 2
Replace noisy pixels with mean of noise-free pixels
Mean = 27

23	25	27
27	29	30
23	27	36

(a) Total number of noise-free pixels is more than 6 in the given filter size.

23	255	27
0	29	0
23	255	36

If the number of noise-free pixels is less than 7, increase the size of the filter by 2 and repeat the process.

(b) Total number of noise-free pixels is less than 7 in the given filter size.

3   Experimental Results

In this section, the qualitative and quantitative performance of the adaptive mean filter for removal of high density salt and pepper (impulse) noise is presented. Results for the five sets of input images containing 10 images each, classified based on varying amounts of objects and increasing entropy are applied for performance evaluation of the adaptive technique. Figure 3 shows the five sets of images selected based on the number of objects with their respective entropy values. Higher entropy indicates more gray levels and complexities in the image.

From the above figure, it is clearly understood that the complexity of the image is directly proportional to the entropy. It is desired to evaluate the performance of the adaptive salt and pepper (impulse) noise removal technique based on evaluation parameters PSNR, MAE and IEF. PSNR, MAE and IEF can be expressed as

$$\mathrm{psnr} = 10 * \log_{10}\left(\frac{255^2}{\frac{1}{m*n} * \mathrm{se}}\right) \tag{3}$$

where m and n are maximum size of the input noisy image.

$$se = \sum_{x=1}^{m} \sum_{y=1}^{n} |I(x, y) - O(x, y)|^2 \tag{4}$$

where $O(x, y)$ is the obtained output/restored image

$$mae = \sum_{x=1}^{m} \sum_{y=1}^{n} |I(x, y) - O(x, y)| \tag{5}$$

$$ief = \frac{\sum_{x=1}^{m} \sum_{y=1}^{n} |In(x, y) - I(x, y)|^2}{\sum_{x=1}^{m} \sum_{y=1}^{n} |I(x, y) - O(x, y)|^2} \tag{6}$$

where $In(x, y)$ is the noisy image and $I(x, y)$ is the original noise-free image.

All the images from the various categories as demonstrated in Fig. 3 were applied to adaptive mean high density salt and pepper noise removal technique. It is required to measure the performance parameters PSNR, MAE and IEF for varying densities of the salt and pepper (impulse) noise. Figure 4 shows the output image for the 0.1–0.9 density of salt and pepper noise. The performance parameters are tabulated in Table 1. Experimental results clearly indicate that the adaptive mean filter technique with variable size filters is effectively applied for high density salt and pepper noise for noise density less than 0.5. The filter size was varied from 3 × 3 to 9 × 9 in steps of 2. Filter size more than 9 × 9 severely degrades the image due to the application of mean filter repeatedly.

4 Conclusion

Adaptive mean filter technique for removal of high density salt and pepper (impulse) noise is presented. The adaptive mean filter technique processes the noisy image with variable filter size that results in better removal of salt and pepper (impulse) noise as compared with fixed filter size. Variable filter size comparatively takes more processing time but results in better evaluation parameters such as signal to noise ratio (SNR), mean absolute error (MBE) and image enhancement factor (IEF). Various experiments were performed on images having different entropy to measure the performance of the presented adaptive mean filter technique. Experimental results clearly indicate that filter size greater than 9 × 9 degrades (blurred) the quality of the output image due to the application of mean filter repeatedly. Also, the adaptive mean filter with variable size filters effectively removes high density salt and pepper (impulse) noise up to density 0.5. Further, adaptive mean filter technique needs significant improvement in case of image corrupted with very high noise density.

Set	Entropy	Objects	Input Image
1.	7.22	1	
2.	7.49	2	
3.	7.39	3	
4.	7.58	4	
5.	7.84	> 5	

Fig. 3 Input images classified based on entropy

Set   Input Image   Output Image
                          Density = 0.1   Density = 0.5   Density = 0.9

1

2

3

4

5

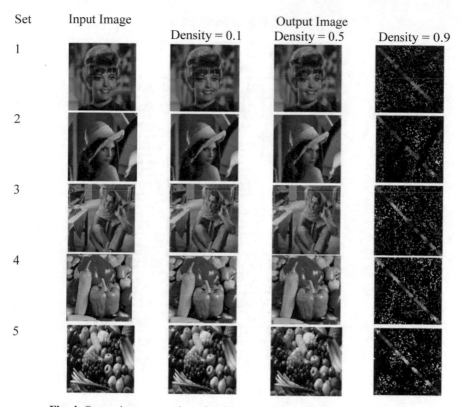

Fig. 4 Output image at various densities of salt and pepper (impulse) noise

Table 1 Performance parameters

Set	Density	PSNR (dB)	MAE	IEF
1	0.1	40.01	0.34	225.21
	0.5	31.92	2.07	169.16
	0.9	34.98	0.92	626.34
2	0.1	37.77	0.43	147.57
	0.5	30.33	2.25	132.29
	0.9	32.06	0.99	350.15
3	0.1	35.87	0.54	69.46
	0.5	28.29	2.98	59.12
	0.9	31.54	1.26	224.16
4	0.1	36.43	0.47	70.47
	0.5	28.11	2.82	52.26
	0.9	30.67	1.28	169.27
5	0.1	31.24	0.91	30.07
	0.5	24.06	4.72	28.11
	0.9	27.35	2.18	108.95

References

1. Chen, T., Wu, H.R.: Adaptive impulse detection using center-weighted median filters. IEEE Signal Process. Lett. **8**, 1–3 (2001)
2. Aiswarya, K., Jayaraj, V., Ebenezer, D.: A new and efficient algorithm for the removal of high density salt and pepper noise in images and videos. In: Second IEEE International Conference on Computer Modeling and Simulation, Sanya, Hainan, China, pp. 409–413 (2010)
3. Gupta, V., Gandhi, D.K., Yadav, P.: Removal of fixed value impulse noise using improved mean filter for image enhancement. In: IEEE International Conference on Engineering, Nirma University, Ahmedabad, India (2012)
4. Talebi, H., Milanfar, P.: Global image denoising. IEEE Trans. Image Process. **23**(2), 755–768 (2014)
5. Wei, Y., Yan, S., Yang, L., et al.: An improved median filter for removing extensive salt and pepper noise. In: IEEE International Conference on Mechatronics and Control, Jinzhou, China, pp. 897–901 (2014)
6. Kumar, R.R., Vasanth, K., Rajesh, V.: Performance of the decision based algorithm for the removal of unequal probability salt and pepper noise in images. In: IEEE International Conference on Circuit Power and Computing Technologies, Nagercoil, India, pp. 1360–1365 (2014)
7. Wang, Y., Wang, J., Song, X., et al.: An efficient adaptive fuzzy switching weighted mean filter for salt-and-pepper noise removal. IEEE Signal Process. Lett. **23**(11), 1582–1586 (2016)
8. Chaitanya, N.K., Sreenivasulu, P.: Removal of salt and pepper noise using advanced modified decision based unsymmetric trimmed median filter. In: IEEE International Conference on Electronics Communication Systems, Coimbatore, India (2014)
9. Wang, X., Shi, G., Zhang, P., et al.: High quality impulse noise removal via non-uniform sampling and autoregressive modelling based super-resolution. IET Image Proc. **10**(4), 304–313 (2016)
10. Bai, T., Tan, J.: Automatic detection and removal of high-density impulse noises. IET Image Process. **9**(2), 162–172 (2015)
11. Roig, B., Estruch, V.D.: Localised rank-ordered differences vector filter for suppression of high-density impulse noise in colour images. IET Image Proc. **10**(1), 24–33 (2016)
12. Kumar, R.R., Vasanth, K., Rajesh, V.: Performance of the decision based algorithm for the removal of unequal probability salt and pepper noise in images. IEEE Int. Conf. Circuit Power and Computing Technologies, Nagercoil, India, pp. 1360–1365 (2014)
13. Dash, A., Sathua, S.K.: High density noise removal by using cascadingalgorithms. In: IEEE Fifth Int. Conf. Advanced Computing and Communication Technologies, Haryana, India, pp. 96–101 (2015)
14. Lin, P.H., Chen, B.H., Cheng, F.C., et al.: A morphological mean filter for impulse noise removal. J. Display Tech. **12**(4), 344–350 (2016)

Robust and Secure Lucas Sequence-Based Video Watermarking

Bhagyashri S. Kapre(✉) and Archana M. Rajurkar

MGM's College of Engineering, Near Airport, Nanded, India
kapre_bs@mgmcen.ac.in1, rajurkar_am@mgmcen.ac.in

Abstract. Currently, video copyright protection has become a challenging issue due to the growth of Internet technology and extensive use of multimedia. Digital watermarking plays an important role in protecting multimedia objects as they become easier to modify, copy and exchange data. Embedding the watermark in the video is done in different ways by many researchers. Most of the existing video watermarking (**V-W**) schemes directly apply image watermarking methods to raw and compressed video. Due to redundancy in the video frames, watermark embedding and extraction become time-consuming. To sort out this problem, the watermark is embedded into the selective key-frames. In this paper we use Lucas sequence (**LS**) for selection of key-frames and the watermark is embedded in selected key-frames only. The proposed scheme is the combination of DWT and DCT transforms that improves the performance. This approach not only saves time required for selecting key-frames but also proves to be better in terms of resistance to various types of attacks like frame dropping, frame averaging and noise addition. The proposed LS-based scheme is compared with [1–3]. It is observed that it outperforms in terms of security, robustness and imperceptibility. The performance of the proposed algorithm is evaluated using commonly used parameters PSNR and NCC. It is found that their average values are improved considerably. The extensive experimental simulations show that the proposed scheme is more efficient and robust against common attacks such as rotation, scaling, cropping, common image processing operations and geometric distortion.

Keywords: Lucas sequence · Key-frame selection · DCT · DWT · Watermark · Embedding · Watermark detection

1 Introduction

Due to rapid advances in computers and Internet technology, it has become very easy to create, store, distribute, enhance, and modify the data such as images and videos in the original data. It is useful for protecting the rightful ownership of multimedia audio. At the same time, there is a demand for protection of the digital contents from unauthorized access. The owners of digital content desire to protect their own copyrights avoid duplication, piracy and distribution of their data. Digital watermarking is a technique which

© Springer Nature Singapore Pte Ltd. 2020
B. Iyer et al. (eds.), *Applied Computer Vision and Image Processing*,
Advances in Intelligent Systems and Computing 1155,
https://doi.org/10.1007/978-981-15-4029-5_30

hides the information data. A large amount of digital data is copied and published without the owner's consent. Digital watermarking has proved to be a good solution to tackle ownership problems. Watermark is inserted into the multimedia contents for copyright protection and may be used to check whether the contents have been legally modified or not [4]. The effectiveness of the watermarking algorithm is measured in terms of three parameters, data payload, fidelity and robustness. Depending on the application in which the watermarking is used, a trade-off among these parameters is done. Most of the existing video watermarking (V-W) schemes are based on the techniques of image watermarking. However, V-W is different from image watermarking since it contains a large volume of data with spatial and temporal redundancy among them.

Watermarking is divided into the frequency domain and spatial domain techniques. In [5] authors have proposed an efficient V-W using lab color space. However, the bit error rate of this method is not up to the mark. In frequency domain methods, Discrete Cosine Transform (DCT), Discrete Fourier Transform (DFT) and Discrete Wavelet Transform are used (DWT) [6]. A robust V-W algorithm was presented by [7] in which the scrambled decomposed watermark image is embedded into the mid-frequency DWT coefficients. This scheme is robust to common image processing attacks like frame dropping, averaging, additive noise and lossy compression. Sujatha and Satyanarayana [8] have proposed a watermarking scheme using DCT, DWT and SVD for improving imperceptibility of watermarking. However, this technique could not resist attacks like frame dropping and compression. All the techniques presented in [9–12] embedded watermark in all the video frames which makes these schemes time-consuming and affect the perceptibility of video. Merits of these algorithms are that they are robust to frame dropping and averaging. To reduce the time complexity and number of computations, Tabassum and Islam [13] proposed a V-W scheme in 2012 using identical frame extraction. In this scheme initially the video is segmented into shots. A frame called identical-frame is selected from each shot for embedding the watermark. In [14] authors have proposed a robust blind and secure watermarking method using integer wavelet transform, wherein, watermark is embedded into low-frequency coefficients of each main frame of video. Authors claim that their method is resistant to various types of attacks. However, limitations of this method are the subjective selection of the main frame of the video. Authors [15] have presented a V-W approach based on shot segmentation. They selected the host frames based on the highest luminance value in every shot. The watermark signal is broken into small parts according to the number of host frames in the host video. These small parts are then embedded into the selected host frames. Also, authors [16] developed a scene change-based V-W scheme which embeds different segments of a watermark into different scenes selected based on scene change detection algorithm. In these two last cited methods, the whole frame sequence is required to retrieve the watermark. In [17] Agilandeeswari and Ganesan proposed an algorithm for V-W using SVD-DWT. In their scheme, a histogram difference-based scene change detection algorithm is used to extract the non-motion frames from the video, and watermark is embedded into those extracted frames. However, the problem in this technique is that only a few frames are used for watermark embedding. The watermarking scheme becomes unreliable if those watermarked frames are lost.

Scene change detection techniques have been used in [1, 18] for V-W. In these methods the authors embedded different parts of watermark in different scenes. These methods are robust against many common attacks, however, their time complexity is quite high. Many researchers [2, 3, 19] have proposed scene change-based watermarking scheme. In all the scene change-based watermarking techniques, identification of the exact scene is an important step. Himeur and Boukabou [20] presented a V-W system in which gradient magnitude similarity deviation (GMSD) is used for key-frame extraction. In this scheme, the watermark is encrypted using a chaotic encryption technique and the encrypted watermark information is embedded into the selected key-frames using DWT and SVD. The limitation of this V-W system is that if a video which contains a key-frame is lost, then the watermark cannot be recovered correctly. In [21] proposed blind watermarking technique QIM is used to embed a watermark into fast motion frames that are extracted from each shot. However, the limitation of this system is that small numbers of frames are considered for watermarking. This technique is not reliable if the watermarked frames are dropped.

In [22] authors have proposed Fibonacci-based watermark embedding in the video. In this method, authors have identified key-frames based on the Fibonacci series. They claim that the proposed method saves time to identify key-frames. However, if the unauthorized user checks for Fibonacci series-based frames in the video, anyone can easily perform any illegal activity on the video. So the motivation of our proposed work is to securely select frames as per the users' choice for embedding watermark. In the proposed LS-based V-W, the first two key-frame numbers are decided by the user and successive key-frames are selected based on LS sequence. Different key-frames are generated by different users depending on the initial first two key-frames. Since the LS-based technique is used, it is difficult for unauthorized persons to find out the key-frames. This paper is organized as follows. In Sect. 2 we describe the proposed video watermark embedding and detection scheme. Experimental results are presented in Sect. 3 and conclusion is provided in Sect. 4.

2 Proposed Scheme

A proposed novel V-W scheme based on LS is presented in this section. The novelty of this scheme is that V-W is done by selecting key-frames using LS. This key-frame selection technique ensures that the unauthorized user cannot find watermarked key-frames easily without the knowledge of keys. To improve the performance of the proposed algorithm with respect to imperceptibility and robustness, a combination of DWT-DCT is used for embedding and extraction.

2.1 Watermark Embedding Techniques

In this technique, the key-frames are selected using the LS. The LS is an integer sequence named after the mathematician François Édouard Anatole Lucas. Lucas number is a sequence of integers having recurrence relation to produce a certain term characterized by two parameters, K1 and K2. LS is computed by taking the modulo of key-frames in the given video.

In LS, the first two key-frame numbers are decided by the user and successive key-frames are selected based on LS. Each Lucas number is defined to be the sum of its two immediately previous terms. The Lucas number is calculated as in Eq. (1):

$$L_n = \begin{cases} x & if\ n = 1 \\ y & if\ n = 2 \\ x + y & if\ n > 2 \end{cases} \qquad (1)$$

In the proposed work, values of x and y are considered to be less than 10, x should not be equal to y, n is the frame number in a video. The first two key-frame numbers act as public keys for embedding, and the next key-frame number is obtained by adding the previous two key-frame's numbers. The intention behind using LS is that the unauthorized users will not be able to identify the frames in which the watermark is embedded. Once the key-frames are selected in this novel manner, the watermark is embedded using DWT-DCT using the following algorithm.

Embedding Algorithm

Step1 :- Decompose each of the selected frame into non-overlapped blocks of size 64 × 64.

Step2 :- Apply DWT on each sub-block, to obtain LL ,HL,LH and HH sub-band.

Step3: - Apply DCT on each chosen sub-band(LH and HL).

Step4 :- Generate Two uncorrelated pseudorandom sequences such as PN_0 and PN_1 using key. PN_0 is used to embed watermark bit 0 and PN_1 sequence used to embed watermark bit 1. Size of each of two PN-Sequence must be equal to the number of mid-frequency elements of 8×8 DCT transformed block.

Step5:- Embed watermark using PN –Sequence with a gain factor α , in the DCT transformed 8×8 blocks using eq (2)

$$I_w = \begin{cases} if\ w = 0 & I_w + \alpha \times PN_0 \\ else & I_w + \alpha \times PN_1 \end{cases} \qquad (2)$$

Step6:-Apply Inverse DCT.

Step7:-Apply inverse DWT to obtain watermarked blocks.

2.2 Watermark Detection Techniques

The proposed watermark detection technique is almost the same process as that of the watermark embedding. In order to recover the watermark, the correlation-based watermark detection technique is used which is the best way for Ownership Proof. In this technique, user has to select exactly the same key-frames as they were selected during embedding. The watermarks are extracted from the selected two frames, compared with the original watermark. If the difference between these two is below the threshold then the algorithm terminates as the watermarks are identified. Then there is no need to find further in the successive key-frames. Due to this, the extraction algorithm becomes more efficient in time and accuracy. The watermark detecting process consists of the following steps: Authentic user will be provided with key-frame sequence in which the watermark is embedded.

Detection Algorithm

Step1: -Select watermarked key frame.

Step2:- Divide Watermarked key-frame into non-overlapped blocks of size 64×64.

Step3:- Apply DWT to decompose each block into four multi-resolution sub-bands LL, HL, LH and HH.

Step4:- Divide selected sub-band into 8×8 blocks.

Step5: Apply DCT on each 8×8 sub-blocks and select watermarked mid-frequency coefficients.

Step6:-Generate PN-sequence PN0 and PN1 using same key used in embedding method.

Step7:- Find the correlation between mid-frequency coefficients and generated two PN-sequences. Calculate correlation between mid-frequencies with $PN_0(corr_{PN_0})$ and mid-frequency with $PN_1(corr_{PN_1})$ .Compare them using following eq(3)

$$\begin{cases} if\ corr_{PN_0} > corr_{PN_1} w' = 0 \\ else \qquad\qquad\qquad w' = 1 \end{cases} \tag{3}$$

---

3 Experimental Results

The experimental results of the proposed V-W scheme are presented in this section. Several experiments have been carried out to evaluate the efficiency of the proposed watermarking scheme. The performance of the proposed technique is tested on a few standard video sequences (e.g., Foreman, Stefan, Coastguard, Flower Garden, Container Ship, Traffic) with the number of frames ranging from 100 to 1000. Three different users were asked to provide the first two key-frame numbers and the next key-frames were generated using LS as shown in Table 1. Different key-frames were generated by different users depending on the initial two frames they select. Since LS is used, it is not possible for unauthorized persons to find out the key-frames in which the watermark was embedded. Hence, the video content remains protected. At the time of detection of a watermark, the owner of the video is required to provide the same key-frame numbers.

The performance of the watermarking technique is measured by robustness and imperceptibility. The quality of the watermarked frame is measured by peak signal to noise ratio (PSNR). The peak signal to noise ratio is defined as in Eq. (4). For comparing the similarities between the original and extracted watermark signals, a normalized cross-correlation function is employed as in Eq. (5)

$$PSNR = 10 \cdot \log_{10}\left[\frac{MAX_1^2}{MSE}\right] \tag{4}$$

$$NCC(W, W') = \sum\sum \frac{W(i, j)W'(i, j)}{|W(i, j)|^2} \tag{5}$$

Table 1 shows the comparison of extracted key-frames and PSNR values of the proposed method and the Fibonacci series-based watermarked embedding method

Foreman.avi				
Sequence method	Users	First two frames	Selected frames	PSNR
Sequence method	Users	First two frames	Selected frames	PSNR
Scheme [22]	USER-A	Default 0, 1	0, 1, 1, 2, 3, 5, 8, 13, 21, 34, 55, 89, 144, 233	61.54
LUCAS sequence	USER-B	2, 3	2, 3, 5, 8, 13, 21, 34, 55, 89, 144, 233	61.86
	USER-C	3, 1	3, 1, 4, 5, 9, 14, 23, 37, 60, 97, 157, 254	60
	USER-D	5, 3	5, 3, 8, 11, 19, 30, 49, 79, 128, 207	61.23

W, W' are Original watermark and Extracted watermark, respectively.

It was found that the proposed V-W system achieves a high imperceptibility with an average value of PSNR of 60 dB. In Fig. 1 it is clearly shown that the PSNR of all watermarked frames is above 58 dB, which ensures a good imperceptibility using the proposed V-W scheme as shown in Fig. 1.

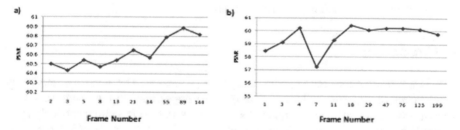

Fig. 1 PSNR (dB) of all selected frames for two watermarked videos: **a** Foreman; **b** Traffic

The proposed V-W scheme is compared with the scheme presented by Sanghavi and Rajurkar [22] in 2013. This technique proves to be the most suitable alternative for key-frame selection-based watermarking scheme. Watermark is embedded in the Fibonacci series frames only. However, this method is not secure since anyone can identify watermarked key-frames easily. The proposed LS-based watermarking technique is more secured than the Fibonacci sequence-based watermarking scheme because LS is not fixed as the Fibonacci series; it changes as the first two frames change as shown in Table 2. To show the robustness of our proposed scheme, the experimental results are conducted with several attacks for the video sequence and the NCC values are compared with the existing method [19, 20, 22]. Experimental results show that the proposed method is more secured and robust than [19, 20, 22]. The proposed key-frame selection technique ensures that unauthorized users cannot find watermarked key-frames easily without the knowledge of keys. Table 1 presents the comparison of extracted key-frames and PSNR values of the proposed method and the Fibonacci series-based watermarked embedding method.

Table 2 PSNR and NCC comparison of different watermarking algorithms for 'Foreman' video sequence

Attacks	Sanghavi and Rajurkar [22]	Scheme in [19]	Scheme in [20]	Proposed
JPEG lossy compression	0.72	0.909	0.99	0.9925
Noise addition	0.69	0.979	1	0.9876
Median filter	0.52	0.633	1	0.9996
Cropping	0.70	0.909	0.99	0.9701
Rescale	0.61	0.636	0.992	1
Frame averaging	0.7	0.818	0.990	1
PSNR	58.26	48.11	49.82	60.52

Table 2 shows the NCC values for different watermarking algorithms against several attacks and PSNR. From this table, the presented scheme can achieve high robustness for almost all attacks used in experimentation.

4 Conclusion

V-W is a powerful tool for ensuring copyright protection. The integrity, quality and transparency of watermarked videos are of great importance to the owner of a video content. It is always desired to embed the watermark in the video frames which no one can predict. In this research, we have presented a key–frames-based V-W. LS-based key-frame selection technique is used that proves to be the most suitable alternative for key-frame-based watermarking methods. This key-frame selection technique ensures that unauthorized user cannot find watermarked key-frames easily without the knowledge of keys. Owner of the video content is prompted to select the first two key-frames which are random. It is not possible for the unauthorized person to predict those key-frames for intended attack in the content. Hence, the proposed method is most secured among existing key-frames-based watermarking methods. A novel watermark detection algorithm is presented that is more efficient in time and accuracy because watermarks are extracted only from selected two key-frames and compared with the original. There is no need to search it in the successive key-frames. The experimental results showed that the proposed V-W system achieves a high imperceptibility with an average value of PSNR and NCC of 60 dB and 0.1, respectively, as well as robustness against common attacks like rotation, scaling, cropping and geometric distortion.

References

1. Junxiaoa, X., Qingbinc, L., Zhi yong, L.: A novel digital video watermarking algorithm. In: Science Direct International Conference on Advances in Engineering, pp. 90–94 (2011)
2. Li, Z., Liu, G.: A novel scene change detection algorithm based on the 3d wavelet transform. In: ICCP IEEE International Conference, pp. 1536–1539 (2008)
3. Xiaona, Z., Guoqin, Q., Qian, W., Tao, Z.: An improved approach of scene change detection in archived films. In: ICSP 2010 Proceeding with IEEE, pp. 825–828 (2010)
4. Patel, K.R., Patel, P.A.: A survey on digital video watermarking. 5(1) (2015). ISSN 2249-555X
5. Mandeep Kaur, Ada: An efficient video watermarking using spatial domain and lab color space. Int. J. Comput. Appl. **146**(5), 11–18 (2016)
6. Manoj Kumar, Dolley Shukla: Review of video watermarking techniques. IJIRS Int. J. Innov. Res. Sci. Technol. **1** (2013)
7. Patel, S.V., Yadav, A.R.: Invisible digital video watermarking using 4 level DWT. In: National Conference on Recent Trends in Engineering & Technology, pp. 1–6 (2011)
8. Sujatha, C.N., Satyanarayana, P.: High capacity video watermarking based on DWT-DCT-SVD. Int. J. Sci. Eng. Technol. Res. (IJSETR) **4**(2), 245–249 (2015)
9. Singh, T.R., Singh, K.M., Roy, S.: Video watermarking scheme based on visual cryptography and scene change detection. AEU Int. J. Electron. Commun. **67**(8), 645–651 (2013)
10. Faragallah, O.S.: Efficient video watermarking based on sin-gular value decomposition in the discrete wavelet transform domain. AEU Int. J. Electron. Commun. **67**(3), 189–196 (2013)
11. Rasti, P., Samiei, S., Agoyi, M., Escalera, S., Anbarjafari, G.: Robust non-blind color video watermarking using QR decomposition and entropy analysis. J. Vis. Commun. Image Represent. **38**, 838–847 (2016)
12. Youssef, S.M., ElFarag, A.A., Ghatwary, N.M.: Adaptive video watermarking integrating a fuzzy wavelet-based human visual system perceptual model. Multimed. Tools Appl. **73**(3), 1545–1573 (2014)
13. Tabassum, T., Islam, S.M.: A digital video watermarking technique based on identical frame extraction in 3-level DWT. In: Proceedings of the 2012 15th International Conference on Computer and Information Technology (ICCIT), Chittagong, Bangladesh, Dec 2012, pp. 101–106
14. Farri, E., Ayubi, P.: A blind and robust video watermarking based on IWT and new 3D generalized chaotic sine map. Int. J. Nonlinear Dyn. Chaos Eng. Syst. **93**(4), 1875–1897 (2018)
15. Jiang, X., Liu, Q., Wu, Q.: A new video watermarking algorithm based on shot segmentation and block classification. Multimed. Tools Appl. **62**(3), 545–560 (2013)
16. Chetan, K., Raghavendra, K.: DWT based blind digital video watermarking scheme for video authentication. Int. J. Comput. Appl. **4**(10), 19–26 (2010)
17. Agilandeeswari, Ganesan, K.: A robust color video watermarking scheme based on hybrid embedding techniques. Multimed. Tools Appl. **75**(14), 8745–8780 (2016)
18. Liu, A., Zhao, J.: A new video watermarking algorithm based on 1D DFT and Radon transform. Elsevier J. Sig. Process. 628–639 (2010)
19. Xiang-yang, W., Yu-nan, L., Shuo, L., Hong-ying, Y., Pan-pan, N., Yan, Z.: A new robust digital watermarking using local polar harmonic transform. Comput. Electr. Eng. **46**, 403–418 (2015)
20. Himeur, Y., Boukabou, A.: A robust and secure key-frames based video watermarking system using chaotic encryption. Springer Science + Business Media, New York (2017)
21. Nouioua, I., Amardjia, N., Belilita, S.: A novel blind and robust video watermarking technique in fast motion frames based on SVD and MR-SVD. Secur. Commun. Netw. (2018)
22. Sanghavi, M., Rajurkar, A.M.: Fibonacci series based watermark embedding in a video. IJCA. In: Proceedings of CRTET'2013

Creating Video Summary Using Speeded Up Robust Features

M. R. Banwaskar$^{(\boxtimes)}$ and A. M. Rajurkar

M. G. M's College of Engineering, Nanded 431605, India
{banwaskar_mr,rajurkar_am}@mgmcen.ac.in

Abstract. There has been unprecedented growth in video data in the last few years due to advances in imaging and video capturing devices. Management of huge video data is a challenging issue these days. Video summarization provides solution to this issue, in which a few representative frames of video are used instead of the whole video. This saves time for the user and also memory space. In this paper, an effective video summarization technique is proposed using preprocessing and Speeded Up Robust Feature (SURF) algorithm. In the preprocessing stage, redundant frames in the video are eliminated using adaptive local thresholding. SURF algorithm is then used to obtain key points in the candidate frames. Key point matching of adjacent frames is done to decide the key frames. Experiments have been conducted on videos from I2V dataset, open-video project dataset and videos downloaded from YouTube. Experimental results demonstrate the superiority of the proposed method against state-of-the-art approaches. Average compression ratio of 0.98 is achieved which is quite high compared to existing techniques. Also, better results are obtained in terms of precision, recall and F-measure. The video summaries produced using the proposed method match the summaries provided by human observers.

Keywords: Key frame extraction · SURF · Summarization · Preprocessing

1 Introduction

There has been explosive growth in video data due to the easy availability of photographic equipment. The increase in digital video poses a great challenge in searching for the required video content. Users of video data want to find the required video quickly without viewing it completely. Also, large memory space is required for the storage of lengthy videos. Video summarization is the solution to this problem wherein the video data is represented using a few frames that carry the main content in the video. In some applications, video skims are used which consist of a collection of video segments, and their corresponding audio. A good video summary should contain frames that capture key content of the video [1–3] and should be visually diverse [1, 4, 5] as well.

Video summarization has applications in video database management, consumer video analysis, surveillance, medical field, sports video, etc. There is a pressing need to propose an effective video summarization method. A video summarization method

© Springer Nature Singapore Pte Ltd. 2020
B. Iyer et al. (eds.), *Applied Computer Vision and Image Processing*,
Advances in Intelligent Systems and Computing 1155,
https://doi.org/10.1007/978-981-15-4029-5_31

using SURF algorithm [6] is proposed in this paper. Redundant frames in the video are removed in the preprocessing step and candidate frames are obtained. SURF algorithm is then used for obtaining key points in those frames. Adjacent frames are compared by matching key points. Frames whose key points do not match are preserved as key frames and rest are eliminated. The key points do not match if the frames are distinct. Such frames are preserved as key frames and rest are eliminated. The organization of the paper is as follows. In Sect. 2 we review the related video summarization methods. The proposed method is illustrated in Sect. 3. Experimental results are presented in Sect. 4 and the conclusion is drawn in Sect. 5.

2 Related Work

Video summarization has become a significant research topic in recent years due to its importance in many video-related applications. A video summary consists of sequential key frames such that adjacent frames in the summary are distinct enough to represent the original video effectively. The majority of the existing works focus on structured video summarization [7], such as the movie or sports videos [8, 9]. On account of the specific characteristics of these videos, good video summaries are obtained. However, these methods usually cannot be applied to other video sequences. Usually, key-frame-based method is used for video summarization. Features such as gradient orientations, color features and a combination of color and texture features are used in [10–12].

Dang and Radha [13] have used a novel image feature named heterogeneity image patch index (HIP) to determine the heterogeneity of patches within video frames. Authors have obtained summaries of consumer videos effectively without much complexity.

To select representative frames, video data is mapped to a high-dimensional space called summary space [14]. A perceptual hash algorithm is then used to determine the similarity of those frames to create a video summary. Video frames with average illuminance below 30% are eliminated in this method. However, some of the important video content may be lost due to this.

Davila and Zanibbi [15] have presented Whiteboard Video Summarization using Spatio-Temporal Conflict Minimization for lecture videos. Authors have generated a spatio-temporal index for the handwritten content in the video and used it for temporal segmentation by detecting and removing conflicts between content regions. Key-frame-based summaries are then obtained from those segments. This approach gives good summaries if there is a single and static video camera but fails when there are multiple cameras and zooming/panning is done. In [16], authors have presented a video summarization approach with an attention mechanism to copy the way of selecting the key shots of humans. They have used Attentive encoder-decoder networks for Video Summarization (AVS). The encoder first reads the sequence of frames and then attention-based decoder generates a sequence of importance scores. Based on the visual sequence and output of the decoder, key shots are selected. Though the concept presented in this paper is novel, it is difficult to implement in real-time videos.

In [17] Zhang et al. have proposed a context-aware surveillance video summarization method which captures important video segments through information about individual local motion regions, as well as the interactions between these motion regions. Authors have used an algorithm to learn and revise dictionaries of video features along with feature correlations. But again this technique is domain specific.

In 2017, Shi et al. have used a two-stage key frame extraction [18]. In this method, candidate key frames are extracted using color histogram difference and then covariance between the frames is used to identify the most dissimilar frames as key frames. Though this method is simple, it suffers a fixed value of thresholds.

de Avila et al. [19] employed the k-means clustering algorithm to decide frames with different visual content. Authors have grouped the frames in sequential order to achieve faster convergence.

Video frames are represented by color histogram features in [20] and principal component analysis is then used to decrease feature dimensionality. The Delaunay Triangulation algorithm is applied to the extracted features to identify the key frames. DT algorithm extracts less number of key frames at the cost of low accuracy. Huang and Wang [21] have proposed a comprehensive video summarization method using key frames for video content summarization and optical flows for video motion summarization. Authors claim this method to be fast, but it may take more time if the video contains many transitions.

It is observed that a substantial amount of work has been done in video summarization but most of the researchers concentrate on various image features and give less importance to the number of computations done for getting video summaries. Rahman et al. [22] have used color moment, color histogram and SURF features for creating video summaries. However, it is computationally expensive and does not provide effective key frames. Compression ratio of this method is also low. Hence, in the proposed work, we focus on reducing the number of computations by avoiding redundant frames in the video and, also achieve a good compression ratio. SURF key points are then determined and the candidate frames are compared to give the key frames. Effective video summaries are thus produced by the proposed method.

3 Proposed Work

Frames are the basic units of a video. A shot consists of frames captured in a single uninterrupted camera. A scene is a series of shots that are coherent from a narrative point of view. There are many redundant frames in a video.

To reduce the number of computations needed for creating a video summary, these redundant frames are removed using preprocessing. SURF [6] algorithm is then used to obtain key points in the candidate frames. Comparison of key points is done to decide the key frames and they are presented to the user as a video summary. Figure 1 shows the architecture of the proposed system which is divided into two phases, viz., preprocessing and SURF key point detection that are explained in the following section in detail.

3.1 Preprocessing

The video sequence is first divided into segments consisting of 20 frames [23]. The pixel-wise distance of the luminance component between the first and last frames is calculated for each segment. Every ten segments are then grouped together to form a basic

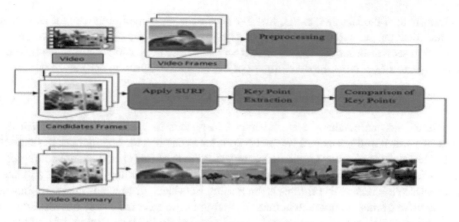

Fig. 1 Architecture of the proposed system

thresholding unit. All the segments within this unit share one threshold. The segments whose distance values are less than the corresponding local thresholds are classified as non-boundary segments. The local threshold for each unit is defined adaptively as

$$T_L = \mu_L + 0.6 * \mu_G \sigma_L / \mu_L \tag{1}$$

where μ_G and μ_L denote the global and local mean values, respectively, σ_L denotes the local standard deviation. The segments whose distance values satisfy the following criterion are classified as candidate segments

$$(d20(n) > 3d20(n-1) \text{ or } d20(n) > 3d20(n+1)) \text{ and } d20(n) > \mu_G$$

For each candidate segment, the distance values between the middle frame (i.e., the tenth frame) and the first as well as the last frames are calculated. Depending on the relationship between distance values, the segments are retained or eliminated. Hence, a large percentage of redundant frames are removed in the preprocessing process.

SURF key point detection

SURF algorithm was proposed by Bay et al. [6]. It is based on sums of 2D Haar wavelet responses. SURF uses the Hessian matrix because of its good performance in computation time and accuracy. Hessian of a pixel is given by

$$H(f(x, y)) = \begin{bmatrix} \frac{\partial^2 f}{\partial x^2} & \frac{\partial^2 f}{\partial x \partial y} \\ \frac{\partial^2 f}{\partial x \partial y} & \frac{\partial^2 f}{\partial y^2} \end{bmatrix} \tag{2}$$

For a point $X = (x, y)$, the Hessian matrix $H(x, \sigma)$ in x at scale σ is defined as

$$H(X, \sigma) = \begin{bmatrix} L_{xx}(X, \sigma) & L_{xy}(X, \sigma) \\ L_{xy}(X, \sigma) & L_{yy}(X, \sigma) \end{bmatrix} \tag{3}$$

where $L_{xx}(x, \sigma)$ is the convolution of the Gaussian second-order derivative with the image I in point x, and similarly for $L_{xy}(x, \sigma)$ and $L_{yy}(x, \sigma)$. In order to calculate the determinant of the Hessian matrix, first, we need to apply convolution with the Gaussian kernel, then the second-order derivative. The 9×9 box filters is an approximation for Gaussian second-order derivatives with $\sigma = 1.2$. These approximations are denoted by D_{xx}, D_{yy}, and D_{xy}. Now the determinant of the Hessian (approximated) is

$$\det(H_{\text{approx}}) = D_{xx} D_{yy} - (w D_{xy})^2 \tag{4}$$

SURF first calculates the Haar wavelet responses in x and y direction. Then the sum of vertical and horizontal wavelet responses is calculated. The number of features extracted from each image is 64 as the image is divided into 4×4 square subregions and 4 features are obtained from each subregion. After determining SURF key points, comparison of consecutive frames is done using Euclidean distance. If the key points in consecutive frames do not match, then those frames are selected as key frames. Frames whose key points match are similar hence discarded. In [22], key points of all frames are determined which is computationally expensive.

4 Experimental Results

We have used 27 videos in our experiment. Ten videos are downloaded from YouTube, 10 videos are from the I2V dataset (https://www.unb.ca/cic/datasets/url-2016.html) and 7 videos are from the open-video dataset. Quantitative and qualitative performance analysis was carried out to check the validity of the proposed method. For qualitative analysis, these videos were presented to 5 human observers. They were asked to view the videos frame by frame and select summary frames. The average number of key frames for each video provided by human observer was noted. Using the proposed method, initially, redundant frames are removed from the video using adaptive local thresholding and candidate frames are selected. In the next step, key points are determined using SURF algorithm. Comparison of key points using Euclidean distance is done for selecting summary frames. These key frames effectively represent the original video. Figure 2a, b shows the video summary of 'Short wildlife video.mp4' and 'I2V dataset vid 6.7.mp4'. It can be seen that the summary is compact and the frames are distinct enough to represent the original video. Such video summaries are useful in applications like movies, lecture videos, surveillance videos, action recognition videos, etc. Compression ratio is an important performance parameter in video summarization which is given by

$$\text{C ratio} = 1 - (N_{kf}/N_{vf}) \tag{5}$$

where N_{kf} is the number of key frames and N_{vf} is the total number of frames in a video. For better compactness of video summary, high compression ratio is desirable. It is also observed that video summaries obtained in the proposed method consist of a negligible amount of duplicate key frames. Table 1 gives the compression ratio and the number of redundant frames obtained for 7 videos in the database. Compression ratio and distinctness of the frames in video summary are two important parameters to decide the effectiveness of a video summary.

Using the method described in [22], many redundant frames are selected for providing a video summary which is not desirable. It has been observed that there is a

(a) Summary of short wild life vid.mp4

(b) Summary of vid6.7.mp4 (I2V dataset)

Fig. 2 Video summary

Table 1 Comparison of the number of key frames

Video	No. of frames	Summary frames [22]	Duplicate frames [22]	Summary frames (human created)	Proposed method summary frames	Compression ratio (proposed method)	Duplicate frames
V1	3569	58	23	35	36	0.99	1
V2	3222	60	30	30	32	0.99	0
V3	3740	60	05	55	52	0.98	0
V4	4875	107	45	62	60	0.99	1
V5	643	20	14	6	5	0.99	0
V6	1085	40	24	16	19	0.98	3
vid 6.7.mp4	2800	80	24	56	70	0.97	4

negligible number of redundant frames in the video summary obtained by the proposed method. The compression ratio of the proposed method is more. User can very quickly get an idea about the original video by observing such a video summary. This reflects the effectiveness of the proposed method. Average compression ratio obtained for 20 videos is 0.98 which is quite high compared with the method in [24].

(a) *Summary of short wildlife vid.mp4*
(b) *Summary of vid 6.7.mp4 (I2V dataset)*

Table 2 Dataset details

Video	Name	#Frames	# Summary frames
V21	The Great Web of Water, segment 01	3,279	15
V22	The Great Web of Water, segment 02	2,118	4
V23	The Great Web of Water, segment 07	1,745	15
V24	A New Horizon, segment 01	1,806	11
V25	A New Horizon, segment 02	1,797	10
V26	A New Horizon, segment 03	6,249	16
V85	Football Game, segment 01	1,577	10

Table 2 provides the details of videos chosen from the open-video dataset.

Summary created by the proposed method for Video v85 from the open-video database shown in Fig. 3 closely matches the ground truth summary in the VSUMM project. (https://sites.google.com/site/vsummsite/download).

Qualitative results are obtained using precision, recall and F-measure.

$$\text{Precision (P)} = \frac{\text{Number of matched frames}}{\text{Number of frames in summary}}$$

Fig. 3 Summary of video v85 (Open-video dataset)

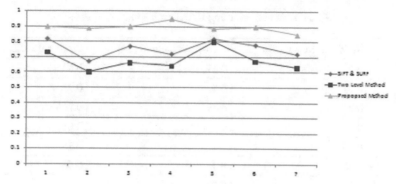

Fig. 4 Comparison of F-measure

Table 3 Results obtained by the proposed method and by [ref. 22 and 24]

Video	SIFT and SURF method [22]			Two-level method [24]			Proposed method		
	P	R	F	P	R	F	P	R	F
V21	0.85	0.8	0.82	0.73	0.73	0.73	0.9	0.9	0.9
V22	0.6	0.75	0.67	0.5	0.75	0.6	0.8	1	0.89
V23	0.75	0.8	0.77	0.61	0.73	0.66	0.9	0.9	0.9
V24	0.64	0.82	0.72	0.7	0.64	0.64	0.9	1	0.95
V25	0.75	0.9	0.82	0.8	0.8	0.8	0.8	1	0.89
V26	0.7	0.88	0.78	0.6	0.75	0.75	0.9	0.9	0.9
V85	0.66	0.8	0.72	0.58	0.7	0.7	0.8	0.9	o.85

$$Recall(R) = \frac{\text{Number of matched frames}}{\text{Number of ground truth frames}}$$

$$\text{F-measure (F)} = \frac{2RP}{R + P}$$

It has been observed that the selected summary frames cover most of the ground truth summary frames in comparison with other techniques. Table 3 gives the performance comparison of proposed method with two other methods in literaturein terms of precision, recall and F measure. Figure 4 shows the performance comparison of F-measure graphically. Also, the deviation from ground truth summary frames is minimum compared with other methods. Table 4 shows this comparison.

Table 4 Deviation from ground truth summary frames

Video	SIFT and SURF [22]	Two-level method [24]	Proposed method
V21	0.2	0.27	0.07
V22	0.25	0.5	0.2
V23	0.2	0.39	0.125
V24	0.19	0.3	0.08
V25	0.17	0.2	0.16
V26	0.125	0.4	0.06
V85	0.2	0.33	0.18

5 Conclusion

The proposed video summarization is based on preprocessing to remove redundant frames from a video. This reduces the number of computations and also saves time. SURF algorithm is used for key point extraction since it is fast and accurate. Comparison of adjacent frames is then done on the basis of key point matching. If the frames are dissimilar, then key points do not match. Such frames are retained as key frames. To decide the effectiveness of the proposed method, the video summaries are compared with summaries provided by human observers. It is observed that there are no duplicate key frames in the video summaries obtained by the proposed method. Users of the video summary obtained by the proposed method strongly agree that they get an exact glimpse of the original video very quickly. Average compression ratio of 0.98 is obtained by the proposed method which is quite high compared with the average compression ratio of 0.80 obtained using the method proposed in [22]. In further work, we plan to use low- and mid-level features along with SURF and propose a summarization method for particular types of videos.

References

1. Gong, B., Chao, W.-L., Grauman, K., Sha, F.: Diverse sequential subset selection for supervised video summarization. In: Advances in Neural Information Processing Systems (2014)
2. Khosla, A., Hamid, R., Lin, C.-J., Sundaresan, N.: Large-scale video summarization using web-image priors. In: IEEE Conference on Computer Vision and Pattern Recognition (2013)
3. Ngo, C.-W., Ma, Y.-F., Zhang, H.-J.: Automatic video summarization by graph modeling. In: IEEE International Conference on Computer Vision (2003)
4. Mahasseni, B., Lam, M., Todorovic, S.: Unsupervised video summarization with adversarial LSTM networks. In: IEEE Conference on Computer Vision and Pattern Recognition (2017)
5. Zhang, K., Chao, W.-L., Sha, F., Grauman, K.: Video summarization with long short-term memory. In: European Conference on Computer Vision (2016)
6. Bay, H., Tuytelaars, T., Gool, L.V.: SURF: speeded up robust features. In: 9th European Conference on Computer Vision, Computer Vision ECCV 2006, Part II, 7–13 May 2006. Springer, pp. 404–417

7. Truong, B.T., Venkatesh, S.: Video abstraction: a systematic review and classification. ACM Trans. Multimed. Comput. Commun. Appl. **3**(1), 1–37 (2007)
8. Cheng, C.-C., Hsu, C.-T.: Fusion of audio and motion information on HMM-based highlight extraction for baseball games. IEEE Trans. Multimed. **8**(3), 585–599 (2006)
9. Li, B., Sezan, I.: Semantic sports video analysis: approaches and new applications. In: Proceedings of International Conference on Image Processing, pp. I-17–I-20 (2013)
10. Pritch, Y., Ratovitch, S., Hendel, A., Peleg, S.: Clustered synopsis of surveillance video. In: Proceedings of International Conference on Advanced Video and Signal-Based Surveillance, pp. 195–200 (2009)
11. Simakov, D., Caspi, Y., Shechtman, E., Irani, M.: Summarizing visual data using bidirectional similarity. In: Proceedings of IEEE Conference on Computer Vision and Pattern Recognition, June 2008, pp. 1–8
12. Zhu, X., Wu, X., Fan, J., Elmagarmid, A.K., Aref, W.G.: Exploring video content structure for hierarchical summarization. Multimed. Syst. **10**(2), 98–115 (2004)
13. Dang, C.T., Radha, H.: Heterogeneity image patch index and its application to consumer video summarization. IEEE Trans. Image Process. **23**(6) (2014)
14. Li, X., Zhao, B., Lu, X.: Key frame extraction in the summary space. IEEE Trans. Cybern. **48**(6) (2018)
15. Davila, K., Zanibbi, R.: Whiteboard video summarization via spatio-temporal conflict minimization. In: 14th IAPR International Conference on Document Analysis and Recognition (2017)
16. Ji, Z., Xiong, K., Pang, Y., Li, X.: Video summarization with attention-based encoder-decoder network. IEEE Trans. Circuits Syst. Video Technol.
17. Zhang, S., Zhu, Y., Roy-Chowdhury, A.K.: Context-aware surveillance video summarization. IEEE Trans. Image Process. **25**(11), 5469 (2016)
18. Shi, Y., Yang, H., Gong, M., Liu, X., Xia, Y.: A fast and robust key frame extraction method for video copyright protection. Journal of Electrical and Computer Engineering, Vol. 2017, Article ID 12317942017
19. de Avila, S.E.F., Lopes, A.P.B., da Luz, Jr., A., de Albuquerque Araújo, A.: VSUMM: a mechanism designed to produce static video summaries and a novel evaluation method. Pattern Recognit. Lett. **32**(1), 56–68 (2011)
20. Mundur, P., Rao, Y., Yesha, Y.: Keyframe-based video summarization using Delaunay clustering. Int. J. Digit. Lib. **6**(2), 219–232
21. Huang, C., Wang, H.: A novel key-frames selection framework for comprehensive video summarization. IEEE Trans. Circuits Syst. Video Technol. (2018). https://doi.org/10.1109/tcsvt.2019.2890899
22. Rahman, Md.A., Hassan, S., Hanque, S.M.: Creation of video summary with the extracted salient frames using color moment, color histogram and speeded up robust features. Int. J. Inf. Technol. Comput. Sci. **7**, 22–30 (2018)
23. Li, Y.-N., Lu, Z.-M., Niu, X.-M.: Fast video shot boundary detection framework employing pre-processing techniques. IET Image Process. **3**(3) (2009)
24. Jahagirdar, A., Nagmode, M.: Two level key frame extraction for action recognition using content based adaptive threshold. Int. J. Intell. Eng. Syst. **12**(5) (2019)

Fingerprint-Based Gender Classification by Using Neural Network Model

Dhanashri Krishnat Deshmukh$^{(\boxtimes)}$ and Sachin Subhash Patil

Department of Computer Science & Engineering, Rajarambapu Institute of Technology, Rajaramnagar, India
dhanashri.deshmukh415@gmail.com, sachin.patil@ritindia.edu

Abstract. A new method of gender classification from a fingerprint is proposed based on the biometric features like minutiae map (MM), orientation collinearity maps (OCM), Gabor feature maps (GFM) and orientation map (OM) for pattern type, 2D wavelet transform (2DWT), principal component analysis (PCA) and Linear Discriminant Analysis (LDA). The classification is performed by using neural networks. For training the neural network in supervised mode, biometric features obtained from fingerprints of 100 males and 100 females are used. For testing classification performance of neural network, 50 male samples and 50 female samples are used. The proposed method achieved an overall classification accuracy of 70%.

Keywords: Minutiae map (MM) · Orientation collinearity maps (OCM) · Gabor feature maps (GFM) · Orientation map (OM) · Discrete wavelet transform (DWT)

1 Introduction

A fingerprint is the epidermis of a finger consisting of the pattern of ridges and valleys. The endpoints and bifurcation points of ridges are called minutiae. Fingerprint minutiae patterns of ridges are determined as unique through the combination of genetic and environmental factors. This is the reason, the fingerprint minutiae patterns of the twins are different. Also, the ridge pattern of each fingertip remains unchanged from birth till death. The gender classification- and identification-based biometric applications are designed for E-commerce, E-Governance, Forensic applications, etc.

Person identification using fingerprint algorithms is well established but a few attempts have been made for gender classification from a fingerprint [1]. To improve the performance of fingerprint gender classification more, the biometric features like the minutiae map (MM), orientation collinearity maps (OCM), Gabor feature maps (GFM) and orientation map (OM) for pattern type, 2D wavelet transform (2DWT), principal component analysis (PCA) and Linear Discriminant Analysis (LDA) features are extracted. The classification is performed using Fuzzy logic–C Means (FCM), and Neural Network (NN) technology. The research work is focused on the gender classification from fingerprint.

© Springer Nature Singapore Pte Ltd. 2020
B. Iyer et al. (eds.), *Applied Computer Vision and Image Processing*,
Advances in Intelligent Systems and Computing 1155,
https://doi.org/10.1007/978-981-15-4029-5_32

2 Related Work

2.1 Gender Classification from Fingerprints

Ridge features have been used for training Support Vector Machine (SVM) [2]. For testing the effectiveness of the method, 200 male fingerprints and 200 female fingerprints were used. The classification accuracy of 91% was observed.

In [3] authors used 57 male and 58 female fingerprints. The ridge features and finger size features were used for gender classification and they obtained classification results of 86% accuracy.

Ridge density differences in two Indian populations were studied. A very sharp differences in ridge density has been found among male and female samples [4].

The research work has found that the males have a slightly higher ridge count compared to females while females have a higher ridge thickness to the valley thickness ratio compared to males [5].

In [6] authors proposed a method for gender classification based on discrete wavelet transform (DWT) and singular value decomposition (SVD). K-Nearest Neighbor (KNN) is used as a classifier. Overall classification accuracy of 88.28% has been obtained.

In [7] authors proposed a method for gender classification based on Fast Fourier transform (FFT), Discrete Cosine Transform (DCT) and Power Spectral Density (PSD). They obtained an overall classification accuracy of 84.53%.

In their work [8] authors proposed a method for gender classification based on SVM which gives more accuracy than existing methods.

In their work [9] FFT, Eccentricity and Major Axis Length were used as features for gender classification and they give overall classification accuracy of 79%.

In [10] authors proposed DWT and SVD-based gender classifier which gives overall classification accuracy of 80%.

A multi-resolution texture approach for age-group estimation of children has been proposed in [11] which gives 80% classification accuracy for children below the age of 14.

In a study focusing on age and aging [12], the authors collected fingerprints of persons from age groups 0–25 and 65–98 and analyzed issues such as permanence and quality of fingerprints.

3 Proposed Work

The proposed research work is aimed at developing the system for accurate classification of gender through fingerprint. The system has the following stages:

3.1 Fingerprint Image Enhancement

To enhance the fingerprint images precisely, we use various preprocessing algorithms like Image Segmentation, Image Normalization, Image Orientation estimation, Image Ridge frequency estimation, Image Binarization and Image Thinning.

3.2 Feature Extraction

In this stage, the features from enhanced fingerprint images—minutiae map (MM), orientation collinearity maps (OCM), Gabor Feature maps (GFM), orientation map (OM) for pattern type, 2D wavelet transform (DWT)—are calculated.

3.3 Gender Classification of Finger Extracted Feature Reading

The classification is performed with the application of Neural Network (NN) technology.

4 Methodology

The thumb fingerprints are acquired from different age group people (150 males and 150 females) from various locations of the country by using an optical fingerprint scanner to form the large database source as shown in Fig. 1. Each fingerprint from the database is passed through various preprocessing stages for image enhancement and image noise removal as shown in Fig. 2.

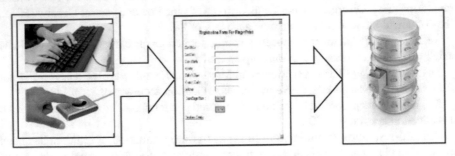

Fig. 1 Fingerprints are acquired in real time

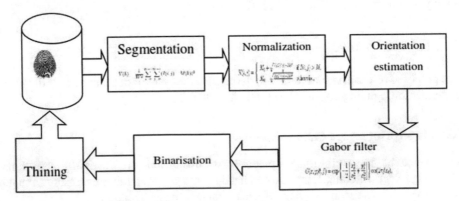

Fig. 2 Preprocessing stages for Image enhancement

After the preprocessing of fingerprints, features are computed in four stages: first is six levels of 2D discrete wavelet transform decomposition, second is PCA, third is the Linear Discriminant Analysis (LDA) and fourth, ridge count, RTVTR and various map readings. The next step is to combine these four feature vectors together into a single combined vector, which will be stored in the database for classification as shown in Fig. 3.

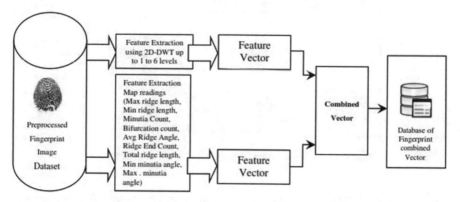

Fig. 3 Feature extraction to form a combined vector database

Then all the combined vectors within the database will be allowed to pass for unsupervised training mode. It uses clustering with neural network technique within the combined vector database. It generates the clusters with unsupervised learning attempts of a neural network model. The neural network model uses similarity measures or minimum distance for all the combined vectors database. This clustering will create the two classes, one male and another female with the calculation of threshold values using any one of the technique, centroid or average. The clustering scheme is illustrated in Fig. 4.

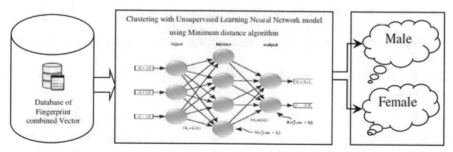

Fig. 4 Unsupervised learning using neural network training model applied to tuples of combined vectors database

This training mode database further will be used for the supervised learning process of the next subsequent new entry of fingerprint-combined vectors for gender classification. The further classification for perception mode or supervised learning will be performed by using Neural Network (NN) technology as shown in Fig. 5.

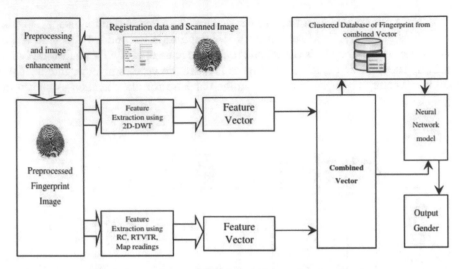

Fig. 5 Perception model of fingerprint gender classification

In this work, all fingerprints were scanned by an optical fingerprint scanner with a size of 512×512. To enhance the fingerprint image quality by reducing noise, the work implements various improved algorithms like Image Segmentation, Image Normalization, Image Orientation estimation, Image Ridge frequency estimation, Image Binarization and Image Thinning. These image enhancement algorithms will issue precise and accurate pixel patterns for further work.

To get the best feature vector readings, the present work implements four advanced feature extraction algorithms like six levels of 2D discrete wavelet transform (DWT) decomposition, second the spatial level undergoing PCA, third the LDA and fourth ridge count, RTVTR and various map readings. These four feature vector readings will be obtained accurately as maximum.

The work is to be performed in two phases:

1. Training mode with unsupervised learning of gender for a training set of fingerprint data.
2. Recognition mode with supervised learning of gender for an eventual fingerprint as an input.

The unsupervised training mode work uses clustering algorithms and is implemented with a neural network model for all combined vectors in the database. The neural network model uses similarity measures or minimum distance for all combined vectors in the database. This clustering will create the two precise classes, one male and another female with the calculation of threshold values using any one of the technique, centroid or average technology.

The supervised learning (Recognition mode) is the process for classifying the next subsequent new entry of fingerprint combined vectors with its gender. The further classification will be implemented with supervised learning that will be performed by using Neural Network (NN) technology.

5 Experimental Results

Table 1 shows the TOTAL RIDGE feature values of one person. Figure 6 shows the total ridge count for male and female fingers. It shows maximum ridge count value and minimum ridge count values for the 50 male and 50 female fingerprints (for 10 fingers) tested. The maximum ridge count for a female is mean of 50 female maximum ridge values. Also, the minimum ridge count for a male is mean of 50 male minimum ridge count values. As compared ridge count of a male is slightly more than female ridge count. For each person, a feature vector of size 160 (10 fingers $\times$ 16 features per finger) is generated. For training, the feedforward backpropagation based neural network with a sigmoid activation function of 100 male and 100 female fingerprints were used. For testing classification accuracy, 50 male and 50 female fingerprints were used. Table 2 shows the classification accuracy of the proposed method for male and female fingerprints. For male and female fingerprints, the classification accuracy of 70% and 72% was observed. An overall classification accuracy of 71% was observed.

6 Conclusion

In this work, a new method for gender classification based on fingerprints is proposed. We used image features computed from minutiae map (MM), orientation collinearity maps (OCM), Gabor feature maps (GFM) and orientation map (OM) for pattern type, 2D discrete wavelet transform (DWT), principal component analysis (PCA) and Linear Discriminant Analysis (LDA). The classification is performed by using neural networks. Our proposed method has achieved a 70% overall classification accuracy.

Our future work will focus in the following directions:

– To investigate the use of deep neural networks in fingerprint-based gender classification.
– To extend our proposed method for fingerprint-based age estimation.
– To improve our proposed method to increase classification accuracy.

Table 1 Total ridge feature values for one person

Feature	Finger No.	New user	Min. male	Max. male	Avg. male	Min. female	Max. female	Avg. female
Total ridge	1	176	155	208	194.34	156	208	190.19
Total ridge	2	165	59	204	176.45	123	194	169.36
Total ridge	3	159	111	204	178.24	77	204	165.30
Total ridge	4	161	131	203	175.8	111	204	159.06
Total ridge	5	152	109	200	160.35	101	176	141.18
Total ridge	6	194	155	208	197.05	156	208	193.47
Total ridge	7	168	72	206	179.25	120	192	166.75
Total ridge	8	181	144	207	182.04	117	201	169.15
Total ridge	9	154	124	204	179.67	116	194	160.18
Total ridge	10	130	91	198	161.32	70	186	143.82

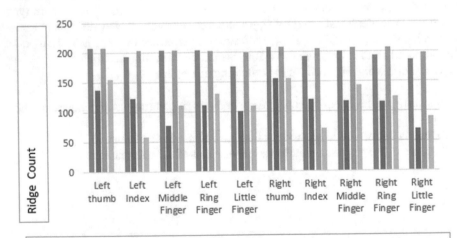

blue -Maximum Average for Female   Gray -Maximum Average for Male
Orange -Minimun Average for Female   Yellow -Maximum Average for Male

Fig. 6 Total ridge count for male and female fingers

Table 2 Classification accuracy of the proposed method for male and female fingerprints

	No. of records to test	No. of accurate result records	Total accuracy
Male	50	35	70
Female	50	36	72

References

1. Verma, M., Verma, M., Agarwal, S.: Fingerprint based male—female classification (2008)
2. Wang, J.F., et al.: Gender determination using fingertip features (2008)
3. Kaur, R., Garg, R.K.: Determination of gender differences from fingerprint ridge density in two Northern Indian populations (2011)
4. Omidiora, E.O., Ojo, O., Yekini, N.A., Tubi, T.O.: Analysis, design and implementation of human fingerprint patterns system towards age & gender determination, ridge thickness to valley thickness ratio (RTVTR) & ridge count on gender detection (2012)
5. Gnanaswami, P.: Gender classification from fingerprint based on discrete wavelet transform (DWT) and singular value decomposition (SVD) (2012)
6. Kaur, R.: Fingerprint based gender identification using frequency domain analysis (2012)
7. Sudha Ponnarasi, S., Rajaram, M.: Gender classification system derived from fingerprint minutiae extraction (2012)
8. Gornale, S.S., Geetha, C.D., Kruthi, R.: Analysis of finger print image for gender classification using spatial and frequency domain analysis (2013)
9. Chand, P., Sarangi, S.K.: A novel method for gender classification using 2D discrete wavelet transform and singular value decomposition techniques (2013)
10. Bharti, P., Lamba, C.S.: DWT-neural network based gender classification (2014)
11. Saxena, A.K., Chaurasiya, V.K.: Multi-resolution texture analysis for fingerprint based age-group estimation (2017)
12. Galbally, J., Haraksim, R.: A study of age and ageing in fingerprint biometrics (2019)

Segmentation of Characters from Degraded Brahmi Script Images

Aniket Suresh Nagane[1]([✉]) and S. M. Mali[2,3]

[1] MIT-Arts, Commerce and Science College, Alandi, Pune, India
asnagane@mitacsc.ac.in, aniket.nagane@gmail.com
[2] Department of Computer Science, Savitribai Phule Pune University, Pune, India
shankarmali007@gmail.com
[3] Dr. Vishwanath Karad MIT World Peace University, Pune, India

Abstract. Segmentation of symbols or characters in the OCR process is a very critical and important phase, as it directly affects the recognition system. If objects in the image are not accurately segmented, then the recognition will also be false and this ultimately affects the performance of the system. Brahmi script contains certain alphabets which comprise some isolated symbols. In this paper, the authors have proposed an innovative approach to segment the lines from the digital Brahmi estampage document, and further segment the alphabets from the line, which are regular characters as well as special characters comprising isolated symbols. The authors have developed an algorithm that effectively segments the lines, regular characters and special characters. Advancements in the algorithm can be made to improve the results for accuracy.

Keywords: Brahmi script · Segmentation · Estampage · Isolated symbols

1 Introduction

The most oldest or ancient scripts which were practiced in Central Asia and the Indian subcontinent are Indus valley script and Brahmi script. Indus valley script was used between BC 3500 and BC 1900. Inscriptions found in Indus script contain very few symbols making it difficult to decode and understand. Indus script is still not deciphered. Brahmi script was used between BC 400 and AD 500 which is a long period of 900 years [1]. Brahmi script became popular during the period of emperor Ashoka from BC 268 to BC 232. As the Brahmi script is practiced for a long span of 900 years, the script is found engraved on the rocks, copper plates (*tampra-patra*) and *bhurja-patra*. The use of Brahmi script spans a vast geographical area. Many variations are observed in the set of symbols used to represent the script because of the long span, diverse mediums used and geographical locations. The period of emperor Ashoka is the time where it has been observed that the script is consistent in its use of character set across the Indian subcontinent. The set of symbols for the Brahmi (Emperor Ashoka's period) script contains 10 vowels and 33 consonants. The Brahmi script also consists of modifiers

© Springer Nature Singapore Pte Ltd. 2020
B. Iyer et al. (eds.), *Applied Computer Vision and Image Processing*,
Advances in Intelligent Systems and Computing 1155,
https://doi.org/10.1007/978-981-15-4029-5_33

such as *anuswar, ukar, velanti*. If the modifier is used with any character from the set, it becomes a conjunct symbol [2–4].

Most commonly, the Brahmi script is found engraved on rocks. These inscriptions of rock-cut edicts are preserved in the form of estampage. Estampage is an impression of rock-cut edicts taken on the special paper inked from one side by pasting that side on the rock-cut edicts and slow hammering is done with the help of a wooden hammer. Digitization of these estampages is done by Archaeological Survey of India, Mysore, and these images are provided to the historians for the purpose of their study.

As the Brahmi script is an ancient script and was practiced many centuries ago, reading it requires expertise, detailed knowledge about the script to understand and interpret it. A large number of documents are available in these forms, many of which are unread [5, 6]. This great valuable content of history in the Brahmi script is not understandable to everyone as it is written in non-functional script (not in use nowadays). If these contents and valuable information is made available in an understandable form (i.e., in functional script of today), historians will get much more information out of it which will help them gather and notify about the history to the society. OCR system can help common people to read and understand the literature available in the Brahmi script [7, 8].

2 Literature Review

OCR is a system that uses document image as input containing some text written in a specific script. This input is further processed and the generated output contains correctly recognized characters from the image [9]. OCR system comprises major steps such as image acquisition, binarization, segmentation and recognition. After image acquisition and binarization (including preprocessing of image and noise removal), segmentation is a very crucial operation [10, 11]. Uneven intensities and noise in the image affect segmentation, if it is drawn on region-based level set models [12]. If the symbols or characters are accurately segmented, recognition will be more accurate. However, if the characters or symbols are not accurately segmented, it will directly affect the recognition [13–15]. Since the input for the recognition phase is output of the segmentation phase, incorrect segmentation will result in incorrect recognition. When the recognition goes wrong, it affects the performance of the system. So correct segmentation is an important aspect of the OCR system [16–18].

A fair amount of work has been already done in OCR for several languages and scripts across the globe. Notable and remarkable success has been achieved in OCR systems for some ancient Indian scripts and most of them are currently in practice [19–23]. Yet, there are some ancient Indian scripts like Brahmi script, in which hardly any work related to OCR has been reported.

Kunchukuttan et al. have proposed a transliteration and script conversion system for the Brahmi script which achieves significant results. Their system supports script conversion between all Brahmi-derived scripts [24].

Gautam and Chai have proposed a geometric method for handwritten and printed Brahmi character recognition. The accuracy of the proposed method is 94.10% and for printed and handwritten Brahmi character recognition, it is 90.62% [23].

Warnajith et al. have proposed the image processing approach. Their study aims at developing Brahmi script fonts so as to provide input for efficient OCR systems for ancient inscriptions [25].

Apart from this work, no considerable work has been found in developing OCR for the Brahmi script. There are different approaches used to segment the alphabets from the image. Based on various approaches several methodologies are also developed [26]. Considering broadly the various techniques of the similar base, segmentation techniques are mainly classified into three major categories which are classical approach or explicit segmentation, recognition-based segmentation or implicit segmentation, and holistic approach [27]. In explicit segmentation, given an image of sentence or words, it is partitioned into sub-images containing only individual characters. In implicit segmentation, recognition is done along with the segmentation. Components in the image are extracted and these are searched for the image class matching the same components. In the holistic approach, template-based matching is used and the entire word is segmented which avoids character segmentation. In this paper, authors have implemented an algorithm based on the explicit segmentation technique.

3 Proposed Method

Line Segmentation. In the proposed algorithm, we first segment all the lines from the document image of Brahmi estampage. For segmentation of a line from a binarized image [28], we perform the following steps:

Step 1: Find row-wise sum intensity values for all the rows in an image [29].

$$\text{sum}(i) = \sum_{j=1}^{n} f(i, j)$$

where $\text{sum}(i)$ is the sum of intensity values from ith row of the image, n is the total number of columns in an image and $f(i, j)$ is an intensity value at location (i, j) in an image f.

Step 2: Find the average value of the row-wise sum using the formula:

$$\text{avgsum} = \left(\sum_{i=1}^{m} \text{sum}(i) \right) / m$$

where m is the total number of rows in an image.

Step 3: Find the parameters $X_{\min}$, $Y_{\min}$, Width, Height for a rectangle to crop the line as mentioned below:

Step 3.1: Let $l = 1$, $k = 1$, $w = n$, $h = 1$, $j = 1$ where n is the number of columns in an image.

Step 3.2: Let $X_{\min} = 1$, $Y_{\min} = k$.

Step 3.3: Check if $\text{sum}(j)$ is greater than avgsum/4, then let $h = h + 1$ and go to step 3.4 otherwise let $h = j - k$ and go to step 3.5.

Step 3.4: $j = j + 1$ and go to step 3.2.
Step 3.5: Crop the rectangle with parameters X_{min}, Y_{min}, Width and Height,
let $k = j$.
Step 3.6: Check if j is less than the number of rows in an image, then go to
step 3.2 otherwise exit from the algorithm.

Once the lines are segmented from the document image, we check for blank lines, i.e., whether the segmented line contains at least one character or not. If the segmented line does not contain a character, then it is considered as a blank line, and further it is deleted. To delete blank lines, we perform the following steps:

Step 1: Read the segmented line image.
Step 2: Count the total number of connected components using 8 connectivity.
Step 3: If count is zero, then delete the segmented line.

After the successful segmentation of lines and removal of blank lines, the segmented line image is given as an input to the next step of the algorithm for segmenting the characters out of it (Fig. 1).

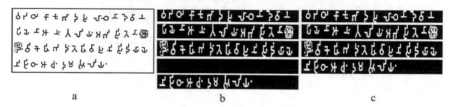

a   b   c

Fig. 1 a Digital estampage image of Brahmi script, **b** line images after successful segmentation, **c** line images after successful removal of blank lines. *Source* ASI, Mysore, India

3.1 Character Segmentation

Character segmentation in the proposed algorithm is carried out based on three types of characters. The size of the character is the parameter considered to categorize them. The size of each character is calculated as area (number of the pixels in the region defining the character). After watchful analysis on 10 different images containing more than 2000 characters, the following observations are noted:

1. All valid characters are found in only two specific sizes.
2. Some valid characters are very small in comparison with the others.
3. Ratio of character sizes is determined and it is found as 3:1, i.e., area/size of three small characters together is equivalent to the area of one normal character. In terms of percentage, it is found that small characters are approximately of size 20–35% with respect to one full character.

On the basis of the above observations, we have categorized characters into three different types as below:

i.   Regular character (RC): it is defined as an alphabet which is a connected component with an area greater than one-third of the average area of a character in an image.
ii. Special character with *anuswar* modifier (SC1): it is defined as a collection of two connected components together, out of which the first component is a regular character (RC) and the second component is *anuswar*, where *anuswar* is defined as a connected component with area less than one-third of the average area of a character in an image.
iii. Special character (b ⸫) (SC2): it is defined as the collection of three *anuswar* together.

For these three types of characters, we are using abbreviations as RC, SC1 and SC2 (Fig. 2).

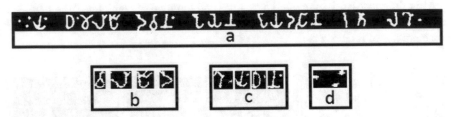

Fig. 2 **a** Segmented line, **b** regular character (RC), **c** special characters with *anuswar* modifier (SC1), **d** special character (∴) (SC2). *Source* ASI, Mysore, India

After successful segmentation of lines, the next step in the proposed algorithm is to segment the characters as per their characteristics which are elaborated with the help of the following steps:

Step 1:   Ordering of characters:

Characters in a line are arranged in chronological order and sequence numbers are assigned to them since it is a necessity for segmenting characters of type SC1 and SC2. The ordering or sequencing of characters is done by calculating the distance of each character from the origin of the image to the left topmost pixel of every character. With respect to the distance of a character from the origin of the image, numbers are assigned to the characters. Shorter the distance, earlier the character. Hence the most distant character is the last character in the sequence [30].

Step 2:   Find the average size of a character from the image.
Step 3:   Segment character as RC if its area is greater than 33% of the average size of the character.
Step 4:   Identify the *anuswar* character and store it in an array.
Step 5:   Find and segment special symbols SC1 and SC2. In the segmentation of these type of characters, multiple cases need to be handled:

Case I: If the first character in the line is *anuswar*, then check the next two consecutive characters. If the next two consecutive characters are *anuswar*, then club these three characters to form special character SC2 (b ⌣).

Case II: If the first character in the line is *anuswar*, then check the next two consecutive characters. If the next two consecutive characters are not *anuswar*, then ignore and do not segment it.

Case III: If the character is not the first character of the line, and it is a regular character RC (not an *anuswar*), then check the following three possibilities:

1. Only the next character is *anuswar*, then combine this *anuswar* with the previous character to form a special character SC1.
2. Next three characters are *anuswar* but not fourth, then club these three characters to form special character SC2 (b ⌣).
3. Next four characters are *anuswar*, then combine the first *anuswar* with the previous character to form a special character SC1 and combine the next three *anuswar* to form special character SC2 (b ⌣).

4 Results

The proposed algorithm for segmentation of lines and character is tested on 24 images. In the first step of the algorithm, we have successfully segmented 262 lines. Table 1 describes the details of line segmentation. With the help of blank line removal, we have achieved 100% successful segmentation.

Table 1 Line segmentation details

Total number of images	Actual number of lines	Number of total lines segmented	Number of blank lines deleted	Number of correctly segmented lines	Percentage of successful segmentation
24	262	268	6	262	100%

The second step of the algorithm, i.e., character segmentation is carried out on 262 images of segmented lines. Character segmentation is based on three different categories. Out of total characters 7689, we have successfully segmented 6771 characters with the accuracy of 88.68%. Table 2 gives the summary of overall character segmentation.

As the character segmentation is based on three categories of characters which are RC, SC1 and SC2, category-wise results are described in Table 3, Table 4 and Table 5, respectively.

Table 2 Category-wise details of segmentation

Type of character	Actual number of characters	Number of correctly segmented characters	Percentage of successful segmentation
RC	6888	6121	91.35
SC1	724	580	81.85
SC2	77	70	92.85
Total	7689	6771	88.68

Table 6 enlists comparative analysis of different segmentation techniques used for various scripts in recent times. It is observed that despite so many techniques being available, the result is yet to reach 100%. Moreover, it is also seen that no definite work has been done for Brahmi script either.

5 Conclusion

In this paper, we have presented a two-step methodology to segment the lines at first, and then the three types of characters (RC, SC1 and SC2). We have implemented this method on 24 Brahmi estampage digital images acquired from ASI, Mysore, India. These images cover all possible variations with respect to the quality of an image. They contain 262 lines and 7689 characters in total. We have achieved 100% accuracy in line segmentation. Whereas for character segmentation of type RC, SC1 and SC2, we have achieved 91.35%, 81.86% and 92.86% accuracy, respectively.

Segmentation accuracy is affected because of the preprocessing algorithm applied in the previous stage of segmentation. Considering the global approach, the same preprocessing algorithm is applied to all images, which are of different quality containing uneven amount of noise and percentage of degradation. One of the steps of erosion in the preprocessing algorithm causes the removal of noise and separation of individual characters from touched characters. But at the same time, it also causes some characters to break down into multiple pieces, and some touched characters don't get separated from each other as it may require more percentage of erosion. So in such a case, if different best suitable preprocessing algorithms are applied to each image separately, it will help significantly to improve the segmentation accuracy.

In this method, line segmentation based on horizontal projection is found to be remarkable. Accuracy of the character segmentation based on three types of characters is significant. Overall enhancement in the preprocessing algorithm will help the segmentation accuracy. Yet, enhancement can be proposed in the segmentation of type SC1.

Table 3 Image-wise segmentation details of RC character

Sr. No.	Image name	Actual RC count	Correctly segmented RC count	Total segmented RC count	Percentage
1	Img 1	107	107	107	100
2	Img 2	137	124	124	90.51
3	Img 3	99	93	94	93.94
4	Img 4	128	124	125	96.88
5	Img 8	195	193	200	98.97
6	Img 10	364	355	368	97.53
7	Img 13	93	88	92	94.62
8	Img 14	154	153	149	99.35
9	Img 17	155	141	153	90.97
10	Img 18	416	401	412	96.39
11	Img 25	594	492	607	82.83
12	Img 26	702	611	701	87.04
13	Img 27	506	476	514	94.07
14	Img 28	525	496	587	94.48
15	Img 37	34	30	34	88.24
16	Img 38	39	30	33	76.92
17	Img 39	46	43	43	93.48
18	Img 40	46	46	47	100
19	Img 41	48	48	48	100
20	Img 47	445	278	355	62.47
21	Img 48	69	63	94	91.3
22	Img 50	600	534	566	89
23	Img 51	784	653	758	83.29
24	Img 52	602	542	574	90.03
		6888	6121	6785	91.35

Table 4 Image-wise segmentation details of SC1 character

Sr. No.	Image name	SC1 count	Correctly segmented	Total segmented	Percentage
1	Img 1	15	15	15	100
2	Img 2	8	5	13	62.5
3	Img 3	15	13	19	86.67
4	Img 4	6	6	7	100
5	Img 8	10	8	10	80
6	Img 10	42	41	48	97.62
7	Img 13	10	9	9	90
8	Img 14	13	12	17	92.31
9	Img 17	25	21	33	84
10	Img 18	47	37	63	78.72
11	Img 25	62	52	79	83.87
12	Img 26	79	72	101	91.14
13	Img 27	74	59	73	79.73
14	Img 28	51	46	66	90.2
15	Img 37	2	1	1	50
16	Img 38	3	3	3	100
17	Img 39	7	6	6	85.71
18	Img 40	6	6	5	100
19	Img 41	5	5	5	100
20	Img 47	29	12	17	41.38
21	Img 48	8	4	14	50
22	Img 50	51	41	41	80.39
23	Img 51	105	67	67	63.81
24	Img 52	51	39	40	76.47
		724	580	752	81.86

Table 5 Image-wise segmentation details of SC2 character

Sr. No.	Image name	SC2 count	Correctly segmented	Total segmented	Percentage
1	Img 1	2	2	2	100
2	Img 2	1	1	2	100
3	Img 3	1	1	2	100
4	Img 4	0	0	0	100
5	Img 8	0	0	0	100
6	Img 10	6	6	6	100
7	Img 13	2	2	2	100
8	Img 14	0	0	0	100
9	Img 17	4	3	3	75
10	Img 18	4	3	4	75
11	Img 25	3	3	6	100
12	Img 26	10	10	26	100
13	Img 27	7	7	8	100
14	Img 28	11	11	11	100
15	Img 37	1	0	0	0
16	Img 38	1	0	0	100
17	Img 39	0	0	1	100
18	Img 40	0	0	0	100
19	Img 41	0	0	0	100
20	Img 47	1	1	1	100
21	Img 48	1	1	1	100
22	Img 50	4	4	4	100
23	Img 51	14	11	11	78.57
24	Img 52	4	4	4	100
		77	70	94	92.86

Table 6 Comparison of results from various segmentation techniques

References	Language/Script	Technique	Result
Chamchong and Fung [7]	Thai Noi, Thom	Line segmentation: partial projection profile Character segmentation: contour tracing algorithm	Line segmentation: 58.46% Character segmentation: 93.99%
Gajjar et al. [9]	Devanagari	Histogram-based approach	92.53%
Ramteke and Rane [14]	Devanagari	Line segmentation: connected component approach Character segmentation: vertical projection profile	Overall segmentation: 97%
Kumar et al. [18]	Gurumukhi	Horizontal and vertical projection profile along with water reservoir technique	92.53%
Mathew et al. [22]	Indic scripts	M-OCR	96.49%
Batuwita and Bandara [26]	English	Skeleton-based approach using Fuzzy features	–
Kumar and Singh [27]	Gurumukhi	Top-down techniques by using features at the highest level structure to features at the lowest level structure	Line segmentation: 92.06% Word segmentation: 91.97% Character segmentation: 89.44%
Patil and Mali [29]	Devanagari	Statistical information and vertical projection	82.17%
Nagane and Mali [30]	Brahmi	Connected component	91.30%

References

1. Ojha, P.G.H.: Bharatiya Prachin Lipimala: The Palaeography of India. Munshiram Manohar-lal, New Delhi (1971)
2. Salomon, R.: Indian Epigraphy: A Guide to the Study of Inscriptions in Sanskrit, Prakrit, and the Other Indo-Aryan Languages. Oxford University Press, New York Oxford (1998)
3. Bandara, D., Warnajith, N., Minato, A., Ozawa, S.: Creation of precise alphabet fonts of early Brahmi script from photographic data of ancient Sri Lankan inscriptions. Can. J. Artif. Intell. Mach. Learn. Pattern Recogn. **3**(3), 33–39 (2012)
4. Kak, S.C.: Indus and Brahmi further connections. Cryptologia **14**(2), 169–183 (1990)
5. https://en.wikipedia.org/wiki/Brahmi_script. Accessed 10 Nov 2019
6. http://www.ancientscripts.com/brahmi.html. Accessed 10 Nov 2019
7. Chamchong, R., Fung, C.C.: Character segmentation from ancient palm leaf manuscripts in Thailand. In: Proceedings of the 2011 Workshop on Historical Document Imaging and Processing, pp. 140–145 (2011)
8. Siromoney, G., Chandrasekaran, R., Chandrasekaran, M.: Machine recognition of Brahmi script. IEEE Trans. Syst. Man Cybern. SMC **13**(4), 648–654 (1983)
9. Gajjar, T., Teraiya, R., Gohil, G., Goyani, M.: Top Down Hierarchical Histogram Based Approach for Printed Devnagri Script Character Isolation. In: Nagamalai, D., Renault, E., Dhanushkodi, M. (eds.). DPPR 2011, CCIS, vol. 205, pp. 55–64 (2011)
10. Saba, T., Rehman, A., Elarbi-Boudihir, M.: Methods and strategies on off-line cursive touched characters segmentation: a directional review. Artif. Intell. Rev. **42**(4), 1047–1066 (2014)
11. Alginahi, Y.M.: A survey on Arabic character segmentation. Int. J. Doc. Anal. Recogn. (IJDAR) **16**(2), 105–126 (2013)
12. Cheng, D., Tian, F., Liu, L., Liu, X., Jin, Y.: Image segmentation based on multi-region multi-scale local binary fitting and Kullback-Leibler divergence. SIViP **12**, 895–903 (2018)
13. Casey, R.G., Lecolinet, E.: A survey of methods for strategies in character segmentation. IEEE Trans. Pattern Anal. Mach. Intell. **18**(7), 690–706 (1996)
14. Ramteke, A.S., Rane, M.E.: Offline handwritten Devanagari script segmentation. Int. J. Sci. Technol. Res. **1**(4), 142–145 (2012)
15. Sinha, R.M.K., Mahabala, H.N.: Machine recognition of Devanagari script. IEEE Trans. Syst. Man Cybern. SMC **9**(8), 435–441 (1979)
16. Pal, U., Chaudhuri, B.B.: Indian script character recognition: a survey. Pattern Recogn. **37**, 1887–1899 (2004)
17. Saba, T., Sulong, G., Rehman, A.: A survey on Methods and strategies on touched characters segmentation. Int. J. Res. Rev. Comput. Sci. (IJRRCS) **1**(2), 103–114 (2010)
18. Kumar, M., Jindal, M.K., Sharma, R.K.: Segmentation of isolated and touching characters in offline handwritten Gurmukhi script recognition. Int. J. Inf. Technol. Comput. Sci. **2**, 58–63 (2014)
19. Shobha Rani, N., Chandan, N., Jain, S.A., Kiran, H.R.: Deformed character recognition using convolutional neural networks. Int. J. Eng. Technol. **7**(3), 1599–1604 (2018)
20. Macwan, J.J., Goswami, M.M., Vyas, A.N.: A survey on offline handwritten North Indian script symbol recognition. In: International Conference on Electrical, Electronics, and Optimization Techniques (ICEEOT). IEEE, pp. 2747–2752 (2016)
21. Kathiriya, H.M., Goswami, M.M.: Word spotting techniques for Indian scripts: a survey. In: International Conference on Innovations in Power and Advanced Computing Technologies [i-PACT 2017], pp. 1–5 (2017)
22. Mathew, M., Singh, A.K., Jawahar, C.V.: Multilingual OCR for Indic scripts. In: 2016 12th IAPR Workshop on Document Analysis Systems (DAS), Santorini, pp. 186–191 (2016)

23. Kunchukuttan, A., Puduppully, R., Bhattacharyya, P.: *Brahmi-Net:* a transliteration and script conversion system for languages of the Indian subcontinent. In: Proceedings of NAACL-HLT, pp. 81–85 (2015)

24. Gautam, N., Chai, S.S.: Optical character recognition for Brahmi script using geometric method. J. Telecommun. Electron. Comput. Eng. **9**, 131–136 (2017)

25. Warnajith, N., Bandara, D., Bandara, N., Minati, A., Ozawa, S.: Image processing approach for ancient Brahmi script analysis (Abstract). University of Kelaniya, Colombo, Sri Lanka, p. 69 (2015)

26. Batuwita, K.B.M.R., Bandara, G.E.M.D.C.: New segmentation algorithm for individual offline handwritten character segmentation. In: Wang, L., Jin, Y. (eds.) Fuzzy Systems and Knowledge Discovery. FSKD. Lecture Notes in Computer Science. Springer, Berlin, Heidelberg, vol. 3614 (2005)

27. Kumar, R., Singh, A.: Detection and segmentation of lines and words in Gurmukhi handwritten text. In: IEEE 2nd International Advance Computing Conference, pp. 353–356 (2010)

28. Anasuya Devi, H.K.: Thresholding: a pixel-level image processing methodology preprocessing technique for an OCR system for the Brahmi script. Anc. Asia **1**, 161–165 (2006)

29. Patil, C.H., Mali, S.M.: Segmentation of isolated handwritten Marathi words. In: National Conference on Digital Image and Signal Processing, pp. 21–26 (2015)

30. Nagane, A.S., Mali, S.M.: Segmentation of special character "∴" from degraded Brahmi script documents. In: Two Days 2nd National Conference on 'Innovations and Developments in Computational & Applied Science' [NCIDCAS-2018], pp. 65–67 (2018)

Sentiment Analysis of Democratic Presidential Primaries Debate Tweets Using Machine Learning Models

Jennifer Andriot[1]([✉]), Baekkwan Park[2], Peter Francia[2], and Venkat N Gudivada[1]([✉])

[1] Department of Computer Science, East Carolina University, Greenville, NC 27855, USA
andriotj17@students.ecu.edu, gudivadav15@ecu.edu
[2] Center for Survey Research, East Carolina University, Greenville, NC 27855, USA

Abstract. Primary election debates play a significant role in the outcome of the general presidential election. This is more so when the candidate from the voter's affiliated party is running against an incumbent. This research focuses on the sentiment analysis of tweets posted during the third Democratic primary debate. The goal of this research is to determine whom the Twitter users believe won or lost the debate. We evaluated the sentiment of tweets by using natural language processing tools and building machine learning models. This paper presents our approach to determining which candidate garnered the most support during the debate.

Keywords: Social media · NLP · Democracy · Political debates · Elections · Politics

1 Introduction

Social Media plays a significant role in determining the outcomes of elections in democratic countries. Manipulating Social Media to a candidate's advantage is currently an unregulated activity. Recently, several consulting businesses have come up that offer ways to influence the electorate using Social Media and other digital tools. For example, Indian Political Action Committee (I-PAC) is one such organization for influencing and manipulating electorates. I-PAC is a misnomer as its goal is to steer election outcomes in favor of clients. In a 2019 election held in the state of Andhra Pradesh, the money paid by the winning party to I-PAC represents 44.4% of the total election expense of the party. This data comes from the Election Expenditure Statement that the party has filed with the Election Commission of India.

B. Iyer et al. (eds.), *Applied Computer Vision and Image Processing*,
Advances in Intelligent Systems and Computing 1155,
https://doi.org/10.1007/978-981-15-4029-5_34

In a related development, the world's oldest democracy—the United States (US) of America—was targeted by adversarial groups and hostile nations to upend US elections in 2016 [22]. As per a report produced for the US Senate Intelligence Committee, the goal of the Russian influence campaign unleashed on social media in the 2016 election was to sway American opinion and divide the country [23]. This campaign made an extraordinary effort to target African–American voters and used an array of tactics whose goal was to suppress turnout among Democratic voters.

The above two scenarios underscore the vulnerability of democratic elections to interference operations through digital tools. Given the recency of this phenomenon, governments do not have oversight on voting systems. A policy paper from the Brennan Center for Justice [17] states that the US government regulates, for example, colored pencils more tightly than it does the country's election infrastructure. These situations demand that academic researchers across the domains rise to address election interference challenges, and conduct research into eliminating or mitigating election interference. A rigorous and data-driven findings from such research will also help in formulating public policy on elections in the era of digital dominance. This is precisely the motivation for the research reported in this paper.

The US is an example of a nation with a two-party political system. The electorate votes largely to only two major political parties—Democratic and Republican. Each party may have multiple aspiring candidates. However, each party nominates only one candidate for the election through a filtering process known as *primary elections* or simply *primaries*. Aspiring candidates of a party go through several public debates in which they articulate their vision for the country and distinguish themselves from the other candidates. During and right after the debates, debate watchers tweet about the debate performance of candidates. The focus of this paper is to analyze the sentiments expressed in the tweets of the 2019 debates associated with the Democratic party primaries.

Following the third and most recent Democratic presidential debate on September 12, 2019, political observers from across the media talked and wrote about which candidates "won" or "lost," but with little consensus emerging. Opinion writers at *The New York Times*, for example, scored the debate performances of each candidate on a scale of 1 to 10, and gave Elizabeth Warren the highest average mark of 7.5 and Andrew Yang the lowest of 3.4 [26]. *CNN*'s Chris Cillizza viewed the debate differently, scoring Joe Biden and Beto O'Rourke as the winners, with Elizabeth Warren, Andrew Yang, and Julian Castro as the losers [8]. While political pundits could not seem to agree on a debate winner or loser, thousands of Americans offered their own opinions on Twitter about the debate. Although certainly not representative of the nation's population as a whole, Twitter users have emerged as a popular target audience for political messaging. Indeed, a few could argue against the fact that Twitter has become highly relevant (for both good and bad) in today's politics. So, how did Twitter users react as they watched the Democratic debate?

In this paper, we draw upon the natural language processing (NLP) tools and machine learning models and examine the sentiments of tweets posted during the hours of the third Democratic presidential debate. Academic research on presidential debates focuses much more heavily on presidential debates during the general election than the primary election or nomination phase [9,16]. It is certainly understandable given that there is greater public interest and attention in the general election. Nevertheless, it leaves a noticeable gap in the literature when one considers that primary presidential debates often provide the first exposure of the candidates to voters. This entry point frames the initial and sometimes lasting public impressions of the candidates, particularly for those voters who begin without much knowledge or information of the candidates [2,25].

Our results show that relying on what political pundits observe about debate "winners" and "losers" may not necessarily reflect popular opinion, or at least at a minimum, what people who use Twitter are thinking in real time as they watch the debate. In Sect. 2, we discuss the existing studies on the topic and methodological approaches. Next, in Sect. 3, we describe the data and our supervised machine learning approach to analyze sentiments of the primary debate. We discuss our key findings in Sect. 4, and Sect. 5 concludes the paper.

2   Related Work

In this section, we discuss the significance of debates in the electoral process and motivate the need for research on election debates of primaries using social media and sentiment analysis.

2.1   Primary Election Debates

Political debates, more specifically, election debates have become an important event not only for the candidates, but also for voters [3]. The events allow voters to compare candidates competing for elections side by side and all at once. Since elections are basically choosing a candidate among other competing options, election debates allow and help audiences to observe the candidates' ideas which can be conflicting and cause some confrontation among candidates. Moreover, the basic rules for election debates are that candidates are not supposed to bring any prepared materials, thus, the audiences can observe how candidates handle unexpected questions or comments from competing candidates. Although a large number of scholars have paid attention to and written about general presidential election debates [9,16], it is remarkable that primary debates have received little attention.

Primary election campaign debates are noteworthy. First, primary debates provide the opportunity for "partisan voters" to determine who is going to represent their political party in the upcoming general election [13]. Second, although general debates usually draw more attention, a lot more campaigns have featured primary debates than general debates [13]. Third, because candidates in primary elections are generally less well known, the primary debates tend to have much higher potential for influencing audiences and voters' attitude and choice than general election debates [3]. Thus, primary debates can shape how voters view a candidate's electability, which can ultimately influence how they cast their ballots [29]. For these important reasons, our analysis focuses on an early primary presidential debate.

2.2   Measuring Tweet Sentiment on Political Debates

Social media such as Twitter contains millions of texts that can be used as an abundant source of emotions, sentiments, and moods [12]. Thus, Twitter has provided audiences (voters) a platform to freely express their feelings, thoughts, and opinions about political candidates. In addition, unlike some political opinion studies that rely on survey data taken days or even weeks after the debate [4], social media based analysis allows us to assess how individuals react in real time. As shown by Srinivasan, Sangwan, Neill, and Zu, sentiment analysis of political tweets proved to be more accurate at predicting how 19 US states would vote in the 2016 general election compared to traditional polling methods. They correctly predicted the outcome of 17 out of 19 states whereas the pollsters correctly predicted 15 [24].

Sentiment analysis and emotion classification have attracted much research during the last decade on the basis of natural language processing and machine learning [1,6,18,19,21,27]. One such research being sentiment analysis is performed on Twitter data by Pinto and Murari using supervised machine learning methods. Using an SVM model with TF-idf weighting scheme, 82% of tweets were correctly classified as positive or negative. [20] Similarly, a study by A. A. Hamoud et al. applied SVMs and Naïve Bayes to classify tweets. In addition, Bag-of-Words (BOW), TF, and TF-idf were used for feature extraction. They found that SVM BOW model performed best at classification with 88% accuracy and SVM TF-idf was the next best performing with 85% accuracy. [11]

Some scholars have paid attention to political elections and tried to measure overall public opinions (or sentiment) on candidates to predict the election results [5,7,10,15,28]. Although, there is little research on primary election debates using social media and sentiment analysis.

3   Our Approach: Data and Methods

We first discuss data collection and labeling, followed by machine learning models for classification and identification of candidates from tweets.

3.1   Data Collection and Labeling

For our analysis, we used the Twitter Streaming API to collect data related to the Democratic Primary Debates on September 12, 2019. To filter and select relevant tweets, we used the key words "#DemocraticDebate" and "#demdebate" during the hours of the debate (i.e., between 8:00 and 11:00 p.m. EST on September 12, 2019).

Supervised machine learning models learn from *training data*. Each training data instance consists of a certain number of *predictive variables* and the associated outcome. For example, for a tweet, the predictive variables represent the textual content of the tweet, and the associated outcome is the sentiment expressed in the tweet—positive or negative. We manually annotated over 4,000 tweets. Five groups of students were assigned a random set of tweets and manually labeled each tweet into one of the two classes: positive or negative. To ensure inter-coder reliability, we examined all the labeled data again.

3.2 Machine Learning Models for Classification

After retrieving and filtering all the tweets about the primary debate, we preprocessed the texts to eliminate noisy data. To determine the positive or negative sentiment of tweets on the debate, we used the following supervised machine learning models: multinomial naive bayes (NB), logistic regression (LR), support vector machine (SVM), and random forest (RF) [14]. We represented the content of tweets using two vectorization schemes—term frequency (TF) and term frequency inverse document frequency (Tf-idf). Under the TF scheme, each word in the tweet is considered as a term. Moreover, a weight is associated with the term to denote its relative importance. The frequency of the term—how many times the term appears in the tweet—is the term weight. In essence, each tweet is represented as an n-dimensional vector and each dimension corresponds to a term. Determining the number of dimensions requires first identifying the set of terms across all the tweets. It should be noted that the n-dimensional vector will have a value zero for those terms that do not appear in a tweet.

The Tf-idf scheme is similar to the TF except that the weight of a term is based on the product of two factors: the *frequency of the term* and the *inverse document frequency*. Given a collection of tweets and a term, the *inverse document frequency* is the reciprocal of the number of tweets in which the term appears.

Next, we consider the variations on the definition of a term. In the simplest case, we consider each word in the tweet as a term. This corresponds to what is called a *unigram* model. For example, in the sentence "machine learning is fun," the unigrams are machine, learning, is, and fun. The bigram model considers two consecutive words as a term. For the above sentence, the bigrams are machine learning, learning is, and is fun. Lastly, the trigram model considers three consecutive words as a term. The trigrams for the above example are machine learning is and learning is fun.

We created 24 models through a combination of four machine learning algorithms (LR, NB, RF, and SVM); two vectorization schemes (TF, Tf-idf); and three term models (unigram, bigram, and trigram). We used fivefold cross validation for evaluating the models. In fivefold cross validation, the training data is divided into five equal parts. During the first iteration, the model is built using the last four parts for building the model and the first part for evaluating the system. In the second iteration, parts 1, 3, 4, and 5 are used for building the model and the second part for evaluating the system. During the fifth iteration, the first four parts are used for building the model and the last part for evaluating the system.

Table 1 shows the classification accuracy of various machine learning models. Classification accuracy is the ratio of the number of correctly classified tweets to the total number of tweets in a class. For example, if the model classified 15 of the 20 tweets correctly for the *positive* class, classification accuracy = 15/20 = 75%. By examining the table, SVM (Tf-idf/bigram) model has the best performance. This model was then used to predict the rest of our debate tweets.

Table 1. Classification accuracy of various machine learning models: it represents the model performance for the different machine learning algorithms used

Model	Vectors	Ngrams	Accuracy	Model	Vectors	Ngrams	Accuracy
LR	TF	Unigram	0.774	RF	TF	Unigram	0.771
LR	TF	Bigram	0.813	RF	TF	Bigram	0.791
LR	TF	Trigram	0.816	RF	TF	Trigram	0.787
LR	Tf-idf	Unigram	0.788	RF	Tf-idf	Unigram	0.796
LR	Tf-idf	Bigram	0.824	RF	Tf-idf	Bigram	0.789
LR	Tf-idf	Trigram	0.819	RF	Tf-idf	Trigram	0.787
NB	TF	Unigram	0.743	SVM	TF	Unigram	0.771
NB	TF	Bigram	0.755	SVM	TF	Bigram	0.813
NB	TF	Trigram	0.732	SVM	TF	Trigram	0.814
NB	Tf-idf	Unigram	0.747	SVM	Tf-idf	Unigram	0.790
NB	Tf-idf	Bigram	0.750	SVM	Tf-idf	Bigram	0.831
NB	Tf-idf	Trigram	0.739	SVM	Tf-idf	Trigram	0.829

3.3   Identification of Candidates

In order to identify candidates in tweets, we built and curated a candidate name dictionary. In addition to each candidate's full name, we also included shortened names (e.g., Liz for Elizabeth), related Twitter hashtags (e.g., #TeamWarren, #feelthebern, #joebiden, etc.), and Twitter handles (e.g., @AndrewYang, @amyklobuchar, @PeteButtigieg, etc.) to identify candidates properly.

3.4   Aggregation and Support Ratio

We calculate the proportion of positive tweets T_{positive} over the total tweets (T_{all}) for each candidate (i).

$$S = T_{i,\text{positive}}/T_{i,\text{all}} \tag{1}$$

4   Findings and Discussion

The results that emerge from the sentiment classification do seem to reveal winners and losers. Figure 1 summarizes the first results. Pete Buttigieg generated the most positive sentiment from Twitter during the Thursday night's debate.

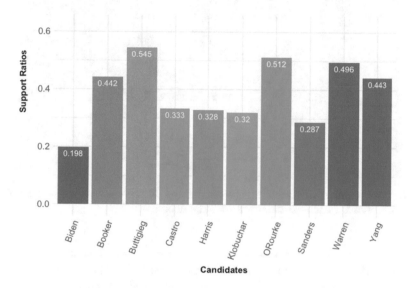

Fig. 1. Support ratios across candidates: it represents the support ratio (the percentage of positive tweets) each candidate received, which is based on the classification results (SVM with TF-idf/bigram)

However, it would be difficult to declare Buttigieg as one of the debate's winners, as he was the subject of the fewest number of tweets, as shown in Fig. 2. Put simply, Buttigieg succeeded in generating positive tweets, but failed in garnering much attention among Twitter users. Similarly, after Buttigieg, Peter O'Rourke generated the highest positive ratio, but people did not tweet about him much, and only a little bit more than Buttigieg.

While the pundits disagreed to some extent over Elizabeth Warren's debate performance (see above), Twitter users scored her positively relative to the other candidates. Warren not only generated significant Twitter activity, but also scored the third highest support ratio. More than 13,000 tweets expressed a sentiment about her debate performance. These results suggest that Elizabeth Warren emerged a debate winner on Twitter. Andrew Yang also generated a high positive ratio. With more than twice the number of tweets than Buttigieg generated, Yang not only received positive reactions from Twitter users, but also a reasonable degree of interest and relevance. In addition to Warren, these results suggest that Andrew Yang also emerged a debate winner on Twitter. These results run counter to the post-debate analysis of most political pundits.

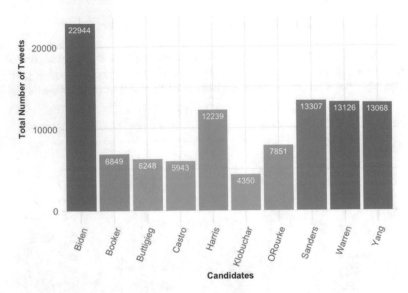

Fig. 2. Total number of tweets across candidates: it represents the total number of tweets collected for each candidate during the debate

If there was an identifiable loser on Twitter, it was Joe Biden and Bernie Sanders, both of whom generated the least positive reaction to their respective debate performances. As the front runner, it is not surprising to see that Biden had the most tweets of any candidate. Interestingly, Sanders generated a little more than Warren or Yang, but scored the second to the bottom, suggesting that he failed on Thursday night.

Twitter's verdict of the debate thus differed from some of the post-debate analysis, particularly with regard to Andrew Yang. If the larger winds of public opinion happen to blow in Twitter's direction, then Warren and Yang will be the beneficiaries. In contrast, Sander's debate performance did little to help his campaign among Twitter users on Thursday night, while Joe Biden struggled as well to impress the "Twitter-verse."

5   Conclusions

Assessing who "won" and who "lost" a presidential debate is a popular topic among political pundits. However, as the analysis in this paper demonstrates, the judgment of the experts is not always consistent with the sentiment of those who are active on Twitter. Given Twitter's growing use as a vehicle for people to communicate and express their political views, and given Twitter's emerging importance in political campaigns and elections, the opinions expressed on Twitter carry increasing significance. This is especially true for major campaign events, such as presidential debates that can have electoral consequences.

Primary presidential debates, in particular, are important because this is when voters are often first exposed to the candidates. These initial impressions ultimately can shape voter behavior. Although Twitter users are not representative of the general population, their opinions still reflect a voice in the larger political chorus. Our results offer an important first step in understanding more about what those Twitter voices are saying during pivotal campaign events, such as primary presidential debates.

We have used random sampling in selecting a subset of the tweets for training the machine learning models. However, random sampling is not necessarily the most appropriate for the problem we analyzed. Given that the President of the United States is chosen by a method that relies upon the *Electoral College* to execute the will of the electorate. To factor this into our analysis, a *stratified sampling* method is more appropriate than the random sampling. We will consider this in our future analysis.

The classification accuracy numbers reported in Table 1 are not particularly stellar. We will explore deep learning based approaches to the tweet classification problem to improve the classification accuracy. Furthermore, manual labeling of tweets data is very expensive in terms of human labor. We plan to explore automated approaches to augmenting the manually generated dataset. Especially for deep learning approaches, the more the training data, the better is the classification accuracy. Lastly, we will replace classification accuracy measure with a more robust one called Area Under ROC curve—AUROC.

References

1. Barbosa, L., Feng, J.: Robust sentiment detection on twitter from biased and noisy data. In: Proceedings of the 23rd International Conference on Computational Linguistics: Posters, pp. 36–44. Association for Computational Linguistics (2010)
2. Benoit, W.L.: Political Election Debates: Informing Voters About Policy and Character. Lexington Books (2013)
3. Benoit, W.L.: Issue ownership in the 2016 presidential debates. Argum. Advocacy **54**(1–2), 95–103 (2018)
4. Benoit, W.L., Hansen, G.J.: Presidential debate watching, issue knowledge, character evaluation, and vote choice. Hum. Commun. Res. **30**(1), 121–144 (2004)
5. Bhatia, S., Mellers, B., Walasek, L.: Affective responses to uncertain real-world outcomes: sentiment change on twitter. PloS one **14**(2), e0212,489 (2019)
6. Brynielsson, J., Johansson, F., Jonsson, C., Westling, A.: Emotion classification of social media posts for estimating people's reactions to communicated alert messages during crises. Secur. Inform. **3**(1), 7 (2014)
7. Burnap, P., Gibson, R., Sloan, L., Southern, R., Williams, M.: 140 characters to victory? using twitter to predict the UK 2015 general election. Elect. Stud. **41**, 230–233 (2016)
8. Cillizza, C.: Winners and losers from the third democratic debate. CNN (2019)
9. Coleman, S., Butler, D.: Televised Election Debates: International Perspectives (2000)
10. Grover, P., Kar, A.K., Dwivedi, Y.K., Janssen, M.: Polarization and acculturation in us election 2016 outcomes-can twitter analytics predict changes in voting preferences. Technol. Forecast. Soc. Chang. **145**, 438–460 (2019)

11. Hamoud, A.A., Alwehaibi, A., Roy, K., Bikdash, M.: Classifying political tweets using naïve bayes and support vector machines. In: Mouhoub, M., Sadaoui, S., Mohamed, O.A., Ali, M. (eds.) Recent Trends and Future Technology in Applied Intelligence, pp. 736–744. Springer International Publishing (2018)
12. Hasan, M., Agu, E., Rundensteiner, E.: Using hashtags as labels for supervised learning of emotions in twitter messages. In: ACM SIGKDD Workshop on Health Informatics, New York, USA (2014)
13. Holbert, R.L., Benoit, W., McKinney, M.: The role of debate viewing in establishing "enlightened preference" in the 2000 presidential election. In: Annual Conference of the International Communication Association, Political Communication Division, Seoul, Korea (2002)
14. Kelleher, J.D., Namee, B.M., D'Arcy, A.: Fundamentals of Machine Learning for Predictive Data Analytics: Algorithms, Worked Examples, and Case Studies. MIT Press, Cambridge, Massachusetts (2015)
15. Kolagani, S.H.D., Negahban, A., Witt, C.: Identifying trending sentiments in the 2016 us presidential election: a case study of twitter analytics. Issues Inf. Syst. **18**(2), 80–86 (2017)
16. Kraus, S.: Televised Presidential Debates and Public Policy. Routledge (2013)
17. Norden, L., Ramachandran, G., Deluzio, C.: A framework for election vendor oversight (2019). https://www.brennancenter.org/our-work/policy-solutions/framework-election-vendor-oversight
18. Pak, A., Paroubek, P.: Twitter based system: using twitter for disambiguating sentiment ambiguous adjectives. In: Proceedings of the 5th International Workshop on Semantic Evaluation, pp. 436–439. Association for Computational Linguistics (2010)
19. Pang, B., Lee, L., Vaithyanathan, S.: Thumbs up?: sentiment classification using machine learning techniques. In: Proceedings of the ACL-02 Conference on Empirical Methods in Natural Language Processing, vol. 10, pp. 79–86. Association for Computational Linguistics (2002)
20. Pinto, J.P., Murari, V.T.: Real time sentiment analysis of political twitter data using machine learning approach. Int. Res. J. Eng. Technol. **6**(4), 4124–4129 (2019)
21. Roberts, K., Roach, M.A., Johnson, J., Guthrie, J., Harabagiu, S.M.: Empatweet: annotating and detecting emotions on twitter. In: Lrec, vol. 12, pp. 3806–3813. Citeseer (2012)
22. Roose, K.: Social media's forever war (2018). https://www.nytimes.com/2018/12/17/technology/social-media-russia-interference.html
23. Shane, S., Frenkel, S.: Russian 2016 influence operation targeted African-Americans on social media (2018). https://www.nytimes.com/2018/12/17/us/politics/russia-2016-influence-campaign.html?module=inline
24. Srinivasan, S.M., Sangwan, R., Neill, C., Zu, T.: Power of predictive analytics: using emotion classification of twitter data for predicting 2016 US presidential elections. J. Soc. Media Soc. **8**(1), 211–230 (2019)
25. Stewart, P.A., Svetieva, E., Eubanks, A., Miller, J.M.: Facing your competition: findings from the 2016 presidential election. In: The Facial Displays of Leaders, pp. 51–72. Springer, Berlin (2018)
26. Team, N.Y.T.O.: Winners and losers of the democratic debate (2019). https://www.nytimes.com/interactive/2019/10/16/opinion/debate-winners.html
27. Thelwall, M., Buckley, K., Paltoglou, G., Cai, D., Kappas, A.: Sentiment strength detection in short informal text. J. Am. Soc. Inf. Sci. Technol. **61**(12), 2544–2558 (2010)

28. Xie, Z., Liu, G., Wu, J., Wang, L., Liu, C.: Wisdom of fusion: prediction of 2016 Taiwan election with heterogeneous big data. In: 2016 13th International Conference on Service Systems and Service Management (ICSSSM), pp. 1–6. IEEE (2016)
29. Yawn, M., Ellsworth, K., Beatty, B., Kahn, K.F.: How a presidential primary debate changed attitudes of audience members. Polit. Behav. **20**(2), 155–181 (1998)

A Computational Linguistics Approach to Preserving and Promoting Natural Languages

Dhana L. Rao[1], Ethan Smith[2], and Venkat N. Gudivada[2(✉)]

[1] Department of Biology, East Carolina University, Greenville, NC 27855, USA
[2] Department of Computer Science, East Carolina University, Greenville, NC 27855, USA
gudivadav15@ecu.edu

Abstract. Natural language is a miracle of human life. They have developed and evolved over thousands of years. Language is more than a medium for communication. Embedded in languages are cultural practices and ancient wisdom. Furthermore, there is cognitive evidence to support that learning in one's mother tongue is easier than it is in a non-native language. Many languages of the world are close to extinction or on a rapid path to extinction. In this paper, we argue that Computational Linguistics offers practical and unprecedented opportunities for preserving and promoting all natural languages. Just as much as biologists care about the extinction of species and their impact on the environmental ecosystem, linguists, computing professionals, and society-at-large should take a similar approach to protecting and celebrating the grandeur of linguistic diversity. We outline an approach to achieving this goal for Indian languages.

Keywords: Mother tongue · Computational linguistics · Natural language processing · Natural languages · Machine learning · Indian languages

1 Introduction

Natural language is one of the miracles of human life. The term *natural language* refers to a language that has naturally developed and evolved in humans. A natural language is in contrast with an *artificial language* such as a computer programming language. For any language, we can find some that speak the language *natively* (called *native speakers*), while others speak the language *non-natively* (called *non-native speakers*). A native speaker of a language is one who has learned and used the language from early childhood. A native speaker's language is her first language and it is the language she uses for thinking. The

© Springer Nature Singapore Pte Ltd. 2020
B. Iyer et al. (eds.), *Applied Computer Vision and Image Processing*,
Advances in Intelligent Systems and Computing 1155,
https://doi.org/10.1007/978-981-15-4029-5_35

language of the native speaker is synonymous with *mother tongue* and *native language*. In this paper, we do not explicitly make these distinctions, and the context should help to infer the intended usage.

Language has multiple purposes, and communication is one of them. It is also used for thinking. Moreover, native languages are also intricately intertwined with cultures and social practices. The literature of the language documents the culture of a group of people and is passed from one generation to the next.

Linguistics is the science of the language, both spoken and written. Linguistics research aims to answer questions such as What is language? How are languages related? How do languages evolve over time? How many languages are there? How do we count languages? Does every human being have a language? Do animals have language? What distinguishes human language from other animal communication systems? More specifically, linguistic researchers study the structure of the language, sounds of the language, syntax, meaning, usage, and evolution [2].

Spoken languages assume more prominence relative to their written counterparts. For all languages, the number of people who speak the language greatly outnumbers those that can also write in the language. Spoken languages came much earlier than the written languages. There are many spoken languages without written counterparts.

Though there are 7,111 known living languages in the world [9], many of them are on the path of decline and eventual extinction. The number of native speakers is also on a rapid decline. This is nowhere more spectacular than India in general and southern India in particular. There are several reasons for this decline including the dominance of English on the World Wide Web (WWW), the colonial past, federal government policies, strong desire of people for migration to other countries, and changing cultural and societal values.

As pointed earlier, language is more than a communication medium. It is intricately woven into the culture of a group of people. It is both an invaluable cultural heritage and a source of ancient wisdom. What would be left in a geographic region if the language and culture are removed from the region? The region becomes simply a piece of land devoid of its unique vibrancy and enchantment.

Computational Linguistics is the application of computational techniques for analyzing and understanding natural languages. Often the terms *Computational Linguistics* and *Natural Language Processing* (NLP) are used synonymously. The term NLP is used in the Computer Science discipline, whereas the *Computational Linguistics* term is used in the Linguistics discipline. Computational Linguistics offers unprecedented opportunities to preserve, promote, and celebrate all languages.

The overarching goal of this paper is to explore this historic opportunity to revive languages whose native speakers are on the decline. Through this position paper, the authors hope to inspire Computational Linguistics researchers across India to embark on a journey to save and actively promote all languages and celebrate the linguistic diversity.

The remainder of the paper is organized as follows. The language diversity in India is presented in Sect. 2. The probable causes for the rapid decline of native speakers in India are discussed in Sect. 3. In Sect. 4, we provide scientific evidence for learning in one's own mother tongue is easier and desirable. Our approach to preserving and promoting natural languages using Computational Linguistics is described in Sect. 5. Section 6 concludes the paper.

2 Spectacular Linguistic Diversity

There are 7,111 known living languages in the world, according to Ethnologue [9]. Languages are as diverse as their commonality. Papua New Guinea offers an example of remarkable language diversity. As per Ethnologue, the number of spoken languages is 853, of which 841 are *living* and 12 are *extinct*. In terms of Ethnologue language taxonomy, of the 841 living languages, 39 are *institutional*, 304 are *developing*, 334 are *vigorous*, 124 are in *trouble*, and 40 are *dying*.

Ethnologue language data for India is 462 listed languages, 448 are living, and 14 are extinct. Of the living languages, 64 are institutional, 125 are developing, 190 are vigorous, 56 are in trouble, and 13 are dying. Even the institutional languages are on a gradual path toward insignificance and eventual extinction.

The English language has deep rooted eminence in India due to British colonial rule and the English-dominated WorldWide Web. English has come to symbolize social status, social mobility, literacy, and economic advancement. English is the de facto common language across India. This created an adverse situation for all major languages in India.

Besides the English dominance, institutional languages are also facingbreak survival challenges from another dimension. The current Indian government equates nationalism with having one language (i.e., Hindi) under the motto "One Nation, One Language." The government aggressively enforces Hindi on non-Hindi speaking population with the eventual goal of replacing English with Hindi. This is clearly a shortsighted and dangerous approach given that most languages spoken in South India predate Hindi and are accorded *classical language* status. Moreover, the linguistic boundaries were the basis for the demarcation of states/provinces. In fact, India is like the European Union. It is apt to call India as the *United States of India* to reflect its federal structure enshrined in the constitution, multiple cultural identities, and vast linguistic diversity.

The institutional languages of India have vast literature dating backbreak thousands of years. As noted earlier, the language and culture are intricately intertwined. Also, the literature in these languages embodies vast ancient knowledge that is unique to the geographic region. For example, consider Telugu, which is the official language of two southern states in India. The Telugu language has a unique literary tradition called *Ashtavadhanam*. It is a public performance in which an *Avadhani* (the performer) demonstrates his/her sharpness of memory and retention, mastery of the language, literature, grammar, and linguistic knowledge. The *Avadhani* is surrounded by eight peers who keep posing questions to the *Avadhani* and also distract her with challenges. The *Avadhani* constructs answers to the questions in a way that the answers adhere to certain grammatical constructions and other linguistic constraints. Advanced versions of *Ashtavadhanam* include *Satavadhanam* and *Sahasravadhanam*, where 100 and 1000 peers, respectively, ask questions and distract the *Avadhani*.

The Telugu language has a vast vocabulary. There are separate words for younger sister and elder sister. One needs to read the literature in the language to appreciate the vastness of its vocabulary. Each word is pronounced exactly as it is spelt. Telugu writing system won the second place in The World Alphabet Olympics held in 2012 (https://languagelog.ldc.upenn.edu/nll/?p=4253). Telugu poetry (called *padyalu*) employs an elaborate set of rules called *chandhas* for defining structural features. *Chandhas* also applies to prose and it generates rhythm to the literature and poetry. Some features of this metrical poetry are unique to Telugu.

From a linguistics perspective, there is so much diversity in Indian languages [3]. Figure 1 shows the seven most-spoken languages in India and the data for this figure comes from the Ethnologue [9]. Shown in Fig. 2 are the linguistic distances between five widely spoken Indian languages and Hindi [6]. The notion of *linguistic distance* is developed by Lewis et al. [9], which is quantified by counting the number of nodes between each pair of languages on the family tree of Indo-European and Dravidian languages. The farther the two languages are in the tree, the greater is the linguistic distance between them. For example, the linguistic distance between Telugu and Hindi is 10, whereas the same between Telugu and Kannada is 6. According to this measure, of the 15 major languages of India, Kannada is the closest language to Telugu with a distance of 6. The linguistic distance can be viewed as a measure of the degree of difficulty of one language speakers to learn another language. Any government language policy that is oblivious to the above findings is not only going to fail, but will also inflict great long-term damage socially, politically, and economically.

3 Why many Indian languages are on the decline?

All languages in India except Hindi are on the path toward a rapid decline in the number of native speakers and eventual extinction. It is projected that by 2050, very few people will be able to read literature and poetry in languages such as Telugu, let alone the ability to generate new literature. Even the number of people who will speak the language natively and fluently will reduce to a single digit by 2100. Many factors are contributing to this dangerous trend. The primary ones are the dominance of English as an international business language from one end, and the shortsighted and misinformed policies related to promoting Hindi at the cost of all other languages from the other end. The major reason for learning English is not the love for the language, but as a means to migrate to well-developed countries. But the truth is that it is impossible for other countries to provide jobs for all the youth of India. Instead, the government should have a

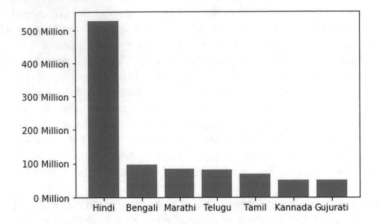

Fig. 1. First language speakers of top seven Indian languages in India   (*Source* Ethnologue)

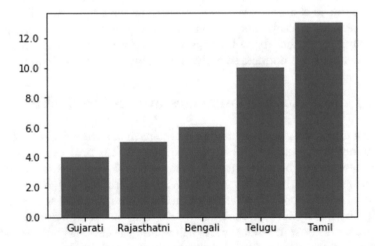

Fig. 2. The linguistic distance between major Indian languages and Hindi

vision and strategy to develop the country and create employment opportunities for its citizens. English can still be learned as an international business language, but this should not come at the cost of native languages.

Secondly, the Government of India relentlessly propagates the myth of a national language, whereas the Constitution of India specifies no national language. The Constitution lists 22 languages which include Telugu, Kannada, Tamil, and Malayalam. Furthermore, the Government of India has designated the following languages as *classical languages*: Tamil (since 2004), Sanskrit (2005), Telugu (2008), Kannada (2008), Malayalam (2013), and Odiya (2014). These languages have vast literature, poetry, plays, and lyrics and have been there for thousands of years. Given the unique traditions, rich literature and linguistic grandeur, and the local knowledge embedded in the literary works, the Government of India should leverage Computational Linguistics to preserve, promote, and celebrate languages and linguistic diversity (see Sect. 5).

The third factor is related to the lack of new literature. For many Indian languages, there is no significant production of new literature. Moreover, most academically talented students are not attracted to study language and literature. The language and linguistics departments end up educating those students who have less than ideal aptitude for linguistics and literature. Based on anecdotal evidence, even the language professors are not proficient to read literary classics.

The fourth factor is related to provincial government policies. The state of Andhra Pradesh is a case in point. In November 2019, the government announced through an executive order that all instruction in elementary and primary schools will be delivered only in English. The students will still learn Telugu as a language, but physical sciences, mathematics, and social sciences all be taught in English. Some claim that this policy is in response to declining enrollments in government-run primary and secondary education schools.

It is said that the executive order was in response to the trend that parents prefer to send their children to private primary and secondary schools where physical sciences, mathematics, and social sciences are taught in English. In such schools, teaching Telugu is optional. The major issue with the private schools is the lack of quality, but an exorbitant cost to the parents. Instead of regulating private schools, the government chose a flawed approach which is bound to fail.

In [7], Myles discusses policy issues related to second language learning of children in the age group 7 to 11 years. She notes research evidence that children are slower at learning a foreign language than adolescents and young adults. Therefore, requiring children to learn multiple languages at a very early age is counterproductive.

4 Why Learn in Mother Tongue?

Setting aside misinformed political policies and economic considerations, learning in the mother tongue is the best way to acquire knowledge and skills for young learners. Mother tongue is also referred to as a first language, native language, or $L1$ in the research literature. In [8], Noormohamadi argues that there is an intimate link between language and cognitive development. Furthermore, the mother tongue helps in the mental, moral, and emotional development of children. There is overwhelming research evidence that shows children who come to school with a solid foundation in their mother tongue develop stronger literacy abilities [8]. Providing a solid foundation in mother tongue requires parents spending time with their children reading stories and discussing issues to enable the development of vocabularies and concepts in the mother tongue.

Jain investigated the impact of official language policies on education using the linguistically demarcated state formation in India [6]. His research notes the detrimental effect of mismatch between the language used for learning and the student's mother tongue. More specifically, linguistically mismatched areas have 18.0% lower literacy rates and 20.1% lower college graduation rates. This undesirable outcome is attributed to the difficulty in acquiring education due to a medium of instruction in schools that differed from students' mother tongue. Surprisingly, this educational achievement gap was closed when linguistically mismatched areas were reorganized on linguistic lines to remove the mismatch.

One should also be wary of business interests that are often masqueraded as research or as an authoritative voice. The position expressed in [1] about the current eminence of English language and its continued dominance will be untenable given the rapid rise of Computational Linguistics techniques and their potential to break language barriers through real-time translation.

5   Computational Linguistics for Preserving and Promoting All Natural Languages

The Association for Computational Linguistics (ACL) defines Computational Linguistics as

> ... the scientific study of language from a computational perspective. Computational linguists are interested in providing computational models of various kinds of linguistic phenomena.

Computational Linguistics is an interdisciplinary domain encompassing linguistics, information retrieval, machine learning, probability, and statistics. It is concerned with analyzing, understanding, and interpreting the written text (aka NLP) and spoken language (aka Speech).

Digital tools for languages have the potential to fundamentally transform lives in rural India. However, the tools must speak the same language as the users. For example, consider ATM machines. They do not provide instructions in native languages, and thus by design, we have excluded millions of native language speakers from using ATM machines. A similar situation exists with air travel. All messaging is done in non-native languages. Imagine the crime that the government would have committed in case of an in-air emergency of a domestic flight and most passengers on board the aircraft do not understand flight crew instructions because they are delivered in a language the passengers do not speak.

Computational Linguistics offers unprecedented opportunities to preserve and promote natural languages. The rapid advances in high-performance cloud computing, machine learning, deep learning, information retrieval, speech and signal processing, and search technologies will provide open-source as well as cost-effective solutions to Computational Linguistics problems. This will also help to create a level playing field for resource-rich and resource-poor languages. Currently, most of the linguistics tools and resources are available for a handful of languages—English, Arabic, Turkish, Japanese, Korean, Spanish, German, among others. However, this resource disparity will change with the emergence of open-source software libraries and application frameworks for the linguistics tasks.

Currently, there are three major approaches to the computational analysis and understanding of human languages: symbolic, statistical learning, and deep learning. Symbolic approaches are the first ones and are designed to reflect the underlying structure of the language. They rely on linguistic principles and hand-crafted rule-based approaches. Such approaches are tied to specific languages and are not transferable from one language to another. The statistical methods, on the other hand, rely on training data and machine learning algorithms. However, creating training data is labor-intensive and hence expensive. In recent years, *deep learning* approaches have become immensely popular and effective [4]. The focus of Computational Linguistics has moved from language analysis to language understanding.

Both machine learning and deep learning algorithms require large training datasets. Some current research efforts in deep learning focus on creating large training datasets incrementally and automatically from a small, manually created dataset. Such data is referred to as *corpus* and is essential for various tasks in Computational Linguistics [5]. Therefore, the first step toward preserving and promoting a language is creating corpora of different genres. In the following, we outline various NLP tasks that can be accomplished with the availability of suitable corpora.

Language Detection is the task of identifying the language of a piece of text. As more and more documents are produced in non-English languages, the ability to perform language detection is crucial. Also, mixed language documents are becoming common.

Sentence Segmentation is concerned with identifying sentence boundaries. For example, a period (i.e., .) does not necessarily imply a sentence ending. For example, the period character is also used in abbreviations such as Ph.D.

Word Segmentation involves identifying individual words in text. For languages based on the Roman or Latin alphabet, word segmentation is trivial as a space character demarcates word boundaries. Word segmentation is a difficult problem for Indian languages as there is no explicit delimiter for words. Often, many words are strung together as one compound word.

Language Modeling involves assigning a probability for every string in the language. In other words, a language model is a probability distribution over all the strings in the language. Language modeling applications include word auto-completion in smartphones and other hand-held devices, spelling correction, and decoding secret codes. It also used in automatic speech recognition systems.

Part-of-Speech (POS) Tagging involves assigning a Part-of-Speech (POS) tag for every word in a piece of text. The POS tags are used in many subsequent activities such as syntactic parsing, word-sense disambiguation, and text-to-speech synthesis.

Named Entity Recognition task identifies the names of people, places, organizations, and other entities of interest in text. Named entities are used in other NLP tasks including word-sense disambiguation, co-reference resolution, semantic parsing, textual entailment, information extraction, information retrieval, text summarization, question-answering, and spoken dialog systems.

Dependency Parsing: Parsing of a sentence involves depicting relationships between words in the sentence. Dependency parsing, on the other hand, is concerned with extracting a dependency parse of a sentence. The latter shows the sentence's grammatical structure as well as the relationships between "head" words and other words in the sentence.

The above are considered as fundamental tasks in Computational Linguistics. They form the foundational basis for advanced NLP applications including machine translation, information extraction, topic modeling, text summarization, document clustering, document classification, question-answering systems, natural language user interfaces, and spoken dialog systems, among others.

6   Conclusions

Natural languages are more than a medium for human communication. They are also tightly integrated with culture and influence our way of thinking. With the popularity of the WWW and the dominance of the English language, many languages are on the path toward extinction. Fortunately, Computational Linguistics techniques can be brought to bear to revive and save the languages. Furthermore, languages cannot simply rest on the past laurels for their existence and growth. New literature needs to be produced, which reflects society's current issues, and to engage with young readers.

Governments that actively seek to destroy native languages through their oppressive policies can learn from the European Union (EU). There is a cultural element that threads through the member countries, yet each country celebrates its uniqueness including language and traditions. The motto "One Country One Language" is a disaster in the making. Instead, the motto should be "One Country Many Languages." Image a situation that a city government decides to replace all trees in the city with just one type of tree. Does this bring beauty and elegance to the city?

To actualize "Unity in Diversity and Diversity in Unity" for nations that have cultural and linguistic diversity, it helps if governments implement policies that require people of one region in a country to learn the language of another region. This policy should not be selective and be implemented uniformly. Such language policies help to understand each other's cultures, promote respect for all citizens, and strive for national unity.

This is an opportune moment to realize this vision given the recent advances in Computational Linguistics featured in products such as Google Translate (https://translate.google.com/) and Microsoft Translator (https://www.microsoft.com/en-us/translator/). Translators address both spoken and written language in real time. Shown in Fig. 3 is Google Translate in action. In the top area of the left pane, we typed a transliteration of a short Telugu language sentence: *yemi chestunnavu?* In this case, transliteration is using the English alphabet for the Telugu sentence, instead of writing the Telugu using the Telugu alphabet. Google Translate correctly reproduced this transliterated sentence in Telugu script (the bottom area of the left pane). Shown in the right pane is the correct translation of the sentence into the English language. Notice the speaker icons in both panes. By clicking on them, you can hear the sentence as spoken in the corresponding language. Compared to the English text-to-speech converter, the Telugu version is rudimentary. However, this is a good beginning. The question then is why only Google and Microsoft are doing this? Why are the computing and engineering researchers in an estimated 751 engineering schools across Andhra Pradesh and Telangana states not doing something to preserve and celebrate their own mother tongue?

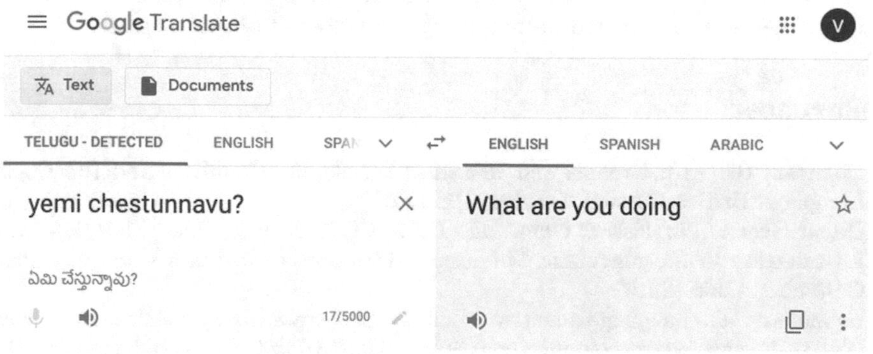

Fig. 3. Google translate: real-time language detection and translation

The argument that science, medicine, and engineering can only be taught using the English language does not hold validity. There are numerous examples of the above disciplines taught in native languages across Europe, South America, and Japan. What is needed is out-of-the-box thinking for the digital world we live in.

Governments should look at language diversity as an asset and provide support for nurturing them. As new body of knowledge is produced in various domains, the language needs to add new words to remain expressive and relevant. Students should be encouraged to contribute to open-source projects related to Computational Linguistics. This is especially so for developing countries as only local students have the knowledge of their language.

In all southern states in India except Tamilnadu, students are required to learn three languages—native language, Hindi, and English. Some schools offer a substitute for the native language. For example, in Andhra Pradesh, a student may study Sanskrit in lieu of Telugu. The motivation for this is not that the student is more interested in Sanskrit over Telugu. Students know very well that relatively Telugu teachers are more proficient than Sanskrit teachers. Therefore, Telugu teachers will grade the exams more thoroughly and make it difficult for students to score high. On the other hand, students can score high in Sanskrit with relatively less effort. This gives them an advantage in securing admissions to professional schools such as medicine and engineering. The government should consider excluding the language exam scores in making admission decisions for professional schools. Government policies are the effective instruments to promote a language as well as to kill the language.

In this digital age, we are blessed with many sophisticated computing and communication tools. Thanks to rapid advances in the computing discipline. These tools eliminated the barriers of geographic space and offer unprecedented opportunities to preserve, promote, and celebrate natural languages. All efforts to kill languages in the name of nationalism and national integration are regressive thinking and amount to a crime against humanity. Therefore, we all need to collectively fight this regressive thinking and pass on the linguistic diversity to future generations. If you do not do it, who else? If not now, when?

References

1. Coleman, H. (ed.): Dreams and Realities: Developing Countries and the English Language. British Council, London, UK (2011)
2. Department of Linguistics: Ohio State University: Language Files: Materials for an Introduction to Language and Linguistics, 12th edn. Ohio State University Press, Columbus, Ohio (2016)
3. Economist, T.: Language identity in India: one state, many worlds, now what? (2013). https://www.economist.com/johnson/2013/06/25/one-state-many-worlds-now-what
4. Goldberg, Y.: Neural network methods for natural language processing. Synth. Lect. Hum. Lang. Technol. **10**(1), 1–309 (2017)
5. Gudivada, V., Rao, D., Raghavan, V.: Big data driven natural language processing research and applications. In: Govindaraju, V., Raghavan, V., Rao, C.R. (eds.) Big Data Analytics, Handbook of Statistics, vol. 33, pp. 203–238. Elsevier, New York, NY (2015)
6. Jain, T.: Common tongue: the impact of language on educational outcomes. J. Econ. Hist. **77**(2), 473–510 (2017). https://doi.org/10.1017/S0022050717000481
7. Myles, F.: Learning foreign languages in primary schools: is younger better? (2017)

8. Noormohamadi, R.: Mother tongue, a necessary step to intellectual development. Pan-Pac. Assoc. Appl. Linguist. **12**(2), 25–36 (2008). https://files.eric.ed.gov/fulltext/EJ921016.pdf

9. Paul, L.M., Simons, G.F., Fennig, C.D. (eds.): Ethnologue: languages of the world, 19th edn., SIL International, Dallas, Texas (2018). http://www.ethnologue.com

Comparative Analysis of Geometric Transformation Effects for Image Annotation Using Various CNN Models

Sangita Nemade[1(✉)] and Shefali Sonavane[2]

[1] Department of Computer Science & Engineering, Walchand College of Engineering, Sangli, India
sangita.nemade@walchandsangli.ac.in
[2] Department of Information Technology, Walchand College of Engineering, Sangli, India
shefali.sonavane@walchandsangli.ac.in

Abstract. For image classification and image annotation applications, a convolutional neural network (CNN) has revealed superior accuracies and computationally efficient methods. Researchers are putting significant efforts across the globe to come up with enhanced deep network architectures. However, the latest CNN models are unable to cope with the geometric transformation of images. In the training stage, the state-of-the-art algorithms mostly make use of data augmentation techniques, that expand training samples but it leads to data overfitting. This paper focuses on how geometric transformation affects the annotation performance of various CNNs such as Alexnet, GoogleNet, ResNet50, and DenseNet201 on the Corel dataset. Comparative analysis of these deep learning models for image annotation is evaluated using the F1 score and is presented with the help of a box plot. Based on the obtained results, it is observed that the DenseNet201 model performs better than other models. It is further observed that the ResNet50 model works better for a specific class of images.

Keywords: CNN · Image annotation · Geometric transformation

1 Introduction

CNN, a deep learning model, is a very prevailing tool, for computer vision applications such as image annotation [1], image retrieval [2, 3], semantic segmentation [4], and video captioning [5]. The basic building blocks of CNN contain a convolutional layer, nonlinear activation function, pooling layer, and fully connected layer. Although CNN exhibits high potential to learn increasingly complex patterns, it has some limitations. Convolutional Layer is the important module of the CNN. In the convolutional layer, filters get convolved with input image of the previous layer's output and produce feature maps which may be basic features or complex features. The mechanism of weight sharing in the CNNs greatly reduces the total number of parameters. However, it can restrict models from learning other types of invariance in the feature map. (ii) The pooling

© Springer Nature Singapore Pte Ltd. 2020
B. Iyer et al. (eds.), *Applied Computer Vision and Image Processing*,
Advances in Intelligent Systems and Computing 1155,
https://doi.org/10.1007/978-981-15-4029-5_36

operation is carried out on each feature map. The pooling operations can be max-pooling (maximum pixel value is selected) or average pooling (where average pixel values are taken) of each feature map. In both pooling, valuable information about the feature map is lost. The aim of the pooling operation is to produce a translation-invariant feature map and to reduce the complexity of the network [6]. This spatial invariance is recognized only through a deep hierarchy of convolution and max-pooling due to small local spatial support for convolution (for example 7×7) and max-pooling (for example 2×2) but intermediate feature representations in a CNN are not truly invariant to the input data's transformations [7].

To get a translation invariance output image, most of the state-of-the-art algorithms have used data augmentation techniques where rotation, flipping, scaling, and crop images are given as training samples to CNN but it increases the training data size and chances of data overfitting. The CNN is generally thought to be invariant to transformations of images mainly due to data augmentation use in training or because of convolutional structure [8]. According to the authors, the convolution architecture disregards the classic sampling theorem, and data augmentation fails to support invariance except for the similar images. Training dataset generated by applying transformations such as centering, translation, elastic deformation, rotation, and their combinations on the MNIST dataset improves the performance of LeNet, DropConnect, and Network3 [9]. Shen et al. [10] have designed an approach to improve CNN's transformation invariance ability by rotating, scaling, and translating feature maps during the training stage. Spatial transformations viz. rotation, translation, scale, and generic warping are performed by a learnable unit placed in the existing network [6]. Vedaldi et al. [7] have studied invariance, equivalence, and equivariance properties of image representation. Rotation invariance and equivalence also improved the performance of the microscopic image classification task [11]. To efficiently learn the parameters, Luan et al. [12] have developed a new Gabor convolutional network that is invariant to rotation and scale.

For this study, CNN models are selected based on their architecture and popularity among the research community. Further studies of image annotation are needed where textual labels are allotted to the image that clearly describes the image content.

(1) The main contribution of this paper is to understand the effect of geometric transformation such as crop, horizontal flipping, vertical flipping, rotation by 45°, rotation by 90°, and rotation by 180° on image annotation performance of different state-of-the-art CNN models on the Corel dataset.

(2) This study provides the comparative analysis of various CNN models such as AlexNet [1], GoogLeNet [13], ResNet50 [14], and DenseNet201 [15] for image annotation.

Section 1 briefed about the CNN and its limitation. Section 2 describes the architecture, models, dataset, and performance measures. Quantitative results are presented in Sect. 3 and lastly, Sect. 4 concludes the paper.

2 Methodology

This section briefs about the architecture of image annotation using CNN models. Further, it explains the four CNN models used for the experimentation analysis and their performance measures.

2.1 Architecture

The architecture of the presented work contains the training and testing phase (Fig. 1). The training phase takes in labeled images of ten distinct categories from the Corel-10 K dataset. These labeled images are fed to the CNN pretrained models. Transfer learning with four pretrained models such as AlexNet, GoogLeNet, ResNet50, and DenseNet201 are used in this work, which are feature extractors that computed the descriptors. In the testing phase, the following geometric transformations have been applied to the same images utilized in the training phase.

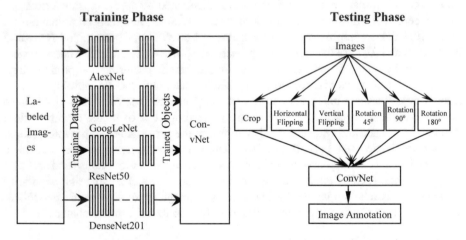

Fig. 1 Architecture of image annotation using CNN models

Geometric operations:

1. **Crop**—Equal sized single crop is taken arbitrarily from all test images.
2. **Horizontal flipping**—All test images are flipped horizontally.
3. **Rotation 45°**—45° rotation is applied to all test images.
4. **Rotation 90°**—90° rotation is applied to all test images.
5. **Rotation 180°**—180° rotation is applied to all test images.
6. **Vertical flipping**—All test images are flipped vertically.

After applying the geometric transformation, labeled images are generated using the trained model.

2.2   CNN Models

AlexNet: The first model used for analysis is the AlexNet [1], which was originally trained on the ImageNet dataset and demonstrated the outstanding performance for image object classification. The eight layers contain five convolutional layers and three fully connected layers. The main feature of this model is the fast downsampling of intermediate feature representation through the convolution layer and max-pooling layer operation. Then, these feature maps are flattened and faded to a fully connected layer which produces a visual descriptor at the output.

GoogLeNet: The second model used for analysis is the 22 layers, GoogLeNet, which is constructed on Inception architecture [16] and trained on the ImageNet dataset. Several inception modules are connected to each other to go deeper. Each inception module comprises various sizes of convolutions and max-pooling. By applying 1×1 convolution, the number of parameters are reduced that are large kernel sizes of convolution ($n \times n$) are decomposed into $n \times 1$ and $1 \times n$ filter sizes. Therefore, the resultant network becomes deeper and wider and also has fewer parameters than AlexNet. This model uses global-average-pooling instead of a fully connected layer that reduces overfitting.

ResNet50 (Residual Network): ResNet50 is of 50 layered residual network. It is trained on the ImageNet dataset. It consists of five stages of convolution layers. At first layer 7×7 convolution matrix is applied with a stride of two, therefore, the input is downsampled by two like pooling layer. Then three identity blocks are followed before downsampling again by two. This process remains the same for multiple layers. Each residual module has a stack of three (1×1, 3×3, and 1×1) convolution layers. The first and last 1×1 convolutions are used for dimensionality reduction and enhancement (restoring). Skip connection is used in it for adding the output of the previous layer to the next layer which helps to reduce the problem of vanishing gradient, as well as helps in recognizing the global features of a network. Average pooling is used in the last layer which produces thousand feature maps.

DenseNet201 (Dense Convolutional Network): The last model of analysis is DenseNet201 which takes fewer parameters for training and achieved good accuracy compared to AlexNet, GoogLeNet, and ResNet50 models. For solving the problem of vanishing the input, as well as gradient information in the deep network, it uses simple connectivity patterns for maximum information flow in forward pass and backward pass computation. Each layer takes input feature maps from all its previous layers and passes output feature maps to all its succeeding layers. Therefore, features are reused all over the network with more compact learning. DenseNet201 is a collection of dense blocks where layers are densely connected to each other. The layers between these blocks are the transition layers that carry out the convolution and pooling operation. Every single layer in a dense block and contains batch normalization, nonlinear activation function ReLu, and 3×3 convolution size.

Table 1 summarizes the CNNs quantitative properties on ImageNet dataset such as a number of parameters, input size of the model, output feature descriptor size, depth of the CNN, Top-5 (with 10-crop testing) validation error, and FLOPs (floating-point operations per second) for forward pass computation.

Table 1 Comparison of computable properties of CNN

Models	Number of parameters (Millions)	Input size of model	Output size of model	No. of layers	Top-5 error rate (%)	FLOPs
AlexNet [1]	62	(227, 227, 3)	1000	08	15.3	1.1×10^9
GoogLeNet [12]	7.0	(224, 224, 3)	1000	22	9.15	5.6×10^9
ResNet50 [13]	25.6	(224, 224, 3)	1000	50	6.71	3.8×10^9
DenseNet201 [14]	20.0	(224, 224, 3)	1000	201	5.54	3.8×10^9

2.3 Datasets

The Corel dataset [17] contains 10,000 images having 100 categories, each of the categories has 100 images. Mostly, this dataset contains images with a single concept. Ten distinct categories, such as Antiques, Aviation, Balloon, Bonsai, Car, Cat, Doll, Gun, Ship, and Ski are used for the training purposes. Single crop, horizontal flipping, vertical flipping, rotation 45°, rotation 90°, and rotation 180° have been performed on the categories selected for the training and utilized as testing dataset. The testing dataset contains 50 images in a category.

2.4 Performance Measures

For classification problems recall, precision, F1 score, and overall accuracy are generally applied to validate the models' performance. F1 score is utilized as a performance metric in this study which is a statistical measure and is a harmonic mean of precision and recall (Eq. 1).

$$\text{F1score} = 2 * \frac{(\text{Precision} * \text{Recall})}{\text{Precision} + \text{Recall}} \tag{1}$$

Precision is a proportion of relevant labels among the total number of retrieved labels by the CNN model and recall is given as a ratio of correctly predicted labels with ground truth labels.

3 Experimental Results

The conducted experimentation deals with the performance investigation of four CNN models on the effect of geometric transformation for the Corel dataset to annotate images. The result of this work is presented and discussed in this section.

Figure 2 depicts comparative performance measure, F1 score, for geometric transformations of ten classes from the Corel dataset using four CNN models.

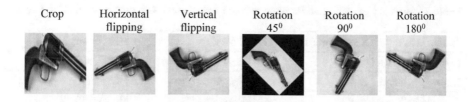

Crop | Horizontal flipping | Vertical flipping | Rotation 45° | Rotation 90° | Rotation 180°

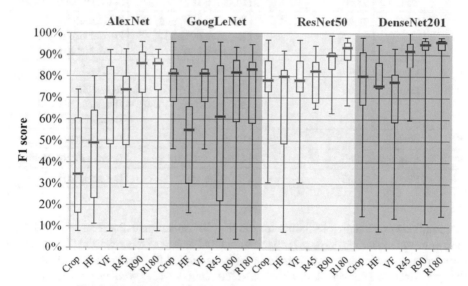

Fig. 2 Impact of geometric transformation on CNNs performance

The experimental results are shown using box plots. The AlexNet performed worst on crop and horizontal flipping operations, however, have a slightly preferred position over the other models with rotation 90° operations. The GoogLeNet performed good than AlexNet for cropping and vertical flipping but performed least well for horizontal flipping and rotation 45° operations. The ResNet50 model is suitable for cropping, vertical flipping, and rotation 180° operations. The DenseNet201 outperformed over other models for horizontal flipping, rotation 45°, rotation 90°, and rotation 180° operations while performed better for cropping.

Table 2 shows maximum F1 score values achieved by various classes for different geometric transformations on CNNs models. Based on the experimentation, a class cat has achieved a maximum F1 score of 98.49% for rotation 90° operation using ResNet50.

4 Conclusion

This experiment gives a comparative analysis of geometric transformations' effects on ten categories from the Corel dataset using AlexNet, GoogLeNet, ResNet50, and DenseNet201. The ResNet50 model, for instance, is the more powerful for cropping, vertical flipping, and rotation 180° operations. The result also shows that the class cat

Table 2 Summary of maximum F1 score gained by each class with operation and CNN model

Sr. No.	Class	Maximum F1 score (%)	Operation	CNN model
1	Antiques	98.03	Rotation 90°	DenseNet201
2	Aviation	84.90	Rotation 180^0	ResNet50
3	Balloon	89.90	Rotation 45°	ResNet50
4	Bonsai	96.00	Rotation 180°	AlexNet, DenseNet201
5	Car	95.83	Rotation 180° and Rotation 90° and Rotation 180°	ResNet50, DenseNet201 and DenseNet201
6	Cat	98.49	Rotation 90°	ResNet50
7	Doll	98.00	Rotation 180°, Rotation 90° and Rotation 180°	ResNet50, DenseNet201 and DenseNet201
8	Gun	98	Crop	DenseNet201
9	Ship	94.62	Crop, vertical flipping	ResNet50, ResNet50
10	Ski	93.33	Rotation 180°	DenseNet201

has a maximum F1 score of 98.49% for rotation 90° operation using ResNet50 as it uses skip connections which helps in recognizing global features. Whereas, the DenseNet201 outperformed over other models for cropping, horizontal flipping, rotation 45°, rotation 90°, and rotation 180° operations. DenseNet201 model works well even when the training dataset is insufficient as it utilizes features of all complexity levels. In the future, it is intended to improve feature representations of CNN in terms of orientation and scaling capabilities.

References

1. Krizhevsky, A., Ilya S., Hinton G.: Imagenet classification with deep convolutional neural networks. In: Advances in Neural Information Processing Systems, pp. 1097–1105 (2012)
2. He, K., Xiangyu Z., Shaoqing R., Jian S.: Deep residual learning for image recognition. In: Proceedings of the IEEE Conference on Computer Vision and Pattern Recognition, pp. 770–778 (2016)
3. Simonyan, K., Andrew, Z.: Very deep convolutional networks for large-scale image recognition 1409–1556. arXiv preprint (2014)
4. Long, J., Evan, S., Trevor, D.: Fully convolutional networks for semantic segmentation. In: Proceedings of the IEEE Conference on Computer Vision and Pattern Recognition, pp. 3431–3440 (2015)
5. Pan, Y., Tao, M., Ting, Y., Houqiang, L., Yong, R.: Jointly modeling embedding and translation to bridge video and language. In: Proceedings of the IEEE Conference on Computer Vision and Pattern Recognition, pp. 4594–4602 (2016)
6. Jaderberg, M., Simonyan, K., Zisserman, A.: Spatial transformer networks. In: Advances in Neural Information Processing Systems, pp. 2017–2025 (2015)

7. Lenc, K., Vedaldi, A.: Understanding image representations by measuring their equivariance and equivalence. In: Proceedings of the IEEE Conference on Computer Vision and Pattern Recognition, pp. 991–999 (2015)
8. Azulay, A., Weiss, Y.: Why do deep convolutional networks generalize so poorly to small image transformations? arXiv preprint. arXiv:1805-12177 (2018)
9. Tabik, S., Peralta, D., Herrera, P.A., Herrera, F.: A snapshot of image pre-processing for convolutional neural networks: case study of MNIST. Int. J. Comput. Intell. Syst. **10**(1), 555–568 (2017)
10. Shen, X., Tian, X., He, A., Sun, S., Tao, D.: Transform-invariant convolutional neural networks for image classification and search. In: Proceedings of the 24th ACM International Conference on Multimedia, pp. 1345–1354 (2016)
11. Chidester, B., Zhou, T., Do, M.N., Ma, J.: Rotation equivariant and invariant neural networks for microscopy image analysis. Bioinformatics **35**(14), 530–537 (2019)
12. Luan, S., Chen, C., Zhang, B., Han, J., Liu, J.: Gabor convolutional networks. IEEE Trans. Image Process. **27**(9), 4357–4366 (2018)
13. Szegedy, C., Liu, W., Jia, Y., Sermanet, P., Reed, S., Anguelov, D., Erhan, D., Vanhoucke, V. Rabinovich, A.: Going deeper with convolutions. In: Proceedings of the IEEE Conference on Computer Vision and Pattern Recognition, pp. 1–9 (2015)
14. He, K., Zhang, X., Ren, S. Sun, J.: Deep residual learning for image recognition. In: Proceedings of the IEEE Conference on Computer Vision and Pattern Recognition, pp. 770–778 (2016)
15. Huang, G., Liu, Z., Maaten, L., Weinberger, K.Q.: Densely connected convolutional networks. In: Proceedings of the IEEE Conference on Computer Vision and Pattern Recognition, pp. 4700–4708 (2017)
16. Lin, M., Chen, Q., Yan, S.: Network in network (2013). arXiv preprint. arXiv:1312.4400
17. Liu, G.H., Yang, J.Y.: Content-based image retrieval using color difference histogram. Pattern Recogn. **46**(1), 188–198 (2013)

Prominent Feature Selection for Sequential Input by Using High Dimensional Biomedical Data set

Archana Kale$^{(\boxtimes)}$ and Shefali Sonavane

Walchand College of Engineering Sangli, Sangli, Maharashtra 416415, India
archana.mahantakale@gmail.com, shefali.sonavane@walchandsangli.ac.in

Abstract. The main aim of feature selection algorithms is to select prominent (optimal) features that are not irrelevant and redundant. Reducing the number of features by keeping classification accuracy the same is one of the critical challenges in Machine Learning. High dimensional data contains thousand of features with the existence of the redundant and irrelevant features which negatively affect the generalization capability of the system. This paper designs the improved genetic-based feature selection (IGA) for Online Sequential—Extreme Learning Machine (IGA-OSELM) algorithm with additive or radial basis function (RBF). Experimental results are calculated for the Extreme Learning Machine (ELM), OSELM, IGA-ELM, and IGA-OSELM. With the result, it is inferred that IGA-OSELM maintains the classification accuracy by minimizing 58.50% features. The proposed algorithm is compared with the other popular existing sequential learning algorithms as the benchmark problem.

Keywords: Sequential problem · Feature subset selection problem · Pattern classification problem

1 Introduction

Machine learning and computational intelligence are the subareas of artificial intelligence. By considering the rapid growth in databases and technologies, Feature subset selection (FSS) and classification problems are important challenges in the field of machine learning. FSS methods are evaluated by using various search methods like sequential search, bidirectional search, rank search, random search, etc., [1]. Generally, for high dimensional biomedical dataset (HDD) random search is suitable as linear search requires more computational time. Genetic algorithm (GA) is one of the examples of random search strategies that is able to select a prominent (nonredundant and relevant) feature subset [2]. In GA, the

© Springer Nature Singapore Pte Ltd. 2020
B. Iyer et al. (eds.), *Applied Computer Vision and Image Processing*,
Advances in Intelligent Systems and Computing 1155,
https://doi.org/10.1007/978-981-15-4029-5_37

prominent feature subset varies as the population size changes. Many researchers choose the size of population randomly or may initially set the population size as 50 or 70 [3,4]. However, the literature survey missed the feature selection by considering the different population size.

In the original ELM, a batch mode learning method is used. ELM considers that the training data is present previously which is not suitable for many real applications where the training data may arrive sequentially. Nan-Ying Liang et al., developed the OSELM algorithm for sequential input [3]. In ELM and OSELM, the input weights are randomly initialized and analytically determined by the output weights. Hence, the overall system performance may degrade due to the random initialization. One of the alternative solutions is to select only the required features (prominent feature subset selection). The combination of irrelevant and redundant features with random initialization is the main cause of degradation.

To address the said problem; in this paper, an improved GA based optimization approach for sequential input by using OSELM classifier (IGA-OSELM) with RBF hidden nodes is proposed. The main contribution of the paper is the variation in initial block training data. In many papers, OSELM is used by differing the number of hidden nodes [3,4] but a vast literature survey limits to identify variation in the initial block of training data.

This paper is organized as follows: Sect. 2 is represented as the related work. The proposed methodology is illustrated in Sect. 3. The discussion on the experimental results of HDD is carried out in Sect. 4. Section 5 is outlined with the conclusion and future research direction.

2 Background and Context

ELM is basically developed for batch learning mode. The batch ELM requires complete data readily available that is not convenient for many applications where the training data reach sequentially [3]. Liang et al., have developed OSELM algorithm for sequential input [3]. OSELM is used for applications like watermarking in DWT domain [5], Hematocrit estimation [6], etc.

Liang et al., developed OSELM algorithm for sequential input [3]. The output weight matrix of ELM is the least square solution and is written as Eq. 1

$$H^\dagger = (H^Y H)^{-1} H^Y \tag{1}$$

where, Y shows the target class and $H^\dagger$ shows the moore-penrose's generalized the inverse of H.

Least square solution to $H\beta = Y$ is in Eq. 2. The sequential implementation of Eq. 2 results in the OSELM. For given a chunk of initial training set,

$$Y_0 = (x_i, y_i)_{i=1}^Y \tag{2}$$

Where, $Y \geq m$ (m—number of hidden nodes), Y_0 is the chunk of initial training data. Consider the new block of data as

$$Y_1 = (x_i, y_i)_{i=Y_0+1}^{Y_0+Y_1} \tag{3}$$

where Y_1 is the number of considerations in the chunk of new observations. The description of the same is detailed in the Sect. 3.

Various algorithms and analytical methods have been designed for input weight assignment [7]. In the literature, the combination of ELM and various optimization algorithms for feature selection have been used. Zhu et al., designed the differential evolution to search for optimal input weights and hidden biases [7], modified ELM [8]. Recently, summation wavelet ELM [9] is used to reduce the impact of random initialization. In many papers, OSELM is evaluated by varying the number of hidden nodes [3,4].

3   Methodology

The proposed Improved Genetic Algorithm for Online Sequential Extreme Learning Machine (IGA-OSELM) algorithm is divided into three subsystems like Input, FSS, and Classification.

Input Dataset As dimension space is used as an evaluation criterion, various datasets with dimensional scope from 2000 to 12600 are used [10] like Lung Cancer-Harvard (LCH-2), DLBCL-Harvard (DH), Colon Tumor (CT), and Nervous System (NS). These datasets have 12534, 7129, 2000, and 7129 attributes, respectively, with two classes. Further, the datasets are divided into a training dataset (70%) and a testing dataset (30%).

Feature Subset Selection (FSS) FSS is a critical step that is used to select nonredundant and relevant features (prominent features) [11] for efficient machine learning [12]. GA uses the fitness function for calculating the fitness value [13]. For evaluation three genetic operators are used like Selection, Crossover, and Mutation. Selection chooses the strings for the next generation. Lost generic material is restored by using mutation where the crossover is used for information exchange. In the proposed model, the experimental observations are evaluated by considering the population size from 10 to 90 with an incremental size of 10 (10:10:90). The IGA-ELM results in different nine subsets. One of the subsets is selected as a prominent feature subset having a topmost percentage of accuracy due to the presence of relevant features as shown in Fig. 1 for further processing. For example, 690 prominent features are selected for the CT dataset from all (2000) features.

Classification Subsystem In this subsystem, ELM and OSELM learning algorithms are used for classification. Initialization and sequential learning are the two phases of OSELM.

In this phase, all required initialization can be done like the number of required data, the chunk size, and the number of hidden nodes. According to

Theorem II.1 [3], the number of training data required can be equal to the number of hidden nodes. The data used once is not reused. A small chunk of initial training data is defined as $Y_0 = (x_i, y_i)_{i=1}^{Y}$ from the training dataset.

In sequential learning phase, get l+1 chunk of new observations as an input.

$$Y_{l+1} = (x_i, y_i)_{i=(\sum_{j=0}^{l} Y_j)+1}^{\sum_{j=0}^{l+1} Y_j} \tag{4}$$

where, Y_{l+1} denotes the number of observations in the $(l+1)$th chunk.

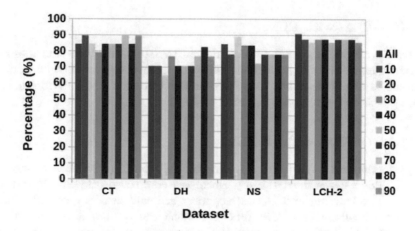

Fig. 1. Prominent FSS by using IGA model

The sigmoidal additive activation function (Sig) and Gaussian radial basis activation function (RBF) have been used in OSELM. In many articles, only the number of hidden nodes is considered for evaluation in OSELM [3,4], but the use of the initial training data in combination with the number of hidden nodes are not considered. The size of the initial training data (number of input neurons) is usually less than the number of hidden neurons. Through experimentation, it is noticed that the initialization of initial training data is as important as the number of hidden nodes. For every iteration of the hidden neuron, the initial training data (n) is also varying from the number of hidden node (j), j to n varying with i $(j : i : n)$ where, i is the incremental value. The results are calculated by considering the various learning modes like a chunk size of 1 and 20 (fixed), as well as by varying the chunk size between 10 to 30 (randomly). The time complexity of the proposed algorithm is $O(n^2)$.

4   Experimental Results and Discussion

For Experimentation, MATLAB©R2014a is used. The evaluation measures (metrics) like accuracy, precision, recall, f-measure, and g-measure are used [14–16].

The experimental results are evaluated for OSELM and ELM with all features and prominent features. The evaluation performance is compared in three different ways—1. IGA-ELM with ELM 2. IGA-OSELM with OSELM 3. IGA-OSELM with an existing sequential algorithm and classifier.

4.1   IGA-ELM with ELM

The results are calculated for four datasets by using all evaluation measures. The performance is compared by considering the average value of all datasets. From the results, it is noted that IGA-ELM improves 11.04% performance by reducing 58.50% features as compared to ELM.

4.2   IGA-OSELM with OSELM

The performance comparison between OSELM and IGA-OSELM is shown in Table 1. Table 2 shows the detailed performance comparison of OSELM and IGA-OSELM. The training and testing accuracy are calculated by considering the average value of Sig and RBF function. With the results, it is inferred that the proposed IGA-OSELM maintains the accuracy by using 41.5% features as compared to OSELM for HDD.

5   Conclusion

The main goal of this paper is to solve the sequential problem by using the proposed Improved Genetic Algorithm for OSELM (IGA-OSELM) algorithm. In order to prove, the significance and the strength of the proposed approach; experimental results for OSELM and IGA-OSELM are analyzed. It is noticed that, IGA-OSELM maintains the accuracy of classification with only 41.5% features as compared to OSELM and IGA-ELM provides on an average 11.04% improvement in the accuracy by reducing almost 58.50% features as compared to ELM for HDD. The performance of IGA-OSELM and IGA-ELM is tested on a new application. With the results, it is noticed that the IGA-ELM provides 9.86% improvement in accuracy and IGA-OSELM maintains the classification accuracy by using 27.27% features as compared to ELM. Presently, the IGA-OSELM approach is used for supervised binary classification problems. In the future, the work will be extended for unsupervised classification.

Table 1. Performance Comparison of OSELM and IGA-OSELM for HDD

Dataset	AF	Mode	Training accuracy (%)		Testing accuracy (%)	
			OSELM	IGA-OSELM	OSELM	IGA-OSELM
CT	Sig	Batch	97.67	97.67	84.21	84.21
		1-by-1	97.67	97.67	78.94	73.68
		20-by-20	97.67	97.67	73.68	84.21
		[10,30]	97.67	97.67	84.21	84.21
	Avg	Sequential	97.67	97.67	78.94	80.7
	RBF	Batch	97.67	97.67	78.94	84.21
		1-by-1	74.41	72.09	63.15	68.42
		20-by-20	72.09	74.41	63.15	63.15
		[10,30]	79.06	83.72	78.94	78.94
	Avg	Sequential	75.19	76.74	68.41	70.17
DH	Sig	Batch	97.75	97.56	70.58	76.47
		1-by-1	97.56	95.12	70.58	70.58
		20-by-20	95.12	97.56	82.35	84.70
		[10,30]	97.56	97.56	82.35	88.23
	Avg	Sequential	96.75	96.75	78.43	81.17
	RBF	Batch	95.12	97.56	94.11	76.47
		1-by-1	29.26	58.53	82.35	82.35
		20-by-20	29.26	31.70	82.35	82.35
		[10,30]	29.26	70.73	82.35	82.35
	Avg	Sequential	29.26	53.65	82.35	82.35
NS	Sig	Batch	97.56	97.61	78.94	77.77
		1-by-1	87.80	87.8	68.42	72.22
		20-by-20	90.24	90.24	68.42	66.66
		[10,30]	92.68	90.47	78.94	77.77
	Avg	Sequential	90.24	89.50	71.93	72.22
	RBF	Batch	97.56	95.23	78.94	83.33
		1-by-1	68.29	69.04	63.15	66.66
		20-by-20	68.29	69.04	68.42	66.66
		[10,30]	68.29	71.42	73.68	72.22
	Avg	Sequential	68.29	69.83	68.42	68.51
LCH-2	Sig	Batch	99.20	99.20	90.90	87.27
		1-by-1	98.41	95.23	90.90	90.90
		20-by-20	99.20	98.41	90.90	90.90
		[10,30]	99.20	99.20	90.90	90.90
	Avg	Sequential	98.94	97.61	90.90	90.90
	RBF	Batch	92.85	96.03	90.90	90.90
		1-by-1	83.53	81.74	85.45	85.45
		20-by-20	82.53	95.16	87.27	87.27
		[10,30]	82.53	87.27	87.27	92.89
	Avg	Sequential	82.86	88.06	86.66	88.54

Table 2. OSELM and IGA-OSELM performance comparison

Data set	Training accuracy (%)		Testing Accuracy (%)	
	OSELM	IGA-OSELM	OSELM	IGA-OSELM
CT	86.42	87.20	73.67	75.43
DH	85.97	86.74	80.39	78.42
NS	79.26	79.66	70.17	70.36
LCH-2	90.9	92.83	88.78	89.72
Avg	85.63	86.60	78.26	78.49

References

1. Wang, G., Song, Q., Sun, H., Zhang, X., Xu, B., Zhou, Y.: A feature subset selection algorithm automatic recommendation method. J. Artif. Intell. Res. **47**, 1–34 (2013)
2. Koller, D., Sahami, M.: Toward optimal feature selection. Technical Report, Stanford InfoLab (1996)
3. Liang, N.-Y., Huang, G.-B., Saratchandran, P., Sundararajan, N.: A fast and accurate online sequential learning algorithm for feedforward networks. IEEE Trans. Neural Netw. **17**(6), 1411–1423 (2006)
4. Lan, Y., Soh, Y.C., Huang, G.-B.: Ensemble of online sequential extreme learning machine. Neurocomputing **72**(13), 3391–3395 (2009)
5. Singh, R.P., Dabas, N., Chaudhary, V.: "Online sequential extreme learning machine for watermarking in DWT domain. Neurocomputing **174**, 238–249 (2016)
6. Huynh, H.T., Won, Y., Kim, J.: Hematocrit estimation using online sequential extreme learning machine. Bio-Med. Mater. Eng. **26**(s1), S2025–S2032 (2015)
7. Zhu, Q.-Y., Qin, A.K., Suganthan, P.N., Huang, G.-B.: Evolutionary extreme learning machine. Pattern Recognit. **38**(10), 1759–1763 (2005)
8. Chen, Z.X., Zhu, H.Y., Wang, Y.G.: A modified extreme learning machine with sigmoidal activation functions. Neural Comput. Appl. **22**(3–4), 541–550 (2013)
9. Nguyen, D., Widrow, B.: Improving the learning speed of 2-layer neural networks by choosing initial values of the adaptive weights. In: 1990 IJCNN International Joint Conference on Neural Networks, 1990, pp. 21–26. IEEE (1990)
10. Li, J., Liu, H.: Bio-medical data set repository. School of Com-puter Engineering, Nanyang Technological University, Singapore. http://datam.i2r.a--star.edu.sg/datasets/krbd/index.html (2004)
11. Dash, M., Liu, H.: Feature selection for classification. Intell. Data Anal. **1**(1–4), 131–156 (1997)
12. Forman, G.: An extensive empirical study of feature selection metrics for text classification. J. Mach. Learn. Res. **3**, 1289–1305 (2003)
13. Zhuo, L., Zheng, J., Li, X., Wang, F., Ai, B., Qian, J.: A genetic algorithm based wrapper feature selection method for classification of hyperspectral images using support vector machine. In: Geoinformatics 2008 and Joint Conference on GIS and Built Environment: Classification of Remote Sensing Images. International Society for Optics and Photonics, pp. 71 471J–71 471J (2008)
14. Han, J., Pei, J., Kamber, M.: Data Mining: Concepts and Techniques. Elsevier (2011)

15. Mahdiyah, U., Irawan, M.I., Imah, E.M.: Integrating data selection and extreme learning machine for imbalanced data. Procedia Comput. Sci. **59**, 221–229 (2015)
16. Parikh, R., Mathai, A., Parikh, S., Sekhar, G.C., Thomas, R.: Understanding and using sensitivity, specificity and predictive values. Indian J. Ophthalmol. **56**(1), 45 (2008)

An Enhanced Stochastic Gradient Descent Variance Reduced Ascension Optimization Algorithm for Deep Neural Networks

Arifa Shikalgar[1]($\boxtimes$) and Shefali Sonavane[2]

[1] Department of Computer Science & Engineering, Walchand College of Engineering, Sangli 416415, Maharashtra, India
`shikalgar.arifa@walchandsangli.ac.in`
[2] Department of Information Technology, Walchand College of Engineering, Sangli 416415, Maharashtra, India

Abstract. The goal of this paper is to compare the most commonly used first-order optimization techniques with proposed enhanced Gradient Descent-Based Optimization. The simplest optimization method is the gradient-based optimization technique. The optimization concerns instigated with Deep Neural Networks (NNs) are unraveled by the rest other techniques. The common technique used in deep neural network architectural setup is Stochastic Gradient Descent (SGD). In SGD there is a raise of variance which leads to slower convergence. This affects the performance of the system. So to address these issues the non-convex optimization technique with faster convergence using an enhanced stochastic variance reduced ascension approach is implemented. It enhances performance in terms of faster convergence.

Keywords: Optimization · Deep learning · Stochastic gradient descent (GD) · Deep neural network (DNN) · Multimodal data

1 Introduction

Hinton investigated a novel layer-wise greedy learning method for training in 2006, which is denoted as the origin of deep learning techniques. With this first deep architecture known as deep belief network (DBN) is introduced by Hinton [1]. In 2012, Hinton led the research group whose working was for ImageNet image classification using deep learning [2]. The cause of the fame of deep learning is due to two reasons: One is the growth of big data analytic techniques shows that in training data the overfitting problem can be partly resolved. Second is the pre-training procedure prior to unsupervised learning allots nonrandom initial values to the network. Hence, after the training process, better local minima and a faster convergence rate can be achieved.

© Springer Nature Singapore Pte Ltd. 2020
B. Iyer et al. (eds.), *Applied Computer Vision and Image Processing*,
Advances in Intelligent Systems and Computing 1155,
https://doi.org/10.1007/978-981-15-4029-5_38

Multiple nonlinear hidden layers enable deep NN to learn the complex relationships that exist in their input and output with inadequate training data. In some cases, the complicated relationships are created which tend to sample noise. Such complexity exists in the training data but not in actual test data even though it is picked from the corresponding distribution. This prompts overfitting and for reducing it numerous techniques have been established [3]. One of the challenging aspects of deep learning is the optimization of the training criterion over millions of parameters [4–6].

Main aim of any optimization technique is to find the minimum or maximum of a generic function (θ^*), that is

$$\theta^* = \text{argmin } J\left(X^{\text{train}}\right), \emptyset \tag{1}$$

Hence, it is extremely necessary to project an effective and enhanced algorithm for faster convergence.

The structure of the paper is as follows: Sect. 2 gives the survey about existing work. The proposed methodology is described in Sect. 3. Section 3.2 discusses the experimental result and analysis of GD, SGD, SVRG, and the proposed technique. Finally, Sect. 4 concludes the paper with future scope.

2 Literature Survey

One of the most well-known and commonly used optimization algorithms is Gradient Descent. Mostly, the whole library of Deep Learning comprises implementations of numerous algorithms that use gradient descent [7]. Gradient descent is mainly having three types depending upon the quantity of data they utilized for calculation of gradient. Based on the quantity of data used, updated parameter accuracy, and time required to updating are analyzed [8].

(a) **Batch gradient descent:**

With respect to the parameters of the complete training data, a gradient of the objective function is calculated. Suppose $\theta_1, \theta_2, \theta_{i3}, \ldots \theta_n$ is a sequence vector function then the goal is to minimize the prediction error

$$E(W) = \frac{1}{n} \sum_{i=1}^{n} \theta_i(w), \tag{2}$$

where $E(W)$ = prediction error, $\frac{1}{n}$ = training examples, $\theta_i(w)$ = objective function (loss function that could be squared loss, logistic regression, etc.).

Then weight update rule for, gradient descent is,

$$W^t = W^{t-1} - \eta * \nabla_\theta E(W)^{t-1} \tag{3}$$

where θ = objective function, η = learning rate, ∇_θ = provides information about which direction to move.

$$W^t = W^{t-1} - \frac{\eta_t}{n} * \sum_{i=1}^{n} \nabla_\theta W^{t-1} \tag{4}$$

From the equation it's clear that to perform just only one update in gradient descent, it is essential to compute the gradients for the entire dataset for every 'n' example. So this gradient descent is very slow. Also, it is inflexible for datasets that don't fit in new examples on-the-fly. Global minima are ensured by batch gradient descent for convex error planes. And it stables to local minima in non-arched planes.

(b) Stochastic gradient descent:

Stochastic gradient descent (SGD) carries out an updating of parameters for respective training examples as given in Eq. (5),

$$W^t = W^{t-1} - \eta_t g_t(W^{t-1}, \xi_t) \tag{5}$$

where ξ_t = random sample, that may depend upon W^{t-1}

Batch gradient descent does repetitive computations for enormous datasets. Before every parameter update, it recomputes slopes for each example without considering similarities [9, 10]. SGD gets rid of this repetition by performing each update in turn. It is consequently much quicker. It is also utilized for online learning. In contrast, batch gradient descent stables to the minima of basin while SGD keeps fluctuating. From one perspective, SGD is capable to find out new and hypothetically well local minima. On other perspectives, batch gradient descent, at last, convolutes convergence to the precise least, while SGD will continue overshooting [11, 12]. Also in SGD randomness introduces variance which is caused by , $g_t(W^{t-1}, \xi_t)$..

(c) Mini-batch gradient descent:

Mini-batch gradient descent takes the middle way of the two optimizers GD and SGD. It updates every training example by considering mini-batches. Due to which variance generated by randomness is reduced and which gives more stable convergence. Still, it does not assure worthy convergence. In fact, it produces some issues like choosing an appropriate learning rate. Slow convergence occurs due to a very lesser learning rate. And a very huge learning rate hampers on the loss function due to swinging that occurs around the base [11].

Learning Scheduling learning rate is one of the options to adjust the learning rate during training, i.e., lessening the learning rate as per predefined schedule or when the objective value is changed below the decided threshold before set epochs. At starting itself these schedules and thresholds are needed to be declared and hence, it is incapable to adapt a dataset's characteristics [12, 13].

Another important challenge for Deep Neural Networks is to reduce the extremely non-convex error functions, and to avoid getting stuck in the several suboptimal spatial minima [14]. Dauphin claimed that the problem arises due to saddle points, not from local minima [6]. Saddle points are those where one dimension slopes are higher than compared to other neighbors. Also, one dimension is having lower slopes as compared to higher slopes. The saddle points are normally enclosed by a plateau. Plateau means, the all neighboring points which have the same error values exhibit that its gradient is nearer to zero in all neighboring aspects. Such plateau makes extremely hard for SGD to escape from it [15].

Thus in order to deal with this non-convex optimization, prior methodologies are utilized such as Gradient Descent (GD) and Stochastic Gradient Descent (SGD) approaches. These approaches maintain adaptable learning rates for different parameters. With this better learning rates are achieved but with increased variance. This will leads to slower convergence. So to overcome this drawback, an enhanced stochastic variance reduced ascension approach is used here and is explained in Sect. 3. It accelerates the convergence of stochastic techniques by lessening the variance of estimated gradient which results in faster convergence.

3   Research Methodology

Enhanced Stochastic Variance Reduced Ascension approach

Stochastic variance reduction gradient is a variance reduction approach that converges faster than the SGD, as well as GD, approaches. Though the stochastic variance reduction gradient exhibits a better variance reduction rate, it is highly sensitive to the learning rate. Hence the proposed approach integrates the properties/features of both the stochastic variance reduction gradient, as well as the conjugate gradient. It is highly efficient in achieving faster convergence by solving the non-convex optimization problem. The algorithm depicts the nature of this Enhanced Stochastic Variance Reduced Ascension approach.

Algorithm: Enhanced Stochastic Variance Reduced Ascension

Initialize $w = w_0$

for $iter = 0 \, to \, k - 1$ do

Sample a mini batch $S \subset \{1, 2, \ldots, N\}, |S| = b;$

Compute the variance reduced Ascension

$$g(w_k) = \frac{1}{|S|} \sum_{k \in S} f_i'(w_k) - \frac{1}{|S|} \sum_{k \in S} f_i'(\chi_i^k) + u_k;$$

Update rule $w_{k+1} = (w_k - \alpha_t g(w_k));$

Update $u_k \leftarrow u_k + \frac{1}{n} \sum_{i \in S} f_i'(w_k) - \frac{1}{n} \sum_{i \in S} f_i'(\chi_i^k);$

end for

Return $w^* = w;$

Here, first, the variance reduced gradient $g(w_k)$ is computed which is then integrated with the conjugate gradient using mini-batches as mentioned in the algorithm. Whereas, $u_k = \frac{1}{n}\sum_{i=1}^{n} f_i'\binom{k}{i}$ stores a scalar for each sample indexed with i and w_k is defined as the iteration in time interval k. Finally, the loss function is minimized and the non-convex optimization problem is tested with reduced variance, thereby it achieves faster convergence. The minimized loss function is given by Eq. (6),

$$\min L(x, z) + \lambda \chi(z) \tag{6}$$

wher regularization parameter is λ and penalty term is $\chi(z)$.

3.1 Data Acquisition and Preprocessing Phase

In order to carry out the experiment, the ROCO dataset is taken. It involves the subsets of "radiology" and "out-of-class". From that 81,825 are true positives and 6,127 are false positives, respectively. The figures in the "out-of-class" is having synthetic radiology images.

3.2 Experimental Results

The experimental setup is as follows: Deep Convolution Network architecture is used which contains a filter size and Maxpooling size of 3×3 and stride size is 1. After that, a fully connected layer is applied. Nonlinear activation function Rectifier Linear Unit (Relu) is also used for the individual convolutional layer.

(a) **Performance metrics:**

To strengthen the efficiency of the proposed framework, it is compared with the existing optimization approaches such as CG, SGD, Stochastic Variance Reduced Gradient Descent (SVRG), and SLBFGS. Conversely to evaluate all these optimizers four loss functions are used as a parameter. They are:

(1) Ridge regression (ridge)

$$\min \frac{1}{n} \sum_{i=1}^{n} (y_i - x_i^T w)^2 + \lambda \, ||w_2^2|| \tag{7}$$

(2) Logistic regression (logistic)

$$\min \frac{1}{n} \sum_{i=1}^{n} \ln(1 + \exp(-y_i x_i^T w)) + \lambda \, ||w_2^2|| \tag{8}$$

(3)   L1 loss L2-regularized SVM (hinge)

$$\min \frac{1}{n} \sum_{i=1}^{n} \ln\left(1 - y_i x_i^T w\right)) + \lambda \, ||w_2^2|| \tag{9}$$

(4)   L2 loss L2-regularized SVM (sqhinge)

$$\min \frac{1}{n} \sum_{i=1}^{n} \ln\left(1 - y_i x_i^T w\right))^2 + \lambda w_2^2 \tag{10}$$

Based on these learning models the effectiveness of our proposed work has been analyzed and is shown by Fig. 1. Here the number of epochs are represented by the x-axis and the loss value is represented by the y-axis.

4   Conclusion

In this research, it is observed that SGD unstably converges. In SGD, inappropriate learning rate effects on fluctuation. Large fluctuations in the loss value occurred due to inappropriate learning rate. Both SLBFGS and SVRG do not show better convergence as it is sensitive to the learning rates. In general, proposed enhanced stochastic variance reduces the ascension approach, converges faster than conjugate gradient, SLBFGS, SGD, and SVRG. Particularly, the chance for the occurrence of non-convex optimization has been reduced with sequent reduction in the variance using an enhanced stochastic variance reduced ascension approach, thereby ensuring faster convergence. In this research work, only one Deep learning model of Deep convolution network is used and one multimodal dataset is considered. But in the future, it would be interesting to use more different types of datasets with different characteristics and analyze the comparative effects of Gradient-based optimizers and it would propose algorithms in different architectural designs and in-depth.

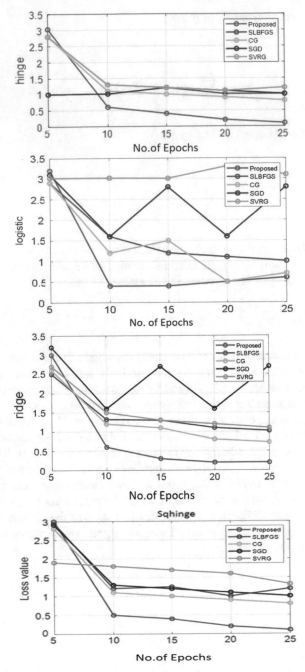

Fig. 1 Performance comparison of proposed optimizer

References

1. Hinton, G.E., Osindero, S., Teh, Y.W.: A fast learning algorithm for deep belief nets. Neural Comput. **18**(7), 1527–1554 (2006)
2. Krizhevsky, A., Sutskever, I., Hinton, G.E.: Imagenet classification with deep convolutional neural networks. Adv. Neural. Inf. Process. Syst. **25**, 1097–1105 (2012)
3. Srivastava, N., Hinton, G., Krizhevsky, A., Sutskever, I., Salakhutdinov, R.: Dropout: a simple way to prevent neural networks from overfitting. J. Mach. Learn. Res. **15**(1), 1929–1958 (2014)
4. Wang, Yu., Yin, W., Zeng, J.: Global convergence of ADMM in nonconvex nonsmooth optimization. J. Sci. Comput. **78**(1), 29–63 (2019)
5. Zhang, Z., et al.: A new finite-time varying-parameter convergent-differential neural-network for solving nonlinear and nonconvex optimization problems. Neurocomputing **319**, 74–83 (2018)
6. Dauphin, Y., Pascanu, R., Gulcehre, C., Cho, K., Ganguli, S., Bengio, Y.: Identifying and attacking the saddle point problem in high-dimensional non-convex optimization. 1–14 (2014)
7. Hardt, M., Ma, T., Recht, B.: Gradient descent learns linear dynamical systems. J. Mach. Learn. Res. **19**(1), 1025–1068 (2018)
8. Ruder, S.: An overview of gradient descent optimization algorithms. http://arxiv.org/abs/1609.04747. Accessed 29 Oct 2018
9. Hallen, R.: A study of gradient-based algorithms. http://lup.lub.lu.se/student-papers/record/8904399. Accessed 29 Oct 2018
10. Shalev-Shwartz, S., Shamir, O., Shammah, S.: Failures of gradient based deep learning (2017). arXiv:1703.07950
11. Papamakarios, G.: Comparison of stochastic optimization algorithms. School of Mathematic, University of Edinburgh. https://www.maths.ed.ac.uk/~prichtar/papers/Papamakarios.pdf. Accessed 26 Oct 2014
12. Darken, C., Chang, J., Moody, J.: Learning rate schedules for faster stochastic gradient search. Neural Networks for Signal Processing II. Proceedings of the IEEE Workshop (September), pp. 1–11 (1992). http://doi.org/10.1109/NNSP.1992.253713
13. Qian, N.: On the momentum term in gradient descent learning algorithms. Neural Netw. Off. J. Int. Neural Netw. Soc. **12**(1), 145–151 (1999). 6080(98)00116-6
14. Nesterov, Y.: A method for unconstrained convex minimization problem with the rate of convergence o(1/k2). Doklady ANSSSR (translated as Soviet. Math. Docl.), **269**, 543–547 (1983)
15. Sutton, R.S.: Two problems with backpropagation and other steepest-descent learning procedures for networks. In: Proceedings of 8th Annual Conference on Cognitive Science Society (1986)

Threat Detection with Facial Expression and Suspicious Weapon

Kiran Kamble(⊠), Swapnil Sontakke, Pooja Mundada, and Apurva Pawar

Walchand College of Engineering, Maharashtra Sangli, India
kirankamble5065@gmail.com

Abstract. Recently criminal activities such as robbery with the threat of life using weapons have increased exponentially. In the past, CCTV cameras were used for providing proof of criminal activities. But due to the technological advancements in areas like deep learning, image processing, etc., various ways are coming into the picture to prevent those criminal activities. Weapon detection was the first criteria introduced to define an activity as "suspicious", but it had many drawbacks, such as dummy weapons were being classified as a threat and also the system failed where there was frequent use of weapons. As the phrase "Suspicious Activity" is a relative entity, human being can identify a friendly environment, where people are carrying weapons but the motto is not harmful. On the other hand, existing technology lacks this context. Here, we have tried to align the existing system of weapon detection and facial expression with this context. This enhances the ability of the system to take decisions as if it is thinking as a human brain. Our CNN model inspired by the VGGNet family named as suspExpCNN model optioned accuracy 66% was yielded for facial expression detection and Faster RCNN and SSD models yielded for weapon detection with accuracy 82.48% and 82.84%, respectively. The work further combined suspExpCNN model with weapon detection model and generated alert if an angry, scared or sad expression is detected.

Keywords: Object detection · Faster RCNN · SSD · Suspicious activity · Human facial expression detection

1 Introduction

For many years, banks are getting robbed at the point of weapons use. Many people have lost their life in an attempt of creating an alert at the time of the incident. With the increase in the development of technologies, every problem keeps on demanding a better solution with the time. CCTV's are being commonly used in our society with the purpose to have proof if something goes wrong. But if a CCTV could have the thinking capability like a brain, it will be able to prevent many such incidents from happening or at least will help in taking appropriate counteraction within time. To take action immediately against a suspicious activity, we first need to be sure that the activity is actually suspicious. For example, children playing with toy weapons are not a suspicious activity and it can be inferred from their joyful facial expressions. This analysis of the environment done

© Springer Nature Singapore Pte Ltd. 2020
B. Iyer et al. (eds.), *Applied Computer Vision and Image Processing*,
Advances in Intelligent Systems and Computing 1155,
https://doi.org/10.1007/978-981-15-4029-5_39

by our human brain by taking into consideration many possible answers and choosing the appropriate one is what we want to embed into our systems. We are not far from the day when drones can be used as safety guards floating around the city and taking appropriate actions if things are seemed to be fishy. It is not possible to have a human as the security guard at every possible place, but machines can be used for serving this purpose if they are as smart as humans. A complex system is needed to serve the purpose of detecting suspicious activity because claiming the activity as suspicious right after detection would be a dumb way in today's world. We have many systems that work as independent solutions to simple problems. In this work, we have tried to combine weapon detection and facial expression detection to solve this complex problem.

2 Literature Study

Object retrieval methods generally have two steps, first is to search for images which contain the query object and the second one is to locate the object in the image which have bounding box. The previous step is important for the classification of image and the use of convolutional neural network has given many high performance results by Hossein Azizpour [1]. But, R-CNN requires a substantial amount of time to process every object proposal without sharing computation.

It has been seen that better accuracy was achieved by using Convolutional Neural Networks with big data, but the difficulty is the unavailability of publicly available datasets for facial expression recognition with deep architectures. Hence, to face the problem, Andre [2] proposed an approach to apply preprocessing techniques which were used to extract features that were only expression specific from a face image and further explored the presentation order of the samples during training.

In Fast R-CNN, a Region of Interest (RoI) pooling layer is designed to obtain faster detection speed by Lina Xun [3]. This approach gave the result which had features extracted only once per image by Hailiang Li [4]. For embedded systems, it was found that the applied approaches were accurate but extremely computationally intensive and showed high execution time for real time application even if high-end hardware was used.

Wei Liu [5] presented the first deep network-based object detector which detected the area of interest without resampling pixels or features for bounding box hypotheses and still gave the same accuracy as given by the approaches which are mentioned. Due to this high-accuracy detection (59 FPS with mAP 74.3% on VOC2007 test, versus Faster R-CNN 7 FPS with mAP 73.2% or YOLO 45 FPS with mAP 63.4%) with improved speed was achieved. By removing the step of building the bounding box and resampling of features, improvement in speed was achieved.

In our study, we have compared both the models, i.e., Faster RCNN and Single Shot Detection for object detection.

3 Proposed Approach

As per the methodology given in Fig. 1, the steps involved are as follows:

Step 1: Video is processed frame by frame.

Step 2: Detection of human object in the frame is done.

Step 3: Obtain the face of a human if present using Haarcascade Frontleface vertical and horizontal features.

Step 4: From the face obtained in the previous step, the facial expressions of the human are extracted.

Step 5: Next step is to check if a weapon is being introduced in the frame with the help of Faster RCNN or SSD model, whichever is suitable for the respected deployable area.

Step 6: If there is a presence of weapon along with humans present in the frame are detected to be scared, angry or sad.

(1) Send a high alert to the respective security administrator, along with the cropped video where the suspicious activity was detected.

(2) The type of alert may depend on the probability of the activity being highly suspicious.

If there is only the presence of weapon and not any objectionable expression detected, keep sending the images captured to the respective security administrator.

4   Dataset Generation

- **Facial Expression**: The dataset consists of images for face recognition. Dataset used for face recognition is Fer2013 [6] provided by Kaggle's facial expression reorganization which consists of 35,887 gray scales, 48 × 48 sized face images with 7 emotions. Emotion labels in the dataset with a count of number of images are depicted in Table 1 below. The dataset is collected by Ian Goodfellow et al. [7]. Samples of Kaggle's facial expression images are shown in Fig. 2.
- **Suspicious Weapons**: The dataset for suspicious weapons were generated by collecting images of guns (majorly pistols) and knives and were labeled using the LabelImg tool [8]. For the weapon detection model, training images are 3187 and for testing 450 images were used.

5   Custom Training of the SSD Architecture

After the generation of the dataset, the model was trained on more than 3000 images depicting suspicious weapons both guns and knives. As depicted in Fig. 3, SSD's architecture [5, 9] is built on the venerable VGG-16 architecture, but does not use it's fully connected layers. VGG-16 was decided to be used as the base network is because it

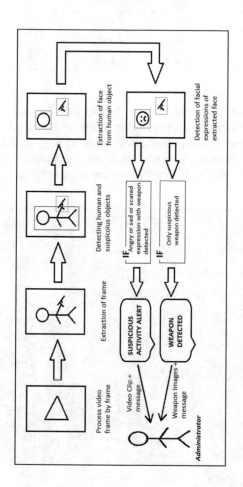

Fig. 1 Schematic diagram of the proposed approach

Table 1 Statistics of FER2013 dataset

Emotion number	Emotion labels	Count of images
1	Angry	4593
2	Disgust	547
3	Fear	5121
4	Happy	8989
5	Sad	6077
6	Surprise	4002
7	Neutral	6198

has shown good performance in high quality image classification tasks and is highly suitable for problems where transfer learning helps in improving the obtained results. A set of auxiliary convolutional layers (from conv6 onwards) were added removing the traditional layers, which thus enables the extraction of features from multiple scales and eventually decreases the input size to each following subsequent layer.

Fig. 2 Sample image with different expression inside the Kaggle

6   Custom Training of the Faster RCNN Architecture

As per Fig. 4, Faster RCNN [4] is one of the powerful models used for object detection and uses convolution neural networks. It consists of three parts:

a. **Convolution layers**: Here appropriate features from the image are extracted by training filters. Convolution networks are composed of the followings:

- Convolution layers
- Pooling layers and
- Fully connected last component or another extended thing which will be used classification or detection.

Fig. 3 Generated sample dataset of for suspicious weapons

This layer does the task of preserving the relationship between pixels using the knowledge of image features.

b. **Pooling layers**: The main focus of pooling is to decrease the quantity of features in the features map by using the method of elimination for pixels which are having low values.
c. **Region Proposal Network (RPN)**: RPN is a small neural network that is implemented on the last feature map of the convolution layers and used to predict the presence of an object along with a bounding box for it.
d. **Classes and Bounding Boxes prediction**: In this step the use of fully connected neural networks is done which has the regions proposed in the previous step as an input and gives the prediction of an object class (classification) and bounding boxes (regression) as an output.

Both SSD and FASTER RCNN models are trained on the same dataset of suspicious weapons to compare results. Faster RCNN is also used for face detection and expression recognition.

7   Training and Testing the Custom Model

Initially, this work was experimented on Fer2013 dataset for the classification of 7 emotions inspired by the family of VGG CNN Model [10]. CNN model's first layer is framed by 32 3 X3 tiny matrices to learn features with activation functions as Exponential Linear Unit (ELU) followed by batch normalization. The second layer also consists of 32 3X3 tiny matrices to learn the next pattern in expression images followed by ELU activation and batch normalization. The next layers are max-pooling with a dropout of 25%. With the use of a stack of network and softmax classifier at the end details architecture of suspExpCNN model shown in Fig. 5. We obtained 66% accuracy. Our focus is on suspicious expressions: scared, angry or sad. If anyone of the expression is detected out of these, then our system triggers an alert. The second phase of work focuses on weapon detection.

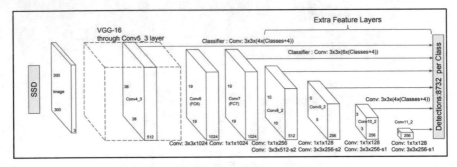

Fig. 4   Architecture of SSD model [5]

The dataset generated consisted of 2627 images of class—"gun" and 800 + images of class "knife". The model built was trained for 10,000 + iterations for SSD and 8000 + iterations for the Faster RCNN model and a loss of about 1.5 and 0.01 was achieved for SSD and Faster RCNN, respectively. The trained model is tested using gun images of count 94; knife sample of 89, and No-weapon sample of 91.

8   Results

The results achieved are given below as a Confusion Matrix. A confusion matrix gives results in tabular form and usually describes the performance of a classification model for test data on the basis of knowledge of the true values for the same. By visualizing the performance of an algorithm, it helps in easy identification of confusion between the classifications of classes. The use of the confusion matrix is popular in the performance measures of algorithms. Tables 2 and 3 show the confusion matrix for suspicious weapon detection for testing samples. Figure 5 shows the result of Weapon detection with softmax classifier confidence is 97% and a threshold of 0.7. Similarly, Fig. 6 shows the result of the subject with a GUN for which softmax classifier confidence is 72% and threshold 0.7. Figure 7 shows the output of weapon detection and alerting on gunpoint. Figure 8 shows result in night vision effectively detecting expression as anger (Figs. 9 and 10).

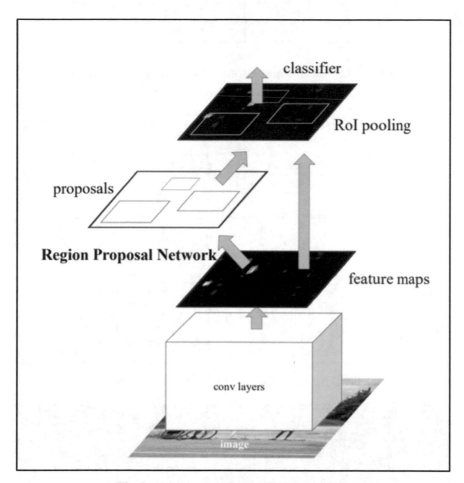

Fig. 5 Architecture of faster RCNN model [11]

The trained SSD model has obtained an accuracy of 82.84% with 86% precision and 83% recall, and the trained FASTER RCNN model has obtained an accuracy of 82.48% with 85% precision and 82% recall.

9 Conclusion

Previously, suspicious activities were detected from the presence of the weapons only, but our work presented a new approach that would use facial expressions of subjects along

Table 2 Confusion matrix of SSD for suspicious weapons

Predicted				
Actual		Gun	Knife	No weapon
	Gun	66	0	28
	Knife	4	74	11
	No weapon	3	1	87

Table 3 Confusion matrix of Faster RCNN for suspicious weapons

Predicted				
Actual		Gun	Knife	No weapon
	Gun	91	0	3
	Knife	16	61	12
	No weapon	17	0	74

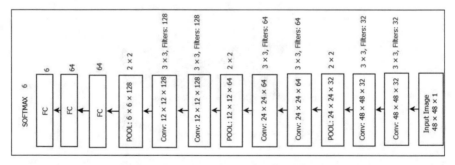

Fig. 6 suspExpCNN model for facial expression

Fig. 7 Results of weapon detection using faster RCNN

Fig. 8 Results of subject with GUN with neutral expressions using suspExpCNN model for expression and faster RCNN for GUN detection

Fig. 9 Results of alerting on Weapon detected and alerting on gunpoint **a** using faster RCNN **b** using SSD

with the suspicious weapons for analyzing the context to better predict the suspicious activities and reduce false predictions. Both SSD and Faster RCNN provided results with 82.84% and 82.48% accuracy. Faster RCNN being good in detecting small guns (91 of 94) whereas (66 of 94) in case of SSD and SSD being good in providing results in less amount of time (1.5 s) to process one frame whereas (2.5 s) required for Faster RCNN. This system can be used to monitor any public place, private shops, can be used by government agencies to detect suspicious activities and prevent it.

(1) If the system is to be used at places where the distance of subjects would be less than 10 m, SSD model can be used for quick response.
(2) If the system has to be used in public places where small and distant objects are of our concern then Faster RCNN model can be used for better precision.

10 Assumptions

1. Images are taken in daylight with an appropriate camera with a high configuration that can capture objects from the required distance.
2. To detect threats at night, a night vision camera is used for the system to work properly.
3. The system fails and acts only as a weapon detection system when the faces of the subjects are not visible.
4. The system works with acceptable performance with camera radius of about 15 m, but can work for more than that if a more highly configured camera is used.

11 Future Scope

1. As the CCTV cameras are fixed, the facial expression of a person facing against the camera cannot be detected or there is a limitation on the area of surveillance, we can also think of using drones to avoid such cases.
2. As the real time system is time-consuming we can improve it by using parallel computing wherein the frames will be processed in parallel which will help in minimizing the time required to process a frame.

References

1. Sharif Razavian, A., Azizpour, H., Sullivan, J., Carlsson, S.: CNN features off-the-shelf: an astounding baseline for recognition. In: Proceedings of the IEEE Conference on Computer Vision and paTtern Recognition Workshops, (2014)
2. Lopes, A.T., de Aguiar, E., De Souza, A.F., Oliveira-Santos, T.: Facial expression recognition with convolutional neural networks: coping with few data and the training sample order. Elsevier, (2016)
3. Li, J., Zhang, D., Zhang, J., Zhang, J., Li, T., Xia, YY., Yan, Q., Xun, L.: Facial expression recognition with faster R-CNN. Procedia Comput. Sci. (2017)
4. Li, H., Huang, Y., Zhang, Z.: An improved faster R-CNN for same object retrieval. IEEE Access (2017)
5. Liu, W., Anguelov, D., Erhan, D., Szegedy, C., Reed, S., Fu, C-Y., Berg, A.C.: SSD: single shot multibox detector. In: European Conference On Computer Vision (ECCV), Springer, Cham, (2016)
6. Kaggle Team.: Challenges in representation learning: facial expression recognition challenge.https://www.kaggle.com/c/challenges-in-representation-learningfacial-expression-recognition-challenge
7. Goodfellow, I.J. et al.: Challenges in representation learning: a report on three machine learning contests. In: Neural Information Processing: 20th International Conference, ICONIP 2013, Daegu, Korea, November 3–7, (2013). Proceedings, Part III. Lee, M. et al. (ed.) Berlin, Heidelberg: Springer Berlin Heidelberg, (2013), pp. 117–124. ISBN: 978-3-642-42051-1. https://doi.org/10.1007/978-3-642-42051-1_16
8. LabelImg.: A Graphical Image annotation tool. https://github.com/tzutalin/labelImg Reference Video Link: https://youtu.be/p0nR2YsCY_U

9. Object detection with deep learning and OpenCV by Adrian Rosebrock. https://www.pyimagesearch.com/2017/09/11/object-detection-with-deep-learning-and-opencv/
10. Simonyan, K., Zisserman, A.: Very deep convolutional networks for large-scale image recognition. arXiv preprint arXiv 1409.1556, (2014)
11. Ren, S., He, K., Girshick, R. Sun, J.: Faster R-CNN: towards real-time object detection with region proposal networks. IEEE Trans. Pattern Anal. Mach. Intell. 39, (2015). https://doi.org/10.1109/TPAMI.2016.2577031

Integrated YOLO Based Object Detection for Semantic Outdoor Natural Scene Classification

C. A. Laulkar[⊠] and P. J. Kulkarni

Computer Science and Engineering, Walchand College of Engineering, Sangli, India
chaitalivs@gmail.com

Abstract. Scene classification facilitates many applications like robotics, surveillance system, image retrieval, etc. Recently, various CNN architectures are designed to attempt the problem of scene classification. This paper proposes two stage model for scene classification using semantic rules. In the first stage, YOLOv2 (You Only Look Once) CNN architecture is trained for seven types of objects: sky, green_land, house, tree, sand_land, water, and windmill. In testing phase, all objects which are present in scene images are detected and recognized through trained network with overall accuracy of 79.35%. Semantic rules are designed using context of objects for the classification of scene image. Objects detected in stage-1 are parsed through the semantic rules to classify the scene image into six classes: barn, beach, desert, windfarm, green_ground, and water_land. Scene classification using semantic rules has achieved overall F-score of 76.81% on scene images from SUN397 dataset. Results of the proposed model using YOLOv2 are compared with results of same model using Faster RCNN which indicate that performance of scene classification depends on performance of object recognition system.

Keywords: Object detection and recognition · YOLOv2 (You only look Once) · Semantic rules · Scene classification

1 Introduction

The human brain is capable to understand the scenes of real-world in a single glance. Latest trends in the art of technology are focusing on capturing features of scenes in machines that can assist human being in dealing with extreme situations of understanding the scene. Scene classification plays an important role to build an artificially intelligent system which is able to act in such situations by understanding the surrounding scenes. It also contributes mainly to other real-life applications like automated surveillance systems, image retrieval, etc.

The presence of multiple objects in a scene and relationship among them assists the system to understand the scene semantically. Many object detection methods based on

© Springer Nature Singapore Pte Ltd. 2020
B. Iyer et al. (eds.), *Applied Computer Vision and Image Processing*,
Advances in Intelligent Systems and Computing 1155,
https://doi.org/10.1007/978-981-15-4029-5_40

CNN architecture like RCNN, SPP-Net, Fast RCNN, Faster RCNN, DPM v5, SDD, and YOLO has shown remarkable performance in detection and recognition of objects [1].

This paper is envisioned to attain the improved performance of our earlier work on outdoor natural scene classification using semantic rules [2]. Performance of semantic rules based scene classification system depends on the accuracy of recognition and localization of objects. The proposed framework works mainly in two phases: object detection and recognition using CNN-based YOLOv2 architecture and scene classification using semantic rules. Images are chosen from six classes of SUN397 dataset [3]: barn, beach, desert, windfarm, green ground, and lakes for this research work. The work revealed in this paper detects and classifies the objects into seven categories, i.e., sky, green_land, house, tree, sand_land, water, and windmill. Semantic rules designed using context of these objects are applied for the classification of scene image into any of the six classes, i.e., barn, beach, desert, windfarm, green_ground, and water land.

The paper is structured as follows. Related work of systems designed for object detection and recognition and scene classification are discussed in Sect. 2. Section 3 describes the working of the proposed system. Section 4 talks about the outcomes accomplished by both modules of the proposed system and Sect. 5 delivers the conclusion based on results attained through applied methods.

2　Related Work

Object detection methods are mainly classified into three types: 1.Sliding window detector with neural network 2. Region-based classification of objects 3. Object detection and classification with localization. Sliding window detector extracts the local features like color, texture, gradients, etc., to generate feature vector which is further used for object classification through machine learning algorithms [4]. Region-based classification begins with the generation of regions using S x S grids [5, 6] or various segmentation [7, 2, 10] methods. Feature vectors created by extracting numerous local features for these regions are used to classify the objects through machine learning algorithms. Various Convolution Neural Network (CNN) architectures like AlexNet, ZFNet, VGG, GoogleNet, ResNet, GoogleNetv4, and SENet have shown their best performances in evaluating object detection and image classification during each year in ImageNet Large Scale Visual Recognition Challenge (ILSVRC) since 2012 till 2017 [8]. CNN architectures are slow and computationally very expensive. For object detection, it is not possible to run CNN on all bounding boxes produced by window detector. Region-based CNN (RCNN) has used a selective search algorithm that lowers the number of bounding boxes around 2000 region proposals. Due to constrain of fixed size input to Fully Connected (FC) layer, RCNN warps each region proposal into the same size that causes loss of data which eventually affects the object recognition accuracy. Spatial Pyramid Pooling-Net (SPP-Net) combines CNN with spatial pyramid pool methodology to dispense with fixed size input. To overcome the slowness of RCNN, Fast RCNN model is developed which uses only one CNN on entire image and substitutes Support Vector Machine (SVM) with softmax layer that extends the neural network to extract probabilities of classes. Selective search is the slowest part of Fast RCNN which is replaced with Region Proposal Network (RPN) to generate Region of Interest (RoI) in Faster RCNN. These region-based methods handled object detection as a classification problem in which object proposal is generated and sent to classification heads. You Only Look Once (YOLO) and Single Shot Detector (SSD) pose object detection as a problem of regression which improves the performance of speed [9, 13, 14].

Sematic scene classification is one of the promising approaches, which can show improvement in performance of the scene classification system. LBP feature based modeling of natural scene has attained overall accuracy of 77.14% [5] in which LBP features and prior information of neighborhood are utilized as texture, and spatial features are utilized to annotate the scene image. Semantic classification of scene images uses the context of objects generated by object detection system for classification of image. Spatial information of the objects is used for scene content understanding and improving region labeling [15, 16]. Semantic rules generated by using context of object, like type of object and its localization, are used for outdoor natural scene image classification [2]. Scene classification using Places-CNN and ImageNet-CNN architecture has attained 54.32% and 42.61% accuracy for SUN397 dataset [17]. The proposed work intends to improve the performance of scene classification by applying semantic information.

3 Scene Classification Using Semantic Rules

Outdoor natural scene images comprise natural and man-made objects. Compared to man-made object, which possesses prominent features, classification of natural objects like sky, green_land, sand_land, and water is a challenging task due to resemblance in their color and texture features. Performance of semantic scene classification system mainly depends on performance of object detection system and semantic rules. Proposed research work aims at the classification of outdoor natural scene images in two stages: i. Object detection and recognition using YOLOv2 and ii. Semantic rules based scene classification. For experimental purpose, images from six different classes of SUN397 dataset are selected and labeled manually for the seven types of objects present in it. Labeled image dataset is used to train the YOLOv2 CNN architecture for all seven types of objects. The trained architecture of YOLOv2 is used for the detection and recognition of the objects in test images as shown in Fig. 1. Each recognized object has contextual information about its type, confidence score, and localization. Semantic rules designed using information about type of objects are applied for classification of scene images.

3.1 Object Detection and Recognition Using YOLOv2

Natural scene image contains multiple objects which assist classification of the image. For the purpose of semantic scene classification, extraction and recognition of individual object in a scene are necessary. You Only Look Once (YOLO) is an object detection and recognition algorithm which detects multiple objects present in an image [11, 12]. It is a single CNN which at the same time predicts multiple bounding boxes with their class probabilities.

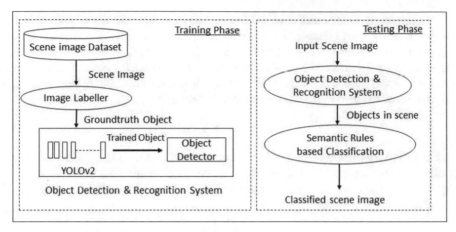

Fig. 1 Architecture of semantic rules based scene classification system

In proposed research work, YOLOv2 model is designed with transfer learning approach in which AlexNet is used as pre-trained network [18, 19]. YOLOv2 model begins with feature extractor network in which the first 15 layers of AlexNet are used for feature extraction. These 15 layers contain two convolution layers, which are trailed by three layers each, i.e., Rectified Linear Unit (ReLU), cross channel normalization, and max pooling layer. Each of the 3rd, 4th, and 5th convolution layer is trailed by ReLU layer. Using transfer learning approach, reorganization layer is added after ReLU5 layer which contains series of convolution, batch normalization, and ReLU layers. Reorganization layer is followed by convolution, transform, and output layer as shown in Fig. 2. The transform layer extracts activations of the last convolutional layer and transforms the bounding box predictions to fall within the bounds of the ground truth.

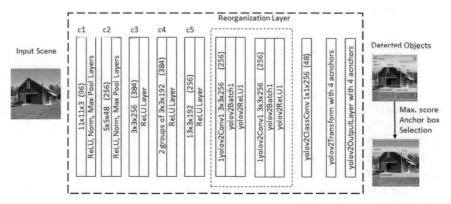

Fig. 2 AlexNet based YOLOv2 architecture for object detection and recognition

The proposed architecture has used image size of 256 × 256 with 5 anchor boxes. Labeled training image dataset is used to train the network for seven types of objects: sky, green_land, house, tree, sand_land, water, and windmill. YOLOv2 divides input image into 14 × 14 grids. The grid cell is responsible to detect the object that comprises the center of that object. Each grid cell predicts 5 bounding boxes along with contextual information of depth "D" which is calculated by using Eq. 1.

$$D = B * (5 + C) \tag{1}$$

Here, B represents the count of bounding box and C is the total number of classes of objects to be detected [19]. Contextual information contains 12 fields as shown in Fig. 3. Tx, Ty represent the center, whereas, Tw and Th are the width and height of the bounding box, respectively.

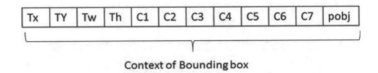

Context of Bounding box

Fig. 3 Bounding box descriptor

C1–C7 represent conditional class probabilities using Eq. 2.

$$C_i = \Pr\left(\frac{the\,object\,belongs\,to\,i - th\,class}{an\,object\,exist\,in\,this\,box}\right) \tag{2}$$

The probability of an object of ith class is given by $C_i P(obj) \leq 1.0$. If this value is greater than threshold value of 0.25, then network has predicted an object of ith class existing in this bounding box.

3.2 Semantic Rules Based Scene Classification

Contextual information of objects in a scene assists the proposed system for the classification of scene image. Classification of each scene requires presence of object which is mandatory to represent that scene, like barn-house, beach-water and sand, desert-sand, windfarm-windmill, green ground-green_land, and water land-water. Proposed research work has used "class of the object" as a semantic information for the generation of semantic rules. Semantic rules, which are parsed for the objects detected by YOLOv2 model, are given in Table 1. Matching of semantic rule classifies the scene image into one of the six classes. Semantic rules fail to match for wrongly recognized or non-recognized objects.

Table 1 Semantic rules for classification of outdoor natural scene

Semantic rules	Explanation
Scene_class → C1|C2|C3|C4|C5|C6|Not_Classified Not_Classified → No match for rule	Rules for the classification of scene image into six classes: barn, beach, desert, windfarm, green_ground, and water_land. The scene will not get classified into any of the proposed class if rules do not match
C1 → House + T1 T1 → T2 + T1 | ε T2 → Sky | Green_land | Tree | ε	Rules to classify the scene image as Barn in which mandatory object House is present exclusively or in various combinations with other three objects Sky, Green_land, and Tree
C2 → Sand_land + Water + T1	Rules to classify scene image as Beach in which mandatory objects Sand_land and Water are present exclusively in various combinations with other three objects Sky, Green_land, and Tree
C3 → Sand_land + T3 T3 → T4 + T3 | ε T4 → Sky |Tree | ε	Rules to classify scene image as Desert in which mandatory object Sand_land is present exclusively or in various combinations with other two objects Sky and Tree
C4 → Windmill + T5 T5 → T6 + T5 | T7 + T5|ε T6 → Sky| Green_land | Sand_Land | Tree | ε T7 → Sky | Water|ε	Rules to classify scene image as Windfarm in which mandatory object Windmill is present exclusively or in various combinations with other objects like Sky, Green_land, Sand_land, Water, and Tree
C5 → Green_land + T8 T8 → T9 + T8 | ε T9 → Sky | Tree | Water | ε	Rules to classify scene image as Green_ground in which mandatory object Green_land is present exclusively or in various combinations with other objects like Sky, Water, and Tree
C6 → Water + T1	Rules to classify scene image as Water_land in which mandatory object Water is present exclusively or in various combinations with other objects like Sky, Green_land, and Tree

4 Results and Discussion

4.1 Experimental Setup

The proposed system works in two stages: i. Object detection and recognition and ii. Semantic rules based scene classification. It is intended to classify the scene images into six different classes: barn, beach, desert, windfarm, green_ground, and water_land. Total 700 color scene images which cover seven types of objects: sky, green_land, house, tree, sand_land, water, and windmill are selected from seven: barn, beach, desert, windfarm, green ground, and lake classes of SUN397 dataset are used to train the YOLOv2 model for object detection. Each input image is resized to 256 × 256. Multiple objects present in training image are labeled by using image labeler tool before submitted for training. Minimum 100 instances of each type of object are used to train the system. It is assumed that input scene image contains maximum five types of objects at a time. The proposed system has been tested against 350 scene images for the detection of seven types of objects and 6 classes of scenes.

4.2 Object Recognition System

For the purpose of experiment, the proposed system has used five anchor boxes of sizes [200 200; 100 200; 150 200; 50 200; 150 50] to train theYOLOv2 model with learning rate of 0.0001 and epochs of 20. Ground truth data of objects in each image is generated manually. Object recognition system is trained for 500 and 700 images. Generated trained objects, i.e., Detector_500 and Detector_700 are tested against 350 images. Classification accuracy calculated by comparing system assigned labels of objects with ground truth labels is as given in Eq. 3.

$$Accuracy = \frac{(label\_assigned == ground\_truth\_labels)}{count(ground\_truth\_labels)} \tag{3}$$

Classification accuracy achieved by the proposed system is shown in Table 2. System trained with 500 images has achieved lower accuracy for green_land, sand_land, andwater objects. To improve the accuracy of these objects, 200 more images containing green_land, sand_land, and water objects are added during training. System trained with 700 images achieved improved accuracy for said objects but accuracy for tree and windmill classes is decreased slightly. From Table 2, it is perceived that windmill object has achieved highest accuracy due to its prominent features whereas green_land and sand_land objects have attained lower values of accuracy due to similarity in texture feature.

Accuracy of YOLOv2 is compared with accuracy of Faster RCNN architecture which shows improved performance of YOLOv2 by 0.39% in overall accuracy of Detector_700 for the same set of training and testing dataset. Accuracy of sky, tree, water, and windmill objects is found improved whereas it is impaired for green_land, house, and sand_land objects.

Table 2 Classification Accuracy of ORS using YOLOv2 and Faster RCNN

Object class	Faster RCNN-classification accuracy in %		YOLOv2-classification accuracy in %	
	Detector 500	Detector 700	Detector_500	Detector_700
Sky	91.37	81.55	91.96	93.75
Green land	87.96	90.74	69.44	69.44
House	98.00	98.00	82.00	86.00
Tree	50.91	50.00	88.18	70.91
Sand Land	78.57	67.35	47.96	55.10
Water	59.87	71.05	57.89	84.21
Windmill	96.00	94.00	100.00	96.00
Average	80.38	78.96	76.78	79.35

4.3 Semantic Rules Based Scene Classification

Outdoor natural scene images containing natural objects like sky, green_land, sand_land, and water are difficult to classify correctly due to resemblance in their color and texture feature whereas man-made objects like house and windmill are classified easily due to their prominent features. For the purpose of experiment, more number of natural object instances are used during training of natural objects compared to man-made objects. Table 3 shows comparison of classification results for the scene images through YOLOv2 and Faster RCNN object detection models. Evaluation metrics TP, FP, and

Table 3 Performance of scene classification for faster RCNN and YOLOv2 model

Method (iteration count)/Object class		Sky	Green_land	House	Tree	Sand_land	Water	Windmill
Accuracy in %	YOLOv2 (10485)	94.94	62.03	76.00	64.55	64.28	80.92	96.00
	YOLOv2 (13980)	93.75	69.44	86.00	70.91	55.10	84.21	96.00
	AlexNet [2]	93.90	96.40	–	–	74.40	77.50	–

FN are calculated from given confusion matrix in Table 3. F-score evaluation metric is harmonic mean of precision and recall which is calculated using Eq. 4. For higher value of F-score, both precision and recall indicate good results.

$$F - score = 2 * \frac{Precision * Recall}{Precision + Recall} \tag{4}$$

It has been observed from Table 3, for YOLOv2 model, windfarm images containing man-made object like windmill has scored highest F-score. For other classes, scene images are occasionally misclassified or not classified in any of the given classes. Misclassification or non-recognition of mandatory objects in scene images results in wrong match or no match of semantic rules which affects performance of scene classification, e.g., Desert and beach images must contain water and water + sand_land objects, respectively. Due to less object detection and recognition accuracy of sand_land object, many times it is not detected in these scene images. Detection of only sky and water objects misclassifies most of the beach images as water_land images. In desert scene images, sand mountains are occasionally recognized as house which misclassify the desert scene as barn image. Similarly, for barn images, the mandatory object is house which is not recognized. Detection of only sky, green_land, and tree objects misclassify the barn scene images as green_ground class. Due to less accuracy of green_land object, only sky object is detected in green ground scene images. When only sky object is detected for given classes, scenes do not get classified in any of the given class as there is no semantic rule that has been designed for only sky object.

Table 3 also presents the results of scene classification for the same dataset under two models: YOLOv2 and Faster RCNN. Comparison shows that F-score for certain categories for Faster RCNN is higher than YOLOv2. YOLOv2 has achieved better performance than Faster RCNN for the detection and recognition of sky, tree, water, and windmill objects which improve the performance of classification of windfarm and water land scene images containing these objects as shown in Table 4. Similarly, Faster RCNN has achieved better performance than YOLOv2 for the detection and recognition of green_land, house, and sand_land objects which improve the performance of classification of barn, beach, desert, and green_land and of images containing these objects. Overall performance of the scene classification mainly depends on the accuracy of object detection system. Objects which share more similarity in features are difficult to detect exclusively. Objects with prominent features are detected with great accuracy.

5   Conclusion

The proposed system has detected and recognized the multiple objects present in scene image using YOLOv2 object detector with the overall accuracy of 79.35% for detection of seven objects. Performance of scene classification mainly depends on accuracy of object recognition system. Objects detected by YOLOv2 CNN model are parsed through semantic rules to identify the class of scene image. Scene classification has achieved highest F-score value for images of windfarm class. The proposed system has achieved minimum F-score of values for images of beach class due to less accuracy of object recognition system for sand_land object. Overall 76.81% F-score is achieved for scene classification model. F-score of scene classification is improved by improving accuracy of object recognition system using YOLOv2 model of CNN.

Table 4 Impact of object recognition system on classification of scene

Scene class	Performance of scene classification for YOLOv2 and faster RCNN													
	Faster RCNN							YOLOv2						
	Barn	Beach	Des	WF	GG	WL	NR	Barn	Beach	Des	WF	GG	WL	NR
Barn	49	0	0	0	1	0	0	38	0	0	1	4	0	7
Beach	0	23	7	0	0	11	9	0	11	3	2	1	26	7
Desert	5	0	38	0	0	0	7	1	1	31	0	0	0	17
Windfarm	1	0	0	45	1	0	3	0	0	0	50	0	0	0
Green_ground	0	0	0	0	50	0	0	0	0	0	0	45	0	5
Water land	1	2	1	0	3	70	23	0	3	0	0	3	80	14
TP, FP and FN														
TP	49	23	38	45	50	70		38	11	31	50	45	80	
FP	1	18	5	2	0	7		5	32	2	0	0	6	
FN	1	27	12	5	0	30		12	39	19	0	5	20	
Precision, recall and F-ratio in %														
Precision	98.00	56.10	88.37	95.74	100.00	90.91		88.37	25.58	93.94	100.00	100.00	93.02	
Recall	98.00	46.00	76.00	90.00	100.00	70.00		76.00	22.00	62.00	100.00	90.00	80.00	
F-score	**98.00**	**50.55**	**81.72**	92.78	**100.00**	79.10		81.72	23.66	74.70	**100.00**	94.74	**86.02**	

References

1. Melek, C.G., Sonmez, E.B., Albayrak, S.: Object detection in shelf images with YOLO. In: IEEE EUROCON 2019-18th International Conference on Smart Technologies, pp. 1–5. IEEE, (2019)
2. Laulkar, C.A., Kulkarni, P.J.: Semantic rules-based Classification of outdoor natural scene images. In: Computing in Engineering and Technology, pp. 543–555. Springer, Singapore, (2020)
3. Xiao, J., Hays, J., Ehinger, K.A., Oliva, A., Torralba, A.: Sun database: large-scale scene recognition from abbey to zoo. In: 2010 IEEE Computer Society Conference on Computer Vision and Pattern Recognition, pp. 3485–3492. IEEE, (2010
4. Laulkar, C.A., Kulkarni, P.J.:Outdoor natural scene object classification using probabilistic neural network, in: 2018. Int. J. Comput. Sci. Eng. (IJCSE), 1(6), 26-31 (2017). E-ISSN:2347–2693
5. Raja, R., Roomi, S.M.M., Kalaiyarasi, D.: Semantic modeling of natural scenes by local binary pattern. In: 2012 International Conference on Machine Vision and Image Processing (MVIP), pp. 169–172. IEEE, (2012)
6. Vogel, Julia, Schiele, Bernt: Semantic modeling of natural scenes for content-based image retrieval. Int. J. Comput. Vision 72(2), 133–157 (2007)
7. Mylonas, Phivos, Spyrou, Evaggelos, Avrithis, Yannis, Kollias, Stefanos: Using visual context and region semantics for high-level concept detection. IEEE Trans. Multimedia 11(2), 229–243 (2009)
8. http://www.image-net.org/challenges/LSVRC/
9. Zhao, Z-Q., Peng, Z., Xu, S., Wu, X.: Object detection with deep learning: a review. IEEE Transactions on Neural Networks Learn. Syst. (2019)
10. Li, L-J., Richard, S., Li, F-F.: Towards total scene understanding: classification, annotation and segmentation in an automatic framework. In: 2009 IEEE Conference on Computer Vision and Pattern Recognition, pp. 2036–2043. IEEE, (2009)
11. Redmon, J., Santosh, D., Ross, G., Ali, F.: You only look once: unified, real-time object detection. In: Proceedings of the IEEE Conference on Computer Vision and Pattern Recognition, pp. 779–788, (2016)
12. Redmon, J., Ali, F.: YOLO9000: better, faster, stronger. In: Proceedings of the IEEE Conference on Computer Vision and Pattern Recognition, pp. 7263–7271 (2017)
13. Lan, W., Jianwu, D., Yangping, W., Song, W.: Pedestrian detection based on YOLO network model. In: 2018 IEEE International Conference on Mechatronics and Automation (ICMA), pp. 1547–1551. IEEE, (2018)
14. Lu, Z., Jia, L., Quanbo, G., Tianming, Z.: Multi-object detection method based on YOLO and resNet hybrid networks. In: 2019 IEEE 4th International Conference on Advanced Robotics and Mechatronics (ICARM), pp. 827–832. IEEE, (2019)
15. Singhal, A., Jiebo, L., Weiyu, Z.: Probabilistic spatial context models for scene content understanding. In: 2003 IEEE Computer Society Conference on Computer Vision and Pattern Recognition, 2003. Proceedings, vol. 1, pp. I–I. IEEE, (2003)
16. Boutell, M.R., Jiebo, L., Christopher, M.B.: Improved semantic region labeling based on scene context. In: 2005 IEEE International Conference on Multimedia and Expo, p. 4. IEEE, (2005)
17. Zhou, B., Agata, L., Jianxiong, X., Antonio, T., Aude, O.: Learning deep features for scene recognition using places database. In: Advances in Neural Information Processing Sstems, pp. 487–495 (2014)
18. https://medium.com/@y1017c121y/how-does-yolov2-work-daaaa967c5f7
19. https://hackernoon.com/understanding-yolo-f5a74bbc7967

Customer Feedback Through Facial Expression Recognition Using Deep Neural Network

Tejashri P. Dandgawhal$^{(\boxtimes)}$ and Bashirahamad F. Momin

Walchand College of Engineering, Sangli, Maharashtra, India
tejashri19dandgawhal@gmail.com,
bashirahamad.momin@walchandsangli.ac.in

Abstract. In day-to-day life, people express their emotions through facial expressions. Human Facial Expression Recognition has been researched for several years. Real time facial identification and perception of various facial expressions such as happy, angry, sad, neutral fear, disgust, surprise, etc. The key elements of the face are considered for the detection of the face and the prediction of facial expressions or emotions. A facial expression recognition system consists of a dual step process, i.e., face detection and emotion detection on the detected face. For detection and classification of such types of facial expressions, machine learning, as well as deep learning algorithms, are used by training of a set of images. This is an architecture that takes an enormous amount of registered face pictures as an input to the model and classifies those pictures into one of the seven basic expressions. SBI has developed a Convolution Neural Network model to detect the type of emotion expressed by the customer. The system currently recognizes three emotions Happy, Unhappy, and Neutral. This study aims to understand the existing CNN architecture and propose a sophisticated model architecture to improve the accuracy and to reduce the FP (False Positive) rate in the existing system. We also propose to use facial landmarks to train our new model. This survey proposes a deep architecture of the neural network to tackle the FER issue across several face datasets.

Keywords: Deep neural network (DNN) · Convolutional neural network (CNN)

1 Introduction

Facial expression recognition is a way of recognizing facial expressions on one's face. Facial expressions are the fastest way of communication in terms of conveying any type of information in effective manner. It is used not only to express each human being's emotions but also to evaluate his/her mental state. To do so effectively, we can use many approaches. Artificial agents promote communication with human beings and make them think differently about how computer systems of their everyday lives can be used. One of them was found primitive: a device that senses facial expressions automatically in real time. The performance evaluation is being measured using their facial expressions on applications such as the Bank customer feedback service. Charles Darwin was one of

© Springer Nature Singapore Pte Ltd. 2020
B. Iyer et al. (eds.), *Applied Computer Vision and Image Processing*,
Advances in Intelligent Systems and Computing 1155,
https://doi.org/10.1007/978-981-15-4029-5_41

the first scientists to identify facial language as one of the strongest and most immediate means to communicate the emotions, intentions, and opinions of people. It operates entirely automatically and with high precision in real time [1]. In order to identify the features of deep convolutionary networks, we suggest an efficient framework for facial expression detection.

(ConvNets). Figure 1 shows a high-level illustration of our learning to transfer recognition of facial expression [2].

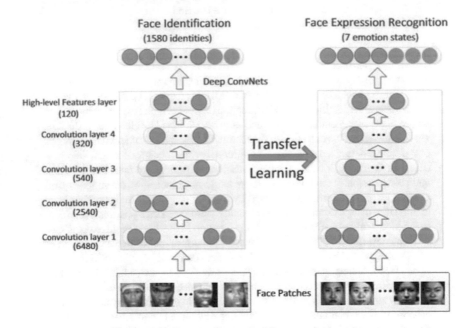

Fig. 1 Facial expression recognition transfer learning

Automated FER approaches attempt to classify faces as one of seven basic emotions in a provided single image or image sequence. While traditional approaches to machine learning such as SVM and Bayesian classifiers have been likely to succeed in classifying posed faces in a controlled environment, but, recent research has shown that such approaches don't have the ability to randomly identify unregulated objects captured. Currently, deep learning technique has seen a rise in popularity due to the availability of more computing power and large training datasets to deal with.

A two-step, fine-tuning driven methodology is used to learn on how to migrate deep CNN architectures. The ImageNet base server with its two distinct CNN systems begins with the simple pretraining protocol. A lower-resolution object database is applied with the support of FER2013 in fine-tuning in the first stage. Step 2 was about the web education EmotiW information set [3].

To obtain precise training data is a much difficult task, especially for facial expressions such as fear or sad which are more difficult for accurate prediction. We explore this approach by training proposed architecture on a subset of the existing training dataset, and then performing different techniques such as cross-validation experiments.

2 Related Work

Deep learning approaches have been very effective in the image-related tasks in recent years with the aid of the convolution neural network, because of its ability of perfect data representation. Predicting a human expression can be even more difficult for people, due to differences in expressions such as fear and sad [3].

There are three basic parts for the development of the facial expression recognition system: face detection, feature extraction, and recognition of facial expression. Many image recognition methods can only identify near-frontal and frontal views of the face in face detection.

This paper attempts to research the aspects of the facial assessments regarded as extremely paranormal identities in depth. We argue that DeepID can be learned effectively, while the need for a multi-class reconciliation mission can expand other features and new identities not found within the training set. Nevertheless, DeepID's capacity to generalize improves with more classes predicted in practice [4].

The component-based, trainable system came in trend for detecting face views in gray images, for both frontal and near-frontal images. This system contains a hierarchy of classifiers for Support Vector Machine (SVM) [5].

Training of large, deep convolution NN to classify the more than one million high-resolution pictures in the 1000 different classes of the ImageNet LSVRC-2010 contest [6].

If we use machine learning for designing such system, then we have limitation in quantity or size of the dataset, which will provide less accurate results. This limitation is overcome by the newly proposed Facial Expression Recognition System which will use Deep learning and Image processing which will overcome the problem of large dataset and accuracy of the system. By doing this, one can easily recognize facial expressions with the highest accuracy and great quality of security than previously designed systems.

3 Methodology/Planning of Study

3.1 Analyzing and Preprocessing of Image Dataset

In this step, we are going to collect datasets from various sources and build a new dataset as we are going to recognize frontal faces. After the collection of datasets, data cleaning and data augmentation is performed with the help of image preprocessing.

We have to separate human images right from the beginning and store them into seven different folders namely: ANGER, FEAR, HAPPY, DISGUST, SURPRISE, SAD, NEUTRAL. Some simple steps for processing of model and preprocessing of the dataset such that they become appropriate for feeding the model for training.

Step 1: Converting images to gray scale (if necessary).
Step 3: Resizing of images into a fixed format for proper alignment.
Step 4: Face detection.
Step 5: Check whether all labels are correct or not.
Step 6: Finally save the image.

3.2 Combining Several Expressions into three Basic Expressions

As we are designing a Facial Expression Recognition system for collecting customers' feedback in the banking sector, we require only three expressions which will help us to improve the banking services, those are HAPPY, SAD, and NEUTRAL. In this step, we are going to combine:

1. Happy, Surprise into HAPPY.
2. Angry, Sad, Disgust, Fear into SAD.
3. Neutral into NEUTRAL.

3.3 Building a Model for Recognize Facial Expression

In this step, our aim to create a model that will recognize the facial expression among the seven expressions mentioned above. For that, we are going to use the Convolution Neural Network for the classification of facial expressions. Working of CNN for classification will be as follows (Fig. 2).

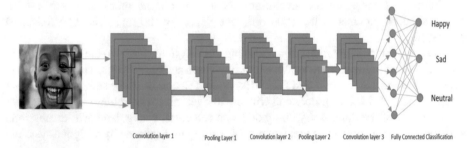

Convolution layer 1   Pooling Layer 1   Convolution layer 2   Pooling Layer 2   Convolution layer 3   Fully Connected Classification

Fig. 2 Working of CNN for classification

4 Results

4.1 Conversion of Colored Image into Gray Scale Image

Colored image consists of a large amount of data that may not be useful in further processing, so we have to convert it into gray scale image which will remove unnecessary information. Figure 3 shows the conversion of a colored image into gray scale image using OpenCV.

4.2 Face Detection and Resizing of Images into 48 * 48

Face detection is the process that should be performed before face recognition. Face detection using HAAR Cascades is a technique that trains a cascade function with a collection of input data. In the preprocessing of the newly added images, the size of the image must be 48 * 48 pixels because of the size of all images in the FER2013 dataset. Figure 4 shows the detection and resizing of the image.

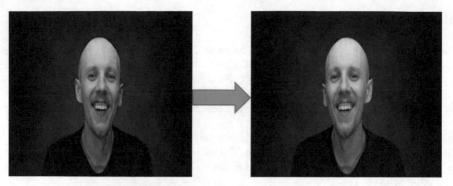

Fig. 3 Colored image to gray scale image

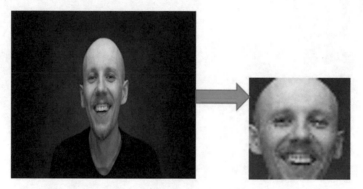

Fig. 4 Face detection and resizing of the image

4.3   CNN for Facial Expression Recognition

The system currently recognizes seven emotions Happy, Sad, Disgust, Surprised, Fear, Angry, and Neutral with an accuracy of 50–55%. This study aims to understand the existing CNN architecture and propose a sophisticated model architecture to improve the accuracy.

5   Conclusion

The study represents a whole DNN architecture for an automatic facial expression recognition system. The proposed architecture contains two convolution layers where each layer contains max pooling layer and Inception layers. The width and depth of the network increases because of Inception layers. The proposed architecture is a one component architecture that takes face images as the input to the model and classifies them into one of seven basic expressions.

Future research topics would include the application of the developed deep learning algorithms to more complex systems with more performance requirements.

References

1. Bartlett, M.S., Littlewort, G., Fasel, I., Movellan, J.R.: Real time face detection and facial expression recognition: development and applications to human computer interaction (2003)
2. Xu, M., Cheng, W., Zhao, Q., Ma, L., Xu, F.: facial expression recognition based on transfer learning from deep convolutional networks (2015)
3. Nguyen, D.V., Vonikakis, V., Winkler, S.: Deep learning for emotion recognition on small datasets using transfer learning
4. Sun, Y., Wang, X., Tang, X.: DeepLearning face representation from predicting 10,000 classes
5. Heisele, B., Serre, T., Pontil, M., Poggio, T.: Component-based face detection (2001)
6. Krizhevsky, A., Sutskever, I., Hinton, G.E.: ImageNet classification with deep convolutional neural networks (2012)

Implementation of E-Health Record with Biometric Authentication

Roshani Ashok Sonar$^{(\boxtimes)}$, Hemraj V. Dhande, Vijay D. Chaudhari, and H. T. Ingale

E & TC Engineering Department (VLSI & Embedded System), GF's Godavari College of
Engineering, Jalgaon 425003, Maharashtra, India
roshanisonar123@gmail.com, hemrajd99@gmail.com,
vinuda.chaudhari@gmail.com, hetui@rediffmaill.com

Abstract. In the health care sector, when the question arises about the patient's safety and security, the most important point on that we have to focus is patient reorganization and patient data truthfulness. Pharmaceutically the fingerprints of any individuals are uncommon and remain unchanged indefinitely. The fingerprint technology used to provide a reliable and accurate method to identify the patients efficiently. The popular point in the fingerprint biometrics technology is that protecting the patient's information, it also preserves against cheating and minimizes human intercession. Using this technology the user enters the information of the patient only one time which minimizes the efforts to add the same data again and again only the further new information related to health, any changes in treatment will get added into the previous record when patient comes into the hospital for the next visit. This helps the doctors to easily study the patient's previous and current health records. The patients get filed quickly at the entrance point simply, by putting down a finger on a self-service terminal or other data collecting device like the emergency department, inpatient areas or outpatient locations.

Keywords: Raspberry pi · Biometrics emergency · Patient's record · E-Health

1 Introduction

The human life is changing nowadays one step ahead due to the IoT, that is, Internet of things. With the new and advanced technology level, IoT is changing the normal simple human life to genius life. In recent years, the use of computer technology strengthening the health care services has received significant outcomes, it also helps to provide online healthcare services. A patient's medical record includes identification, history of medication diagnosis, previous treatment history, patient's body temperature, blood pressure, pulse rate, allergy, dietary habits, genetic information, psychological problem, etc. The above security problems are overcome by using biometrics technology. Our study analyzes the safeguarding and privacy issues that occurred in e-Health. The quality and service of healthcare is boosted by e-Health with the help of making patients' health information easily available to patients' relatives, as well as doctors, improving efficiency and reducing the cost of health service handing over. There are good reasons for

© Springer Nature Singapore Pte Ltd. 2020
B. Iyer et al. (eds.), *Applied Computer Vision and Image Processing*,
Advances in Intelligent Systems and Computing 1155,
https://doi.org/10.1007/978-981-15-4029-5_42

keeping the records private and limiting the access to only required minimum necessary information. According to some behavioral aspects (signature and voice) or physical characteristics (fingerprint or face), a unique identity to an individual is assigned which covers under the biometrics activity. Hence, we use biometrics technology in place of traditional authentication approaches that differentiate between an authorized person and an exploiter. Biometrics refers to technologies used to measure and analyze individual physical or behavioral characteristics to automate the authentication process of the user. There is no chance to lost the Biometric tricks or unremembered, that tricks are tough to copy, share, or circulate among the unknown person.

In this paper, we explore the security and privacy issues in e-Health. Our research pointed out biometrics implementation in user demonstration and health care data encrypting. We conclude that this study will provide a good basic fundamental for further research work in the direction of healthcare data security and patient privacy protection. We are also using the python software. It is a very simple programming language. The errors are easily removed. We are performing many functions, so the code is complex. Python programming language is useful for it. The programming goes easy through this language.

2   Related Work

So many researchers gave their methods and new inventions to the healthcare industry. Wcislik made an android app in an android phone to study patient health continuously. He uses the Bluetooth connection to connect the controller and android phone. The system made by him monitors the patient's pulse rate, respiration rate, ECG wave and patient's body movements, body temperature with the help of Raspberry Pi board and respective sensors. All these parameters are supported only in an android phone. He used the ARM cortex M4F microcontroller for his project. We know that the Bluetooth is a very short distance communication device it works only within 100 m. The IP address is created to monitor patients' health status anytime and anywhere in the world with the help of the webpage [1].

Panda board is used by Amir-Mohammad Rahmani to monitor ECG wave. For connection of internet with the panda board, the Ethernet connection is used in this technique. All the patient's parameters related to health are monitored using the Raspberry Pi board. Ethernet media also operates at very short distance like Bluetooth connection. So, the internet and the Raspberry Pi board is connected by the USB modem [2].

The another technique to monitor the patient's health is invented by using the DRZHG microcontroller. It is a Dual Radio ZigBee Homecare Gateway (DRZHG) which is used to present and impose the support to remote patient monitoring. By using mobile medical sensors the status of long-term patients at home is concurrently tracked. The various sensors measures all the parameters related to patients' health and it provides to the doctor and the users. In this technique, the Zigbee module and the microcontroller are connected to each other [3].

Joao Martinhoa implemented a remotely operated physiological monitoring device. The three types of sensors used for physical measurements of patients are electrocardiography, finger photoplethysmography, and blood pressure plethysmography. Atmega

328 microcontroller is used for connecting these types of sensors. In this technique, the wifi connection is used. The internet and microcontroller are connected [4].

In [5], authors proposed a system that will help to store the patient's each and every detail information in the hospital with the help of the server. The advantage of this is that it helps the doctor to know the history of the patient in few seconds which saves the time of both patient and doctor.

3   Methodology

3.1   Block Diagram Description

The above Fig. 1 shows the generalized system block diagram. The block diagram consists of the block like a patient, fingerprint sensor, Raspberry Pi 3, internet, data server to store data, patient's physical, as well as the psychological parameters related to health and hospital. In that, at the very first stage, the patient fingerprint is taken with the help of a fingerprint sensor. We implement such a simple system in which if the patient is new then we manually enter all the patient's information related to health like patient's pulse rate, respiration rate, ECG wave and patient's body movements, body temperature with the help of Raspberry Pi board and is measured using respective instruments in hospital and it can be monitored in the monitor screen of a computer using Raspberry Pi, as well as their personal information, like name, address, and aadhar number is also entered in the system. On the other hand if the patient is old and we have already implemented this system then all the data is automatically fetched through the system after entering only one single thumb of the patient. After collecting the health information of the patient the system is ready to provide it to the further stage which is the raspberry pi board. Raspberry pi is an operating system that is based on Linux then uploads all this information on the internet automatically because as we connect the internet source to the raspberry pi board it works like a server. This data is also available on the website. The server is used for storing all this data collected by the system. From the server doctors in the hospital, as well as the relatives of the patient are able to monitor that information anywhere in the world using laptops, tablets, and smartphones using the particular IP address. By knowing the patient status a required decision at the critical moment is easily taken by relatives or doctors.

Up till now, the researchers are using so many sensors to measure the body or health parameters of the patient's which increases the complexity, as well as the heaviness of the system. In the hospital already all the measuring devices are available so that's why we are not using so many sensors in our system. Which reduces the cost of the system, space to implement the system, as well as weight of the system. Due to this, we mount this type of system in small hospitals or clinics also. If this system is implemented in every hospital and in any case the patient is changing their place or hospital then also the new doctor will also know the previous history of the patient with entering only a thumb. The advantage of this is that the headache of the doctor and patient is reduced. Some points are missed orally but if that are available in the written form then there is no chance to miss those points, then accordingly the appropriate treatment is given by the doctors to the patient. This process makes the patient happy.

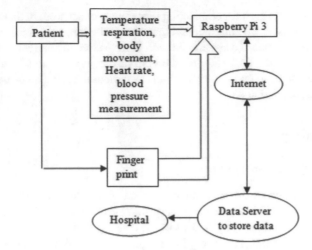

Fig. 1 Generalized system block diagram

This system can also be used in the ambulance. If an emergency occurs and the patient is in the ambulance then at that time if this system is available in the ambulance we can quickly get all the information and upload it on the server and the emergency notification or message is sent to the hospital before the arrival of the patient in the hospital. Due to this, the doctor may get help to attain the patient immediately as he enters in the hospital. The notification message is in the form of mail. In this way implementing simple and easy e-health system.

3.2 Hardware Description

Fingerprint Sensor: A Fingerprint sensor used to connect with Raspberry Pi. Apart from the voltage, the models do not differ much. It is a serial USB converter with 3.3 V and 5 V connection. Female-Female Jumper wires are used, if not included with the USB converter.

USB to TTL Converter: This is a small cable used to connect the raspberry pi and fingerprint sensor. It is a serial tool, uses the PL2303 chip. Some serial devices are connected with the help of this USB port to our PC. It requires the external 5 V power supply. Data transfer status in the real time domain can monitor using two data transmission indicator.

Power Supply: The power requirement of the hardware in the biometric e-health system using raspberry pi is the power supply of 5 V, 3 A.

Raspberry Pi: One can use the raspberry pi as it is open source platform. It can perform the operations that we need. Raspberry Pi has several unique features such as it provides a port for different connections like to connect with PC, sensors, etc., which is the need in our system.

4 Software

The system will perform in the following steps (Fig. 2).

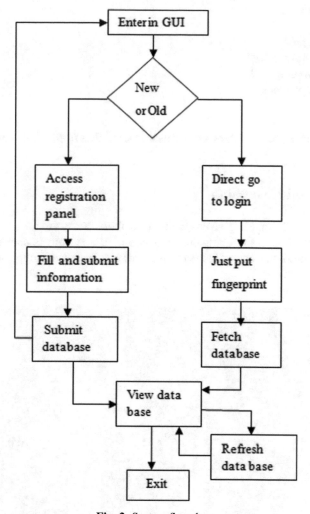

Fig. 2 System flowchart

Algorithm steps for the above flowchart is as follows:

I. Run the program.
    If the user is new then

   i. Access registration panel.
   ii. fill all details of user.

    iii.   submit fingerprint.
    iv.   submit full data.
    v.   exit.

else old user

    i.   Login panel.
    ii.   Enter fingerprint.
    iii.   Information fetched.
    iv.   Displayed on screen.
    v.   Exit.

II.   View database panel to show proper user information from whole database.

5 Results and Discussion

The following result images give us a detail idea about our system.

In the above Fig. 3 the complete hardware of the system is shown. In that our Raspberry Pi 3 module along with the input source that is a power supply and fingerprint sensor is shown in (Fig. 4).

Fig. 3 Complete hardware of the system

In this figure simultaneously we able to see the two hardware components of the system. First is USB to TTL converter cable along with the fingerprint sensor. Which is used to access every user's fingerprint.

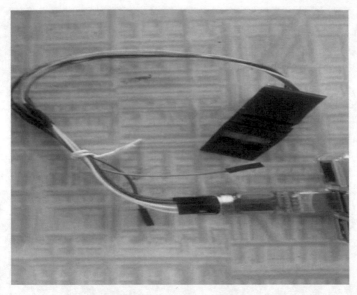

Fig. 4 Fingerprint sensor

The Fig. 5 shows the patient record which is available globally anywhere in the world. Once the information is collected from the user it is saved permanently on the server which will be used in the future for the patient safety and emergency conditions. The Table 1 explains the patient information we can add more detailed information of the patient in that.

The non-invasive techniques and systems are getting popular to gather patents information. These methods are attractive as the cooperation from human subjects is not required during monitoring and recording the data [6, 7]. Further, IoT enabled techniques helped such devices to become more handy, sophisticated, and cost-effective [8–10]. Hence, in the future, the proposed methodology will be extended in terms of its measuring range using RF technology.

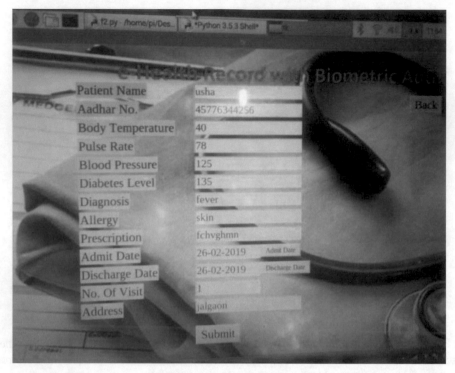

Fig. 5 User/patient record

Table 1 Patient's information

Sr. no.	Patient name	Adhar no.	Body temperature	Pulse rate	Blood pressure	Diabetes level	Diagnosis	Allergy	No. of visits
1.	Usha Sonar	439012349087	40	80	125	160	Fever	Skin	1
2.	Ashok Patel	908745673241	50	66	100	150	Cold	No	3

6 Conclusion

In our project, fingerprint verification is considered to protect the medical information and the confidentiality of data. Patient data can be stored and retrieved by connecting to the hospital database and thus it can be accessed globally by using IoT. The main advantage of our project is the online fetching of patient information. On the another hand it is helpful during emergency conditions. Medical record errors can be reduced by using fingerprint techniques. This system provides a number of benefits to the user.

The system used to capture data, store, view, add, and delete the records entered. In our system at the right stage, we are implementing only one hospital. In the future, if we implement this system in every hospital then the patient's data will be shared globally everywhere in the world with the help of the server.

References

1. Wcislik, M., Pozoga, M., Smerdzynski, P.: Wireless Health Monitoring System. In: IFAC (International Federation of Automatic Control), pp. 312–317. Hosting by Elsevier Ltd. (2015)
2. Rahmani, A.-M., Thanigaivelan, N.K., Gia, T.N., Granados, J., Negash, B., Liljeberg, P., Tenhunen, H.: Smart e-Health gateway: bringing intelligence to internet of-things based ubiquitous healthcare systems (2015)
3. Tung, H.Y., Tsang, K.F., Tung, H.C., Chui, K.T., Chi, H.R.: The design of dual radio Zig-Bee homecare gateway for remote patient monitoring. IEEE Trans. Consum. Electron. **59**(4) (2013)
4. Martinhoa, J., Pratesa, L., Costaa, J., Isel, A., Navarro 1, R.C.E., Portugal, L.: Design and implementation of a wireless multi parameter patient monitoring system. In: Conference on Electronics, Telecommunications and Computers—CETC, pp. 542–549. Hosting by Elsevier Ltd (2014)
5. Sonar, R.A., Dhande, H.V., Ingale, H.: Implementation of E-health record with biometric authentication. Int. J. Innovat. Eng. Sci. (IJIES) **4**(10), 32–37 (2019). ISSN: 2456-3463
6. Iyer, B., Pathak, N.P., Ghosh, D.: Dual-input dual-output RF sensor for indoor human occupancy and position monitoring. IEEE Sens. J. **15**(7), 3959–3966 (2015)
7. Iyer, B., Pathak, N.P., Ghosh, D.: RF sensor for smart home application. Int. J. Syst. Assur. Eng. Manag. **9**, 52–57 (2018). https://doi.org/10.1007/s13198-016-0468-5
8. Deshpande, P., Iyer, B.: Research directions in the Internet of Every Things (IoET). In: International Conference on Computing, Communication and Automation (ICCCA), pp. 1353–1357 (2017)
9. Iyer, B., Patil, N.: IoT enabled tracking and monitoring sensor for military applications. Int. J. Syst. Assur. Eng. Manag. **9**, 1294–1301 (2018). https://doi.org/10.1007/s13198-018-0727-8
10. Patil, N., Iyer, B.: Health monitoring and tracking system for soldiers using internet of things (IoT). In: International conference on computing, communication and automation, pp. 1347–1352 (2017)

Modeling Groundwater Spring Potential of Selected Geographical Area Using Machine Learning Algorithms

S. Husen[⊠], S. Khamitkar, P. Bhalchandra, S. Lohkhande, and P. Tamsekar

School of Computational Sciences, S.R.T.M.University, Nanded, Maharashtra 431606, India
husen09@gmail.com, s_khamitkar@yahoo.com, srtmun.parag@gmail.com,
lokhande_sana@rediffmail.com, pritam.tamsekar@gmail.com

Abstract. The major objective of this paper was to produce groundwater spring potential maps for the selected geographical area using machine learning models. Total seven ML algorithms, viz Logistic Regression, Random Forest, SVM, Gradient Boosting Classifier, XG Boost, KNN, and Decision Tree were deployed. Further modeling was done using six hydrological-geological aspects that control the site of springs during the course of this research. Finally, groundwater spring potential was modeled and planning using fusion of these technologies.

Keywords: Water management · GIS · AHP · DSS · ML

1 Introduction

Groundwater is supposed to be one of the mainly priceless natural treasures. Due to a number of factors such as steady temperature, extensive accessibility, incomplete susceptibility to contamination, short growth cost, and lack of consistency, this treasure is shrinking down [1]. Further, the speedy raise in human populace has tremendously raised the requirement for groundwater provisions for drinking, agricultural, and industrial reasons [2]. Thus, the demand and supply equation is severely damaged leading to massive scarcity. To cater demands for clean groundwater rose in recent years, the demarcation of groundwater spring potential area develop into an important solution. The implementation of this solution needs resolve safety, and good management structure [3]. This is addressed by the creation of groundwater maps. A large number of researchers used integrated GIS, RS, and geo-statistics techniques for groundwater planning [4]. A lot of studies have also examined groundwater viable using the probabilistic model [5]. Additionally, a lot of current research studies use GIS methods build in with frequency ratio (FR), logistic regression (LR), models [6]. After careful investigations of the highlighted references and contemporary research, our work has used Logistic Regression, Random Forest, SVM, Gradient Boosting Classifier, XG Boost, KNN, and Decision Tree methods, included in a GIS to forecast groundwater spring locations for selected the study area [7]. Machine learning (ML) algorithms is a quickly rising area of prognostic modeling, where it is difficult to recognizes structure in complex, frequently

© Springer Nature Singapore Pte Ltd. 2020
B. Iyer et al. (eds.), *Applied Computer Vision and Image Processing*,
Advances in Intelligent Systems and Computing 1155,
https://doi.org/10.1007/978-981-15-4029-5_43

nonlinear data and generates correct prognostic models [8]. The ML approaches showed better authority for deciding complex relationships as ML approaches are not limited to the conventional hypothesis which is usually used along with predictable and parametric approaches. On the backdrop of these discussions, the main objective of this study was to predict groundwater spring potential maps using Logistic Regression, Random Forest, SVM, Gradient Boosting Classifier, XG Boost, KNN, and Decision Tree methods. The selected geographical area is the Kalmnoori taluka, Maharashtra state, India. The result of this research will afford a methodology to expand SPM that can be used by not only government departments, but also by the private sector for groundwater examination, estimation, and fortification.

2 Methodology

Kalmnoori taluka is taken as the study area. The Geographical location of Kalmnoori taluka lies within the coordinate of latitudes between 19.67° North latitude and 77.33° east longitudes. It is in the Hingoli district, Maharashtra state, India and its area is approximately 51.76 km^2. It has an average elevation of 480 m. The Kayadhu is the main river which flows from the study area. The study area is taken out from the top sheet collected from the certified government agency, and it is geo-referenced. In order to generate the DSS for recognition of the Artificial Water Recharge Site (AWRS) the following process has been adopted.

2.1 Spatial Data Collection

Remote sensing data of CartoDEM of 2.5 m resolution and 3.5 m resolution satellite data of IRS P5 LISS-III, which are taken from the Bhuvan portal [9]. Geomorphology and Lineament data is taken out from the WMS layer of the Bhuvan portal. The particulars of the hydrological soil group map are taken from the National Bureau of Soil Survey.

2.2 Formulation of Criteria for Artificial Water Recharge Site

For the selection of Artificial Water Recharge Site, we have selected the below listed six parameters namely Geomorphology, Slope, Drainage density, Lineament density, LU/LC, and Soil. Geomorphology must be pediment-pediplain complex, anthropogenic terrain and water body should be good. The slope required is 0°–5°. Drainage density is 0 to 3.5/km^2. Lineament density is 0–1.5 km^2. LU/LC must be a cropland, water body or scrubland. The soil must be sand or loam.

2.3 Generation of Criteria Map Using GIS

For AWRS the formulation of criteria is completed by using parameters like Geomorphology, soil, slope, LULC, lineament density, and drainage density. After that reclassification process is done for the same unit of all layers. And finally, all vector layers are converted into the raster layer.

2.4 Deriving the Weights Using AHP

Literature reviews, expert advises and from local field experience for the comparative significance to assigning weight age for every parameter and classes. The aim of using an analytic hierarchy process (AHP) is to recognize the ideal option and also conclude a position of the alternatives when all the decision criteria are measured concurrently [10].

2.5 Weighted Overlay Analysis

After giving the weightage of every major parameter has been determined, the weightage for the associate class of major parameters have been assigned. "Weighted Overlay" method is used for overlaying all the layers according to their importance. "Weighted Overlay" is a sub method of Spatial Analyst Tools in ArcGIS. Further, ranking is given from 1 to 5 according to their weight. Where 5 represents excellent prospects and 1 represents a poor prospect of groundwater. Lastly according to importance and rank of layers its entity features are given to each layer [11].

2.6 Data Set

The data set was extracted from the inventory map of artificial groundwater zones in such a manner that it was balanced. The Random Point Selection function of ArcGIS was used for data preparation. The data set consists of 500 samples which consist of 7 parameters namely Geomorphology, soil, slope, LULC, lineament density, drainage density, and result map. Among this, the result map is the target variable and the remaining 6 parameters are the response variable. The sample points used to build the data set from the study area is shown in Fig. 1a.

2.7 Exporting Data in Python

The data set is imported in Python and data exploration is done. In Python, we have done one hot encoding on data. The dataset consists of continuous variable in order to apply classification algorithms the data should be in categorical form, therefore One hot encoding is used to transform the continues variable to categorical. Finally, the dataset is split into training (70%) and testing (30%) dataset.

After the above analysis, efforts are taken for training the Model on Data Set. Following classification algorithms are used to train the model on a dataset.

KNN: The Fig. 1b shows the map of the Artificial Ground Water Recharge Site (AGWRS) predicted by K-Nearest Neighbor (KNN) model of the machine learning algorithm. In this map classes of artificial groundwater recharge site such as very poor, poor, moderated, good, very good are shown along with theses classes Artificial Ground Water Recharge Point predicted by KNN algorithm are also shown. In k-Nearest Neighbors where the k is the number of observations are tuned to 1, 3, 5, 7, 9 and found that $k = 5$ gives the utmost results. By applying k-NN with the value $k = 5$, the confusion matrix for the applied model was created. Total 105 artificial groundwater recharge site (AGRW) points ware correctly classified as good followed by 21 as moderate, and 1,

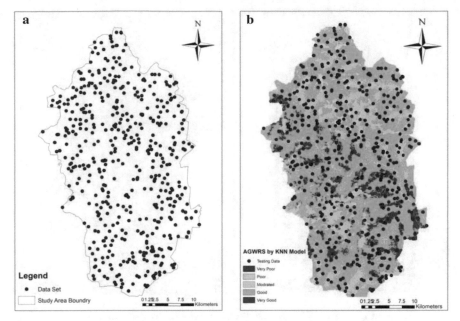

Fig. 1 **a** Data set, **b** AGWRS by KNN model

4, 3 as very poor, poor, and very good, respectively. The misclassifications for each class were 3, 1, 9, 13, and 5 for very poor, poor, moderate, good, and very good class, respectively. The overall accuracy of the model is 0.81% and the error rate is 0.19%.

Random Forest: The Fig. 2a shows the map of the artificial groundwater recharge site (AGWRS) predicted by the Random Forest (RF) model of the machine learning algorithm. This map shows the artificial groundwater recharge point predicted by the RF model and classes of artificial groundwater recharge sites such as very poor, poor, moderated, good, very good. The model is giving relatively good prediction results. This model gave 100% accuracy while predicting very poor class. In poor class, only 1 instance is misclassified as moderate. In moderate, out of 27 instances, 20 were correctly classified as moderate and 2, 7 were misclassified as good and poor, respectively. Out of 165 instances, 108 were classified as good and 2, 9, 5 were misclassified as very poor, moderate, and very good, respectively. In class very good, 3 instances were classified correctly and 3 were misclassified as good. Random forest algorithm has given 0.82 accuracy.

Support Vector Machine: Artificial groundwater recharge site (AGWRS) predicted by Support Vector Machine (SVM) model of machine learning algorithm map is shown in Fig. 2b. This map shows the artificial groundwater recharge point that is testing data and classes of artificial groundwater recharge site such as very poor, poor, moderated, good, very good. In parameter tuning, we have used a linear kernel. Using this approach, the trained model achieved good accuracy, while predicting class very poor. In poor class, 3 instances were correctly classified as poor class and 1 was misclassified as moderate. In class good, out of 121 instances, 108 were correctly classified as good and 1, 7, 5

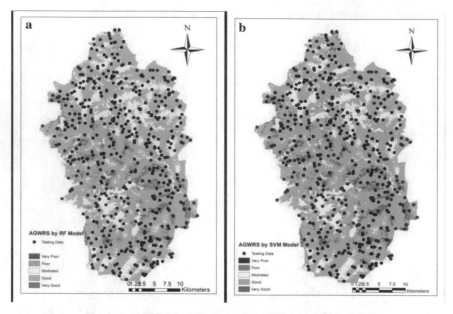

Fig. 2 a AGWRS by RF model, b AGWRS by SVM model

instances were misclassified as very poor, moderate, and very good, respectively. In class very good, 3 instances were classified correctly and 3 misclassified as good. Accuracy of SVM algorithm is 0.83.

XG Boost: In Fig. 3a shows the map of the artificial groundwater recharge site (AGWRS) predicted by the XG Boost (XGB) model of machine learning algorithm. The xg boost's model is a linear combination of decision trees. To make a prediction xg boost calculates predictions of individual trees and adds them. The trained model achieved better accuracy while predicting class very poor. In poor class, 3 instances were correctly classified as poor class and 1 is misclassified as moderate. In moderate, 21 instances out of 32 were correctly classified as moderate and 2 and 9 were misclassified as poor and good. In good class, out of 121 instances, 106 were correctly classified as good and 2, 8, 5 instances were misclassified as very poor, moderate, and very good, respectively. In class very good, 3 instances classified correctly and 3 misclassified as good. Accuracy of XG Boost algorithm is of 0.81.

Logistic Regression: The Fig. 3a shows the map of the Artificial Ground Water Recharge Site (AGWRS) predicted by the Logistic Regression (LR) model of machine learning algorithm. In this map classes of artificial ground water recharge site such as very poor, poor, moderated, good, very good, along with theses classes Artificial Ground Water Recharge Point data is shown on map which is predicted by LR algorithm. The related confusion matrix of the trained logistic regression model has achieved better accuracy while predicting class very poor. In poor class, 1 instance was correctly classified as poor class and 1 was misclassified as moderate. In moderate, 11 instances out of 20 were correctly classified as moderate and 2, 4, and 3 were misclassified as very poor, poor, and good, respectively. In class good, out of 142 instances, 115 were correctly

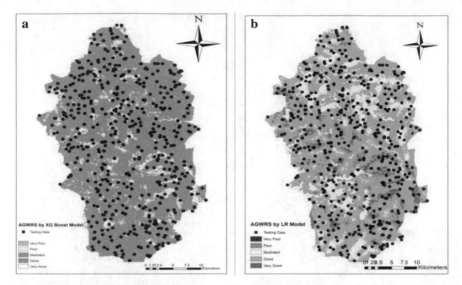

Fig. 3 **a** AGWRS by XG Boost model, **b** AGWRS by LR model

classified as good and 1, 18, and 8 instances were misclassified as very poor, moderate, and very good, respectively. Overall accuracy of LR algorithm is 0.77.

Gradient Boosting Classifier: In Fig. 4a shows the map of the Artificial Ground Water Recharge Site (AGWRS) predicted by the Gradient Booster Classifier (GBC) model of machine learning algorithm. This map shows that, the artificial groundwater recharge point is testing data and classes of artificial groundwater recharge site such as very poor, poor, moderated, good, very good. The related confusion matrix of the trained gradient boosting classifier model was obtained. While predicting class very poor out of 3 instances 2 were correctly classified as very poor and 1 instance was misclassified as good. In poor class, 2 instances were correctly classified as poor and 2 were misclassified as moderate. In moderate, 18 instances out of 29 were correctly classified as moderate and 2, 9 were misclassified as poor and good, respectively. In class good, out of 121 instances, 104 were correctly classified as good and 2, 10, and 5 instances were misclassified as very poor, moderate, and very good, respectively. Overall accuracy of GBC algorithm is 0.78.

Decision Tree: The artificial groundwater recharge site (AGWRS) predicted by the Decision Tree (DT) model of machine learning algorithm map is shown in Fig. 4b. This map shows that, the artificial groundwater recharge point is testing data and classes of artificial groundwater recharge site such as very poor, poor, moderated, good, very good. While predicting class very poor out of 2 instances 2 were correctly classified as very poor. In poor class, 4 instances were correctly classified as poor and 1 was misclassified as moderate. In moderate, 22 instances out of 37 were correctly classified as moderate and 1, 14 were misclassified as poor and good, respectively. In class good, out of 114 instances, 100 were correctly classified as good and 2, 7, and 5 instances were misclassified as very poor, moderate, and very good, respectively. Overall accuracy of Decision tree algorithm is 0.78.

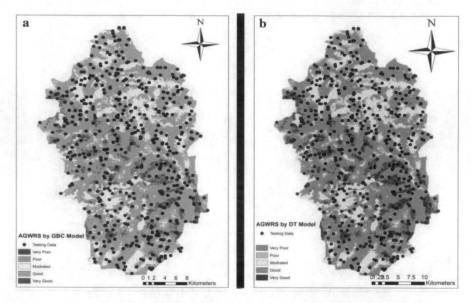

Fig. 4 **a** AGWRS by GBC model, **b** AGWRS by DT model

The performance of machine learning methods for artificial groundwater recharge sites prediction model depends on the data set and machine learning algorithms proper utilization. Picking the correct classification ML technique for the defined crisis is the basic necessity to get the most excellent possible result. Though, the proper selection of algorithms won't guarantee the best possible results. The input dataset served to build the ML model is also a vital issue and for getting the best possible results to feature engineering, the process of modifying the data for machine learning is also an important thing to be considered. Comparing the results of ML techniques applied to the testing data Table 9 shows the prediction results obtained by applying the ML techniques LR, RF, SVM, Gradient Boosting Classifier, XG Boost, KNN, and Decision Tree on the test dataset extracted from geo-data layers. The results are shown in Fig. 5 and Table 1.

3 Conclusions

The aim of the research work was to develop an artificial groundwater recharge site maps (AGWRS), which involves the model comparison of the raw dataset and engineered dataset, in order to improve the result prediction capability of the classification model. The datasets were obtained from the study area. The geo-data layers imported in Python programming language were processed using raster package and information was extracted. The selected six ML algorithms were implemented and their confusion matrix were created which served as input to the prediction model to get the best algorithm for classification. Their outcomes were compared using bench mark evaluation measures. It was observed that the best prediction results from SVM, with a prediction accuracy of 0.83% followed by Logistic Regression, Random Forest, Gradient Boosting

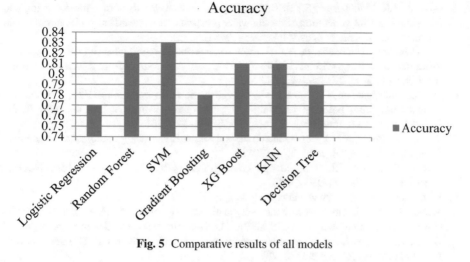

Fig. 5 Comparative results of all models

Table 1 Tabulation of efficiencies of all models

Algorithms	Results
Logistic regression	0.77
Random forest	0.82
SVM	0.83
Gradient boosting classifier	0.78
XG Boost	0.81
KNN	0.81
Decision tree	0.79

Classifier, XG Boost, KNN, and Decision Tree prediction accuracy results are 0.77, 0.82, 0.78, 0.81, 0.81, 0.79, respectively.

References

1. Jha, M.K., Chowdhury, A., Chowdary, V.M., Peiffer, S.: Groundwater management and development by integrated remote sensing and geographic information systems: prospects and constraints. Water Resour. Manage **21**, 427–467 (2007)
2. Lee, S., Song, K.Y., Kim, Y., Park, I.: Regional groundwater productivity potential mapping using a geographic information system (GIS) based artificial neural network model. Hydrogeol. J. **20**(8), 1511–1527 (2012)
3. Ozdemir, A.: Using a binary logistic regression method and GIS for evaluating and mapping the groundwater spring potential in the Sultan Mountains (Aksehir, Turkey). J. Hydrol. **405**, 123–136 (2011)

4. Jaiswal, R.K., Mukherjee, S., Krishnamurthy, J., Saxena, R.: Role of remote sen ing and GIS techniques for generation of groundwater prospect zones towards rural development–an approach. Int. J. Remote Sens. **24**(5), 993–1008 (2003)
5. Srivastava, P.K., Bhattacharya, A.K.: Groundwater assessment through an integrated approach using remote sensing, GIS and resistivity techniques: a case study from a hard rock terrain. Int. J. Remote Sens. **27**, 4599–4620 (2006)
6. Davoodi Moghaddam, D., Rezaei, M., Pourghasemi, H.R., Pourtaghie, Z.S., Pradhan, B.: Groundwaterspring potential mapping using bivariate statistical model and GIS in the Taleghan watershed. Iran. Arab. J. Geosci. **8**(2), 913–929 (2015)
7. Stumpf, A., Kerle, N.: Object-oriented mapping of landslides using random forests. Remote Sens. Environ. **115**(10), 2564–2577 (2011)
8. Olden, J.D., Lawler, J.J., Poff, N.L.: Machinelearning without tears: a primer for ecologists. Q. Rev. Biol. **83**(2), 171–193 (2008)
9. Web resource at http://www.bhuvan.nrsc.gov.in
10. Saaty, T.L.: Decision making with the analytic hierarchy process. Int. J. Serv. Sci. **1**(1) (2008)
11. Husen, S. et al.: Integrated use of GIS AHP and GIS techniques for selection of artificial ground water recharge site, information and communication technology for sustainable development. Adv. Intell. Syst. Comput. **933** (2020)

Printed in the United States
By Bookmasters